CW01023937

Adaptive Control

Stability, Convergence and Robustness

Adaptive Control

Stability, Convergence and Robustness

Shankar Sastry
University of California at Berkeley

Marc Bodson
University of Utah

Dover Publications
Garden City, New York

Bibliographical Note

This Dover edition, first published in 2011, is an unabridged republication of the work originally published in 1989 by Prentice-Hall, Inc., Englewood Cliffs, New Jersey. An errata list has been added to this edition.

Library of Congress Cataloging-in-Publication Data

Sastry, Shankar.
Adaptive control : stability, convergence and robustness / Shankar Sastry and Marc Bodson.
p. cm.
Includes bibliographical references and index.
ISBN-13: 978-0-486-48202-6
ISBN-10: 0-486-48202-2 (pbk.)
1. Adaptive control systems. I. Bodson, Marc. II. Title.

TJ217.S27 2011
629.8'36—dc22

2011010871

Manufactured in the United States of America
48202203 2022
www.doverpublications.com

CONTENTS

Preface xiii

Chapter 0 Introduction 1
- 0.1 Identification and Adaptive Control 1
- 0.2 Approaches to Adaptive Control 4
  - 0.2.1 Gain Scheduling 4
  - 0.2.2 Model Reference Adaptive Systems 5
  - 0.2.3 Self Tuning Regulators 8
  - 0.2.4 Stochastic Adaptive Control 9
- 0.3 A Simple Example 11

Chapter 1 Preliminaries 17
- 1.1 Notation 17
- 1.2 L_p Spaces, Norms 18
- 1.3 Positive Definite Matrices 19
- 1.4 Stability of Dynamic Systems 20
  - 1.4.1 Differential Equations 20
  - 1.4.2 Stability Definitions 23
  - 1.4.3 Lyapunov Stability Theory 25
- 1.5 Exponential Stability Theorems 28
  - 1.5.1 Exponential Stability of Nonlinear Systems 28
  - 1.5.2 Exponential Stability of Linear Time-Varying Systems 32
  - 1.5.3 Exponential Stability of Linear Time Invariant Systems 38
- 1.6 Generalized Harmonic Analysis 39

Chapter 2 Identification 45
2.0 Introduction 45
2.1 Identification Problem 52
2.2 Identifier Structure 53
2.3 Linear Error Equation and Identification Algorithms 57
2.3.1 Gradient Algorithms 58
2.3.2 Least-Squares Algorithms 61
2.4 Properties of the Identification Algorithms—Identifier Stability 63
2.4.1 Gradient Algorithms 63
2.4.2 Least-Squares Algorithms 66
2.4.3 Stability of the Identifier 69
2.5 Persistent Excitation and Exponential Parameter Convergence 71
2.6 Model Reference Identifiers—SPR Error Equation 76
2.6.1 Model Reference Identifiers 76
2.6.2 Strictly Positive Real Error Equation and Identification Algorithms 82
2.6.3 Exponential Convergence of the Gradient Algorithms with SPR Error Equations 85
2.7 Frequency Domain Conditions for Parameter Convergence 90
2.7.1 Parameter Convergence 90
2.7.2 Partial Parameter Convergence 95
2.8 Conclusions 98
Chapter 3 Adaptive Control 99
3.0 Introduction 99
3.1 Model Reference Adaptive Control Problem 103
3.2 Controller Structure 104
3.3 Adaptive Control Schemes 110
3.3.1 Input Error Direct Adaptive Control 111
3.3.2 Output Error Direct Adaptive Control 118
3.3.3 Indirect Adaptive Control 123
3.3.4 Alternate Model Reference Schemes 127
3.3.5 Adaptive Pole Placement Control 129
3.4 The Stability Problem in Adaptive Control 130
3.5 Analysis of the Model Reference Adaptive Control System 132
3.6 Useful Lemmas 138
3.7 Stability Proofs 142
3.7.1 Stability—Input Error Direct Adaptive Control 142
3.7.2 Stability—Output Error Direct Adaptive Control 149
3.7.3 Stability—Indirect Adaptive Control 151

3.8 Exponential Parameter Convergence 154
3.9 Conclusions 156

Chapter 4 Parameter Convergence Using Averaging Techniques 158
4.0 Introduction 158
4.1 Examples of Averaging Analysis 159
4.2 Averaging Theory–One-Time Scale 166
4.3 Application to Identification 175
4.4 Averaging Theory–Two-Time Scales 179
4.4.1 Separated Time Scales 183
4.4.2 Mixed Time Scales 186
4.5 Applications to Adaptive Control 187
4.5.1 Output Error Scheme–Linearized Equations 188
4.5.2 Output Error Scheme–Nonlinear Equations 192
4.5.3 Input Error Scheme 202
4.6 Conclusions 207

Chapter 5 Robustness 209
5.1 Structured and Unstructured Uncertainty 209
5.2 The Rohrs Examples 215
5.3 Robustness of Adaptive Algorithms with Persistency of Excitation 219
5.3.1 Exponential Convergence and Robustness 221
5.3.2 Robustness of an Adaptive Control Scheme 225
5.4 Heuristic Analysis of the Rohrs Examples 231
5.5 Averaging Analysis of Slow Drift Instability 236
5.5.1 Instability Theorems Using Averaging 236
5.5.2 Application to the Output Error Scheme 241
5.6 Methods for Improving Robustness–Qualitative Discussion 248
5.6.1 Robust Identification Schemes 248
5.6.2 Specification of the Closed Loop Control Objective–Choice of Control Model and of Reference Input 250
5.6.3 The Usage of Prior Information 250
5.6.4 Time Variation of Parameters 251
5.7 Robustness via Update Law Modifications 251
5.7.1 Deadzone and Relative Deadzone 251
5.7.2 Leakage Term (σ–Modification) 253
5.7.3 Regressor Vector Filtering 253
5.7.4 Slow Adaptation, Averaging and Hybrid Update Laws 254
5.8 Conclusions 254

Chapter 6 Advanced Topics in Identification and Adaptive Control 257
6.1 Use of Prior Information 257
6.1.1 Identification of Partially Known Systems 257
6.1.2 Effect of Unmodeled Dynamics 263
6.2 Global Stability of Indirect Adaptive Control Schemes 266
6.2.1 Indirect Adaptive Control Scheme 267
6.2.2 Indirect Adaptive Pole Placement 272
6.2.3 Indirect Adaptive Stabilization—The Factorization Approach 274
6.3 Multivariable Adaptive Control 277
6.3.1 Introduction 277
6.3.2 Preliminaries 278
6.3.2.1 Factorization of Transfer Function Matrices 278
6.3.2.2 Interactor Matrix and Hermite Form 282
6.3.3 Model Reference Adaptive Control—Controller Structure 286
6.3.4 Model Reference Adaptive Control—Input Error Scheme 290
6.3.5 Alternate Schemes 292
6.4 Conclusions 293

Chapter 7 Adaptive Control of a Class of Nonlinear Systems 294
7.1 Introduction 294
7.2 Linearizing Control for a Class of Nonlinear Systems—A Review 295
7.2.1 Basic Theory 295
7.2.2 Minimum Phase Nonlinear Systems 299
7.2.2.1 The Single-Input Single-Output Case 300
7.2.2.2 The Multi-Input Multi-Output Case 305
7.2.3 Model Reference Control for Nonlinear Systems 307
7.3 Adaptive Control of Linearizable Minimum Phase Systems 309
7.3.1 Single-Input Single-Output, Relative Degree One Case 309
7.3.2 Extensions to Higher Relative Degree SISO Systems 312
7.3.3 Adaptive Control of MIMO Systems Decouplable by Static State Feedback 320
7.4 Conclusions 322

Chapter 8 Conclusions 324
8.1 General Conclusions 324
8.2 Future Research 325

Appendix 331

References 359

Index 373

Errata 379

PREFACE

The objective of this book is to give, in a concise and unified fashion, the major results, techniques of analysis and new directions of research in adaptive systems. Such a treatment is particularly timely, given the rapid advances in microprocessor and multi-processor technology which make it possible to implement the fairly complicated nonlinear and time varying control laws associated with adaptive control. Indeed, limitations to future growth can hardly be expected to be computational, but rather from a lack of a fundamental understanding of the methodologies for the design, evaluation and testing of the algorithms. Our objective has been to give a clear, conceptual presentation of adaptive methods, to enable a critical evaluation of these techniques and suggest avenues of further development.

Adaptive control has been the subject of active research for over three decades now. There have been many theoretical successes, including the development of rigorous proofs of stability and an understanding of the dynamical properties of adaptive schemes. Several successful applications have been reported and the last ten years have seen an impressive growth in the availability of commercial adaptive controllers.

In this book, we present the *deterministic* theory of identification and adaptive control. For the most part the focus is on linear, continuous time, single-input single-output systems. The presentation includes the algorithms, their dynamical properties and tools for analysis—including the recently introduced averaging techniques. Current research in the adaptive control of multi-input, multi-output linear systems and a class of nonlinear systems is also covered. Although continuous time algorithms occupy the bulk of our interest, they are presented in such a way as to enable their transcription to the discrete time case.

A brief outline of the book is as follows: Chapter 0 is a brief historical overview of adaptive control and identification, and an introduction to various approaches. Chapter 1 is a chapter of mathematical preliminaries containing most of the key stability results used later in the book. In Chapter 2, we develop several adaptive identification algorithms along with their stability and convergence properties. Chapter 3 is a corresponding development for model reference adaptive control. In Chapter 4, we give a self contained presentation of averaging techniques and we analyze the rates of convergence of the schemes of Chapters 2 and 3. Chapter 5 deals with robustness properties of the adaptive schemes, how to analyze their potential instability using averaging techniques and how to make the schemes more robust. Chapter 6 covers some advanced topics: the use of prior information in adaptive identification schemes, indirect adaptive control as an extension of robust non-adaptive control and multivariable adaptive control. Chapter 7 gives a brief introduction to the control of a class of nonlinear systems, explicitly linearizable by state feedback and their adaptive control using the techniques of Chapter 3. Chapter 8 concludes with some of our suggestions about the areas of future exploration.

This book is intended to introduce researchers and practitioners to the current theory of adaptive control. We have used the book as a text several times for a one-semester graduate course at the University of California at Berkeley and at Carnegie-Mellon University. Some background in basic control systems and in linear systems theory at the graduate level is assumed. Background in stability theory for nonlinear systems is desirable, but the presentation is mostly self-contained.

Acknowledgments

It is a pleasure to acknowledge the contributions of the people who helped us in the writing of this book. We are especially appreciative of the detailed and thoughtful reviews given by Charles Desoer of the original Ph.D. dissertation of the second author on which this book is based. His advice and support from the beginning are gratefully acknowledged. Brian Anderson and Petar Kokotovic offered excellent critical comments that were extremely helpful both in our research, and in the revisions of the manuscript. We also thank Karl Astrom and Steve Morse for their enthusiasm about adaptive control, and for fruitful interactions.

The persistently exciting inputs of students at Berkeley and Carnegie Mellon have helped us refine much of the material of this book. We are especially thankful to those who collaborated with us in research, and contributed to this work: Erwei Bai, Stephen Boyd, Michel de Mathelin, Li-Chen Fu, Ping Hsu, Jeff Mason, Niklas Nordstrom, Andy Packard, Brad Paden and Tim Salcudean. Many of them have now adapted to new environments, and we wish them good luck.

We are indebted to many colleagues for stimulating discussions at conferences, workshops and meetings. They have helped us broaden our view and understanding of the field. We would particularly like to mention Anu Annaswamy, Michael Athans, Bob Bitmead, Soura Dasgupta, Graham Goodwin, Petros Ioannou, Alberto Isidori, Rick Johnson, Ed Kamen, Bob Kosut, Jim Krause, Rogelio Lozano-Leal, Iven Mareels, Sanjoy Mitter, Bob Narendra, Dorothee Normand-Cyrot, Romeo Ortega, Laurent Praly, Brad Riedle, Charles Rohrs, Fathi Salam and Lena Valavani.

We acknowledge the support of several organizations, including NASA (Grant NAG-243), the Army Research Office (Grant DAAG 29-85-K0072), the IBM Corporation (Faculty Development Award), and the National Science Foundation (Grant ECS-8810145). Special thanks are due to George Meyer and Jagdish Chandra: their continuous support of our research made this book possible.

We are also grateful for the logistical support received from the administration of the Department of Electrical Engineering and Computer Sciences at the University of California at Berkeley and of the Electrical and Computer Engineering Department of Carnegie Mellon University. Part of our work was also done at the Laboratory for Information and Decision Systems, in the Massachussetts Institute of Technology, thanks to the hospitality of Sanjoy Mitter.

We wish to express our gratitude to Carol Block and Tim Burns, for their diligent typing and layout of the manuscript in the presence of uncertainty. The figures were drafted by Osvaldo Garcia, Cynthia Bilbrey and Craig Louis, at the Electronics Research Laboratory at Berkeley. Simulations were executed using the package SIMNON, and we thank Karl Astrom for providing us with a copy of this software package. We also acknowledge Bernard Goodwin of Prentice Hall for his friendly management of this enterprise and Elaine Lynch for coordinating production matters.

Last, but not least, we would like to express our sincere appreciation to Nirmala and Cecilia for their patience and encouragement. Despite distance, our families have been a source of continuous support, and deserve our deepest gratitude.

Shankar Sastry
Marc Bodson

Berkeley, California

CHAPTER 0
INTRODUCTION

0.1 IDENTIFICATION AND ADAPTIVE CONTROL

Most current techniques for designing control systems are based on a good understanding of the plant under study and its environment. However, in a number of instances, the plant to be controlled is too complex and the basic physical processes in it are not fully understood. Control design techniques then need to be augmented with an identification technique aimed at obtaining a progressively better understanding of the plant to be controlled. It is thus intuitive to aggregate system identification and control. Often, the two steps will be taken separately. If the system identification is recursive—that is the plant model is periodically updated on the basis of previous estimates and new data—identification and control may be performed concurrently. We will see *adaptive control*, pragmatically, as *a direct aggregation of a (non-adaptive) control methodology with some form of recursive system identification.*

Abstractly, system identification could be aimed at determining if the plant to be controlled is linear or nonlinear, finite or infinite dimensional, and has continuous or discrete event dynamics. Here we will restrict our attention to finite dimensional, single-input single-output linear plants, and some classes of multivariable and nonlinear plants. Then, the primary step of system identification (structural identification) has already been taken, and only parameters of a fixed type of model need to be determined. Implicitly, we will thus be limiting ourselves to *parametric system identification*, and *parametric adaptive control.*

Applications of such systems arise in several contexts: advanced flight control systems for aircraft or spacecraft, robot manipulators, process control, power systems, and others.

Adaptive control, then, is a technique of applying some system identification technique to obtain a model of the process and its environment from input-output experiments and using this model to design a controller. The parameters of the controller are adjusted during the operation of the plant as the amount of data available for plant identification increases. For a number of simple PID (proportional + integral + derivative) controllers in process control, this is often done manually. However, when the number of parameters is larger than three or four, and they vary with time, automatic adjustment is needed. The design techniques for adaptive systems are studied and analyzed in theory for *unknown* but *fixed* (that is, time invariant) plants. In practice, they are applied to *slowly time-varying* and *unknown* plants.

Overview of the Literature

Research in adaptive control has a long and vigorous history. In the 1950s, it was motivated by the problem of designing autopilots for aircraft operating at a wide range of speeds and altitudes. While the object of a good fixed-gain controller was to build an autopilot which was insensitive to these (large) parameter variations, it was frequently observed that a single constant gain controller would not suffice. Consequently, gain scheduling based on some auxiliary measurements of airspeed was adopted. With this scheme in place several rudimentary model reference schemes were also attempted—the goal in this scheme was to build a self-adjusting controller which yielded a closed loop transfer function matching a prescribed reference model. Several schemes of self-adjustment of the controller parameters were proposed, such as the sensitivity rules and the so-called M.I.T. rule, and were verified to perform well under certain conditions. Finally, Kalman [1958] put on a firm analytical footing the concept of a general self-tuning controller with explicit identification of the parameters of a linear, single-input, single-output plant and the usage of these parameter estimates to update an optimal linear quadratic controller.

The 1960s marked an important time in the development of control theory and adaptive control in particular. Lyapunov's stability theory was firmly established as a tool for proving convergence in adaptive control schemes. Stochastic control made giant strides with the understanding of dynamic programming, due to Bellman and others. Learning schemes proposed by Tsypkin, Feldbaum and others (see Tsypkin [1971] and [1973]) were shown to have roots in a single unified framework of recursive equations. System identification (off-line) was thoroughly researched and understood. Further, Parks [1966] found a way of redesigning the update laws proposed in the 1950s for model reference schemes so as to be able to prove convergence of his controller.

In the 1970s, owing to the culmination of determined efforts by several teams of researchers, complete proofs of stability for several adaptive schemes appeared. State space (Lyapunov based) proofs of stability for model reference adaptive schemes appeared in the work of Narendra, Lin, & Valavani [1980] and Morse [1980]. In the late 1970s, input output (Popov hyperstability based) proofs appeared in Egardt [1979] and Landau [1979]. Stability proofs in the discrete time deterministic and stochastic case (due to Goodwin, Ramadge, & Caines [1980]) also appeared at this time, and are contained in the textbook by Goodwin & Sin [1984]. Thus, this period was marked by the culmination of the analytical efforts of the past twenty years.

Given the firm, analytical footing of the work to this point, the 1980s have proven to be a time of critical examination and evaluation of the accomplishments to date. It was first pointed out by Rohrs and coworkers [1982] that the assumptions under which stability of adaptive schemes had been proven were very sensitive to the presence of unmodeled dynamics, typically high-frequency parasitic modes that were neglected to limit the complexity of the controller. This sparked a flood of research into the robustness of adaptive algorithms: a re-examination of whether or not adaptive controllers were at least as good as fixed gain controllers, the development of tools for the analysis of the transient behavior of the adaptive algorithms and attempts at implementing the algorithms on practical systems (reactors, robot manipulators, and ship steering systems to mention only a few). The implementation of the complicated nonlinear laws inherent in adaptive control has been greatly facilitated by the boom in microelectronics and today, one can talk in terms of custom adaptive controller chips. All this flood of research and development is bearing fruit and the industrial use of adaptive control is growing.

Adaptive control has a rich and varied literature and it is impossible to do justice to all the manifold publications on the subject. It is a tribute to the vitality of the field that there are a large number of fairly recent books and monographs. Some recent books on recursive estimation, which is an important part of adaptive control are by Eykhoff [1974], Goodwin & Payne [1977], Ljung & Soderstrom [1983] and Ljung [1987]. Recent books dealing with the theory of adaptive control are by Landau [1979], Egardt [1979], Ioannou & Kokotovic [1984], Goodwin & Sin [1984], Anderson, Bitmead, Johnson, Kokotovic, Kosut, Mareels, Praly, & Riedle [1986], Kumar and Varaiya [1986], Polderman [1988] and Caines [1988]. An attempt to link the signal processing viewpoint with the adaptive control viewpoint is made in Johnson [1988]. Surveys of the applications of adaptive control are given in a book by Harris & Billings [1981], and in books edited by Narendra & Monopoli [1980] and Unbehauen [1980]. As of the writing of this book, two other books on adaptive control by Astrom & Wittenmark and Narendra & Annaswamy are also nearing completion.

In spite of the great wealth of literature, we feel that there is a need for a "toolkit" of methods of analysis comparable to non-adaptive linear time invariant systems. Further, many of the existing results concern either algorithms, structures or specific applications, and a great deal more needs to be understood about the dynamic behavior of adaptive systems. This, we believe, has limited practical applications more than it should have. Consequently, our objective in this book is to address fundamental issues of stability, convergence and robustness. Also, we hope to communicate our excitement about the problems and potential of adaptive control. In the remainder of the introduction, we will review some common approaches to adaptive control systems and introduce the basic issues studied in this book with a simple example.

0.2 APPROACHES TO ADAPTIVE CONTROL

0.2.1 Gain Scheduling

One of the earliest and most intuitive approaches to adaptive control is gain scheduling. It was introduced in particular in the context of flight control systems in the 1950s and 1960s. The idea is to find auxiliary process variables (other than the plant outputs used for feedback) that correlate well with the changes in process dynamics. It is then possible to compensate for plant parameter variations by changing the parameters of the regulator as functions of the auxiliary variables. This is illustrated in Figure 0.1.

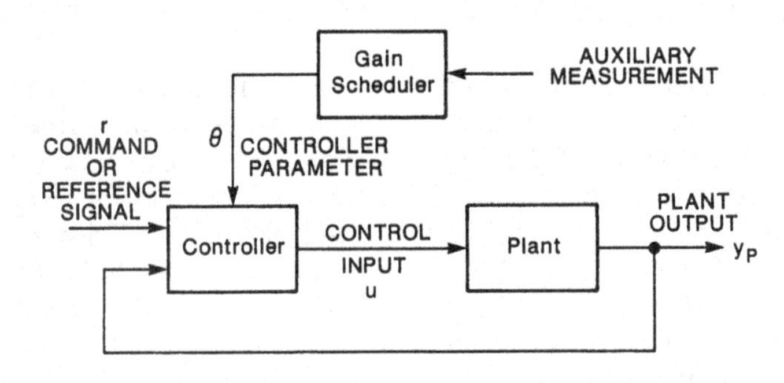

Figure 0.1: Gain Scheduling Controller

The advantage of gain scheduling is that the parameters can be changed quickly (as quickly as the auxiliary measurement) in response to changes in the plant dynamics. It is convenient especially if the plant dynamics depend in a well-known fashion on a relatively few easily measurable variables. In the example of flight control systems, the dynamics depend in relatively simple fashion on the readily available dynamic pressure—that is the product of the air density and the relative velocity of the aircraft squared.

Although gain scheduling is extremely popular in practice, the disadvantage of gain scheduling is that it is an *open-loop* adaptation scheme, with no real "learning" or intelligence. Further, the extent of design required for its implementation can be enormous, as was illustrated by the flight control system implemented on a CH-47 helicopter. The flight envelope of the helicopter was divided into *ninety* flight conditions corresponding to thirty discretized horizontal flight velocities and three vertical velocities. Ninety controllers were designed, corresponding to each flight condition, and a linear interpolation between these controllers (linear in the horizontal and vertical flight velocities) was programmed onto a flight computer. Airspeed sensors modified the control scheme of the helicopter in flight, and the effectiveness of the design was corroborated by simulation.

0.2.2 Model Reference Adaptive Systems

Again in the context of flight control systems, two adaptive control schemes other than gain scheduling were proposed to compensate for changes in aircraft dynamics: a series, high-gain scheme, and a parallel scheme.

Series High-Gain Scheme

Figure 0.2 shows a schematic of the series high-gain scheme.

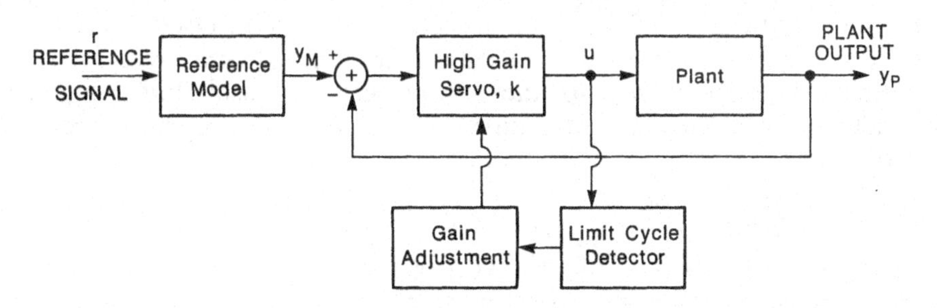

Figure 0.2: Model Reference Adaptive Control–Series, High-Gain Scheme

The reference model represents a pilot's desired command-response characteristic. It is thus desired that the aircraft response, that is, the output y_p, matches the output of the reference model, that is, y_m.

The simple analysis that goes into the scheme is as follows: consider $\hat{P}(s)$ to be the transfer function of the linear, time invariant plant and k the constant gain of the servo. The transfer function from y_m to y_p is $k\hat{P}(s)/1+k\hat{P}(s)$. When the gain k is sufficiently large, the transfer function is approximately 1 over the frequencies of interest, so that $y_m \sim y_p$.

The aim of the scheme is to let the gain k be as high as possible, so that the closed-loop transfer function becomes close to 1, until the onset of instability (a limit cycle) is detected. If the limit cycle oscillations exceed some level, the gain is decreased. Below this level, the gain is increased. The limit cycle detector is typically just a rectifier and low-pass filter.

The series high-gain scheme is intuitive and simple: only one parameter is updated. However, it has the following problems

a) Oscillations are constantly present in the system.

b) Noise in the frequency band of the limit cycle detector causes the gain to decrease well below the critical value.

c) Reference inputs may cause saturation due to the high-gain.

d) Saturation may mask limit cycle oscillations, allowing the gain to increase above the critical value, and leading to instability.

Indeed, tragically, an experimental X-15 aircraft flying this control system crashed in 1966 (cf. Staff of the Flight Research Center [1971]), owing partially to the saturation problems occurring in the high-gain scheme. The roll and pitch axes were controlled by the right and left rear ailerons, using differential and identical commands respectively. The two axes were assumed decoupled for the purpose of control design. However, saturation of the actuators in the pitch axis caused the aircraft to lose controllability in the roll axis (since the ailerons were at maximum deflection). Due to the saturation, the instability remained undetected, and created aerodynamic forces too great for the aircraft to withstand.

Parallel Scheme

As in the series scheme, the desired performance of the closed-loop system is specified through a reference model, and the adaptive system attempts to make the plant output match the reference model output asymptotically. An early reference to this scheme is Osburn, Whitaker, & Kezer [1961]. A block diagram is shown in Figure 0.3. The controller

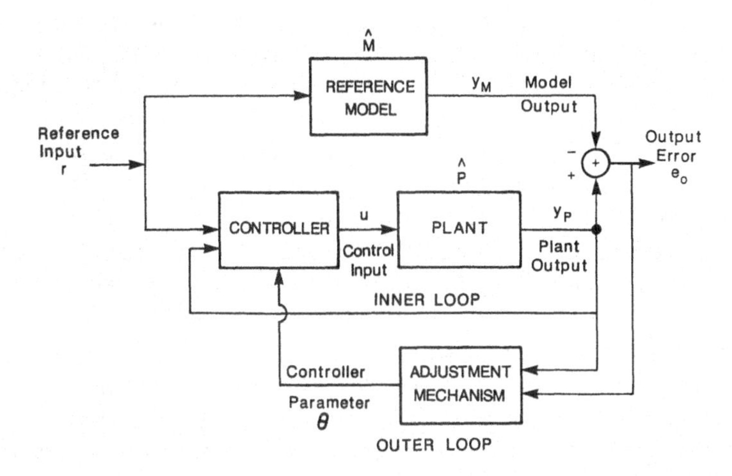

Figure 0.3: Model Reference Adaptive Control—Parallel Scheme

can be thought of as having two loops: an inner or regulator loop that is an ordinary control loop consisting of the plant and regulator, and an outer or adaptation loop that adjusts the parameters of the regulator in such a way as to drive the error between the model output and plant output to zero.

The key problem in the scheme is to obtain an adjustment mechanism that drives the output error $e_0 = y_p - y_m$ to zero. In the earliest applications of this scheme, the following update, called the *gradient update*, was used. Let the vector θ contain the adjustable parameters of the controller. The idea behind the gradient update is to reduce $e_0^2(\theta)$ by adjusting θ along the direction of steepest descent, that is

$$\frac{d\theta}{dt} = -g\frac{\partial}{\partial\theta}(e_0^2(\theta)) \tag{0.2.1}$$

$$= -2g\,e_0(\theta)\frac{\partial}{\partial\theta}(e_0(\theta)) = -2g\,e_0(\theta)\frac{\partial}{\partial\theta}(y_p(\theta)) \tag{0.2.2}$$

where g is a positive constant called the *adaptation gain.*

The interpretation of $e_0(\theta)$ is as follows: it is the output error (also a function of time) obtained by freezing the controller parameter at θ. The gradient of $e_0(\theta)$ with respect to θ is equal to the gradient of y_p with respect to θ, since y_m is independent of θ, and represents the *sensitivity* of the output error to variations in the controller parameter θ.

Several problems were encountered in the usage of the gradient update. The sensitivity function $\partial y_p(\theta) / \partial \theta$ usually depends on the unknown plant parameters, and is consequently unavailable. At this point the so-called *M.I.T. rule*, which replaced the unknown parameters by their estimates at time t, was proposed. Unfortunately, for schemes based on the M.I.T. rule, it is not possible in general to prove closed-loop stability, or convergence of the output error to zero. Empirically, it was observed that the M.I.T. rule performed well when the adaptation gain g and the magnitude of the reference input were small (a conclusion later confirmed analytically by Mareels *et al* [1986]). However, examples of instability could be obtained otherwise (cf. James [1971]).

Parks [1966] found a way of redesigning adaptive systems using Lyapunov theory, so that stable and provably convergent model reference schemes were obtained. The update laws were similar to (0.2.2), with the sensitivity $\partial y_p(\theta) / \partial \theta$ replaced by other functions. The stability and convergence properties of model reference adaptive systems make them particularly attractive and will occupy a lot of our interest in this book.

0.2.3 Self Tuning Regulators

In this technique of adaptive control, one starts from a control design method for known plants. This design method is summarized by a controller structure, and a relationship between plant parameters and controller parameters. Since the plant parameters are in fact unknown, they are obtained using a recursive parameter identification algorithm. The controller parameters are then obtained from the estimates of the plant parameters, in the same way *as if these were the true parameters.* This is usually called a *certainty equivalence principle.*

The resulting scheme is represented on Figure 0.4. An explicit separation between identification and control is assumed, in contrast to the model reference schemes above, where the parameters of the controller are updated directly to achieve the goal of model following. The self tuning approach was originally proposed by Kalman [1958] and clarified by Astrom & Wittenmark [1973]. The controller is called *self tuning*, since it has the ability to tune its own parameters. Again, it can be thought of as having two loops: an inner loop consisting of a conventional controller, but with varying parameters, and an outer loop consisting of an identifier and a design box (representing an on-line solution to a design problem for a system with known parameters) which adjust these controller parameters.

The self tuning regulator is very flexible with respect to its choice of controller design methodology (linear quadratic, minimum variance, gain-phase margin design, ...), and to the choice of identification scheme

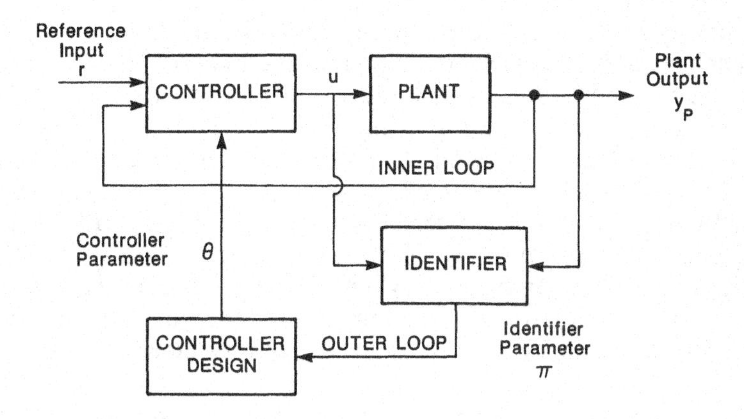

Figure 0.4: Self-tuning Controller

(least squares, maximum likelihood, extended Kalman filtering, ...). The analysis of self tuning adaptive systems is however more complex than the analysis of model reference schemes, due primarily to the (usually nonlinear) transformation from identifier parameters to controller parameters.

Direct and Indirect Adaptive Control

While model reference adaptive controllers and self tuning regulators were introduced as different approaches, the only real difference between them is that model reference schemes are *direct* adaptive control schemes, whereas self tuning regulators are *indirect*. The self tuning regulator first identifies the plant parameters recursively, and then uses these estimates to update the controller parameters through some fixed transformation. The model reference adaptive schemes update the controller parameters directly (no explicit estimate or identification of the plant parameters is made). It is easy to see that the inner or control loop of a self tuning regulator could be the same as the inner loop of a model reference design. Or, in other words, the model reference adaptive schemes can be seen as a special case of the self tuning regulators, with an identity transformation between updated parameters and controller parameters. Through this book, we will distinguish between direct and indirect schemes rather than between model reference and self tuning algorithms.

0.2.4 Stochastic Control Approach

Adaptive controller structures based on model reference or self tuning approaches are based on heuristic arguments. Yet, it would be appealing to obtain such structures from a unified theoretical framework. This can be done (in principle, at least) using stochastic control. The system and its environment are described by a stochastic model, and a criterion is formulated to minimize the expected value of a loss function, which is a scalar function of states and controls. It is usually very difficult to solve stochastic optimal control problems (a notable exception is the linear quadratic gaussian problem). When indeed they can be solved, the optimal controllers have the structure shown in Figure 0.5: an identifier (estimator) followed by a nonlinear feedback regulator.

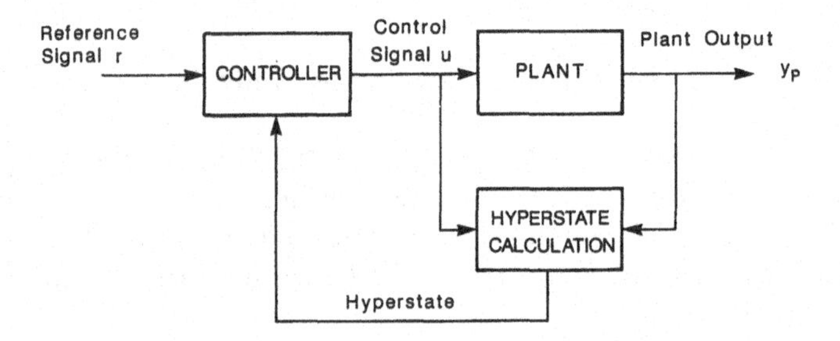

Figure 0.5: "Generic" Stochastic Controller

The estimator generates the conditional probability distribution of the state from the measurements: this distribution is called the *hyperstate* (usually belonging to an infinite dimensional vector space). The self tuner may be thought of as an approximation of this controller, with the hyperstate approximated by the process state and process parameters estimate.

From some limited experience with stochastic control, the following interesting observations can be made of the optimal control law: in addition to driving the plant output to its desired value, the controller introduces *probing* signals which improve the identification and, therefore future control. This, however, represents some cost in terms of control activity. The optimal regulator maintains a balance between the control activity for *learning* about the plant it is controlling and the activity for *controlling* the plant output to its desired value. This property is referred to as *dual control.* While we will not explicitly study stochastic control in this book, the foregoing trade-off will be seen repeatedly: good adaptive control requires correct identification, and for the identification to be complete, the controller signal has to be sufficiently rich to allow for the excitation of the plant dynamics. The presence of this rich enough excitation may result in poor transient performance of the scheme, displaying the trade-off between learning and control performance.

0.3 A SIMPLE EXAMPLE

Adaptive control systems are difficult to analyze because they are nonlinear, time varying systems, even if the plant that they are controlling is linear, time invariant. This leads to interesting and delicate technical problems. In this section, we will introduce some of these problems with a simple example. We also discuss some of the adaptive schemes of the previous section in this context.

We consider a first order, time invariant, linear system with transfer function

$$\hat{P}(s) = \frac{k_p}{s+a} \tag{0.3.1}$$

where $a > 0$ is known. The gain k_p of the plant is unknown, but its sign is known (say $k_p > 0$). The control objective is to get the plant output to match a model output, where the reference model transfer function is

$$\hat{M}(s) = \frac{1}{s+a} \tag{0.3.2}$$

Only gain compensation—or feedforward control—is necessary, namely a gain θ at the plant input, as is shown on Figure 0.6.

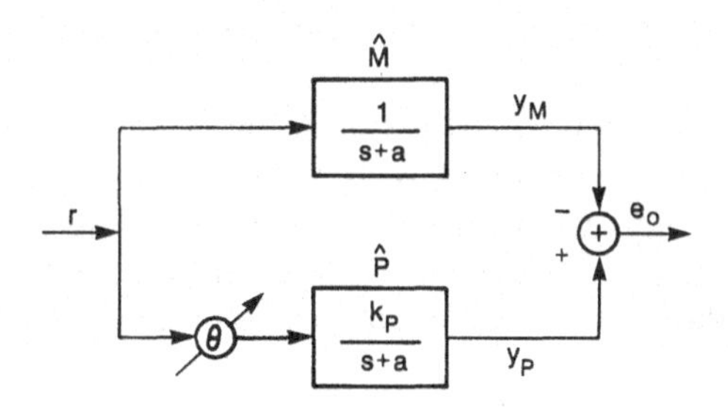

Figure 0.6: Simple Feedforward Controller

Note that, if k_p were known, θ would logically be chosen to be $1/k_p$. We will call

$$\theta^* = \frac{1}{k_p} \tag{0.3.3}$$

the *nominal* value of the parameter θ, that is the value which realizes the output matching objective for all inputs. The design of the various adaptive schemes proceeds as follows.

Gain Scheduling

Let $v(t) \in \mathbb{R}$ be some auxiliary measurement that correlates in known fashion with k_p, say $k_p(t) = f(v(t))$. Then, the gain scheduler chooses at time t

$$\theta(t) = \frac{1}{f(v(t))} \tag{0.3.4}$$

Model Reference Adaptive Control Using the M.I.T. Rule

To apply the M.I.T. rule, we need to obtain $\partial e_0(\theta)/\partial\theta = \partial y_p(\theta)/\partial\theta$, with the understanding that θ is frozen. From Figure 0.6, it is easy to see that

$$\frac{\partial y_p(\theta)}{\partial\theta} = \frac{k_p}{s+a}(\theta r) = k_p y_m \tag{0.3.5}$$

We see immediately that the sensitivity function in (0.3.5) depends on the parameter k_p which is unknown, so that $\partial y_p/\partial\theta$ is not available. However, the sign of k_p is known ($k_p > 0$), so that we may merge the constant k_p with the adaptation gain. The M.I.T rule becomes

$$\dot{\theta} = -g\, e_0\, y_m \qquad g > 0 \tag{0.3.6}$$

Note that (0.3.6) prescribes an update of the parameter θ in the direction opposite to the "correlation" product of e_0 and the model output y_m.

Model Reference Adaptive Control Using the Lyapunov Redesign

The control scheme is exactly as before, but the parameter update law is chosen to make a Lyapunov function decrease along the trajectories of the adaptive system (see Chapter 1 for an introduction to Lyapunov analysis). The plant and reference model are described by

$$\dot{y}_p = -a\, y_p + k_p\theta r \tag{0.3.7}$$

$$\dot{y}_m = -a\, y_m + r = -a\, y_m + k_p\theta^* r \tag{0.3.8}$$

Subtracting (0.3.8) from (0.3.7), we get, with $e_0 = y_p - y_m$

$$\dot{e}_0 = -a\, e_0 + k_p(\theta - \theta^*)r \tag{0.3.9}$$

Since we would like θ to converge to the nominal value $\theta^* = 1/k_p$, we define the *parameter error* as

$$\phi = \theta - \theta^* \tag{0.3.10}$$

Note that since θ^* is fixed (though unknown), $\dot{\phi} = \dot{\theta}$.

The Lyapunov redesign approach consists in finding an update law so that the Lyapunov function

$$v(e_0, \phi) = e_0^2 + k_p \phi^2 \tag{0.3.11}$$

is decreasing along trajectories of the error system

$$\begin{aligned} \dot{e}_0 &= -a\, e_0 + k_p \phi\, r \\ \dot{\phi} &= \text{update law to be defined} \end{aligned} \tag{0.3.12}$$

Note that since $k_p > 0$, the function $v(e_0, \phi)$ is a positive definite function. The derivative of v along the trajectories of the error system (0.3.12) is given by

$$\dot{v}(e_0, \phi)\Big|_{(0.3.12)} = -2a\, e_0^2 + 2k_p\, e_0 \phi\, r + 2k_p \phi \dot{\phi} \tag{0.3.13}$$

Choosing the update law

$$\dot{\theta} = \dot{\phi} = -e_0 r \tag{0.3.14}$$

yields

$$\dot{v}(e_0, \phi) = -2a\, e_0^2 \leq 0 \tag{0.3.15}$$

thereby guaranteeing that $e_0^2 + k_p \phi^2$ is decreasing along the trajectories of (0.3.12), (0.3.14) and that e_0, and ϕ are bounded. Note that (0.3.15) is similar in form to (0.3.6), with the difference that e_0 is correlated with r rather that y_m. An adaptation gain g may also be included in (0.3.14).

Since $v(e_0, \phi)$ is decreasing and bounded below, it would appear that $e_0 \to 0$ as $t \to \infty$. This actually follows from further analysis, provided that r is bounded (cf. Barbalat's lemma 1.2.1).

Having concluded that $e_0 \to 0$ as $t \to \infty$, what can we say about θ? Does it indeed converge to $\theta^* = 1/k_p$? The answer is that one can not conclude anything about the convergence of θ to θ^* without extra conditions on the reference input. Indeed, if the reference input was a constant zero signal, there would be no reason to expect θ to converge to θ^*. Conditions for parameter convergence are important in adaptive control and will be studied in great detail. An answer to this question for the simple example will be given for the following indirect adaptive control scheme.

Indirect Adaptive Control (Self Tuning)

To be able to effectively compare the indirect scheme with the direct schemes given before, we will assume that the control objective is still model matching, with the same model as above. Figure 0.7 shows an indirect or self tuning type of model reference adaptive controller.

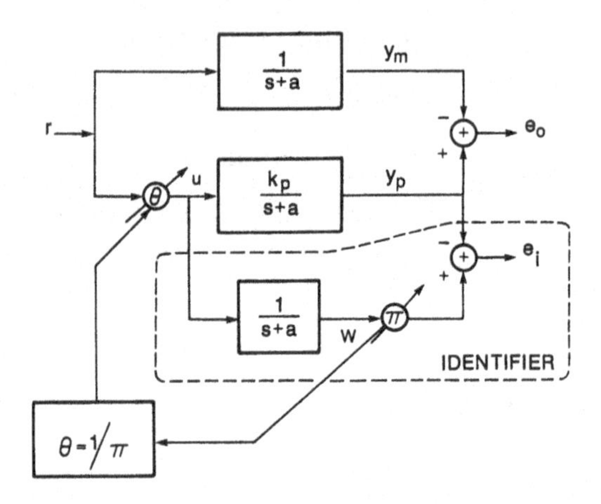

Figure 0.7: A Simple Indirect Controller

The identifier contains an identifier parameter $\pi(t)$ that is an estimate of the unknown plant parameter k_p. Therefore, we define $\pi^* = k_p$. The controller parameter is chosen following the certainty equivalence principle: since $\theta^* = 1 / k_p$ and $\pi^* = k_p$, we let $\theta(t) = 1 / \pi(t)$. The hope is that, as $t \to \infty$, $\pi(t) \to k_p$, so that $\theta(t) \to 1/k_p$.

The update law now is an update law for the identifier parameter $\pi(t)$. There are several possibilities at this point, and we proceed to derive one of them. Define the identifier parameter error

$$\psi(t) := \pi(t) - \pi^* \tag{0.3.16}$$

and let

$$w = \frac{1}{s+a}(\theta r) = \frac{1}{s+a}(u) \tag{0.3.17}$$

The signal w may be obtained by stable filtering of the input u, since $a > 0$ is known. The update law is based on the identifier error

$$e_i := \pi w - y_p \tag{0.3.18}$$

Equation (0.3.18) is used in the actual implementation of the algorithm. For the analysis, note that

$$y_p = \frac{k_p}{s+a}(\theta r) = \pi^* w \tag{0.3.19}$$

so that

$$e_i = \psi w \tag{0.3.20}$$

Consider the update law

$$\dot{\pi} = \dot{\psi} = -g e_i w \qquad g > 0 \tag{0.3.21}$$

and let the Lyapunov function

$$v = \psi^2 \tag{0.3.22}$$

This Lyapunov function has the special form of the norm square of the identifier parameter error. Its derivative along the trajectories of the adaptive system is

$$\dot{v} = -g \psi^2 w^2 \tag{0.3.23}$$

Therefore, the update law causes a decreasing parameter error and all signals remain bounded.

The question of parameter convergence can be answered quite simply in this case. Note that (0.3.20), (0.3.21) represent the first order linear time varying system

$$\dot{\psi} = -g w^2 \psi \tag{0.3.24}$$

which may be explicitly integrated to get

$$\psi(t) = \psi(0) \exp\left(-g \int_0^t w^2(\tau)\, d\tau\right) \tag{0.3.25}$$

It is now easy to see that if

$$\int_0^t w^2(\tau)\, d\tau \to \infty \qquad \text{as } t \to \infty \tag{0.3.26}$$

then $\psi(t) \to 0$, so that $\pi(t) \to \pi^*$ and $\theta(t) \to 1/k_p$, yielding the desired controller. The condition (0.3.26) is referred to as an *identifiability* condition and is much related to the so-called *persistency of excitation* that will be discussed in Chapter 2. It is easily seen that, in particular, it excludes signals which tend to zero as $t \to \infty$.

The difficulty with (0.3.26) is that it depends on w, which in turn depends on u and therefore on both θ and r. Converting it into a condition on the exogenous reference input $r(t)$ only is another of the problems which we will discuss in the following chapters.

The foregoing simple example showed that even when simple feed-forward control of a linear, time invariant, first order plant was involved, the analysis of the resulting closed-loop dynamics could be involved: the equations were *time-varying, linear* equations. Once feed-back control is involved, the equations become nonlinear and time varying.

CHAPTER 1
PRELIMINARIES

This chapter introduces the notation used in this book, as well as some basic definitions and results. The material is provided mostly for reference. It may be skipped in a first reading, or by the reader familiar with the results.

The notation used in the adaptive systems literature varies widely. We elected to use a notation close to that of Narendra & Valavani [1978], and Narendra, Lin, & Valavani [1980], since many connections exist between this work and their results. We will refer to texts such as Desoer & Vidyasagar [1975], and Vidyasagar [1978] for standard results, and this chapter will concentrate on the definitions used most often, and on nonstandard results.

1.1 NOTATION

Lower case letters are used to denote scalars or vectors. Upper case letters are used to denote matrices, operators, or sets. When $u(t)$ is a function of time, $\hat{u}(s)$ denotes its Laplace transform. Without ambiguity, we will drop the arguments, and simply write u and $\hat{u}$. Rational transfer functions of linear time invariant (LTI) systems will be denoted using upper case letters, for example, $\hat{H}(s)$ or $\hat{H}$. Polynomials in s will be denoted using lower case letters, for example, $\hat{n}(s)$ or simply $\hat{n}$. Thus, we may have $\hat{H} = \hat{n}/\hat{d}$, where $\hat{H}$ is both the ratios of polynomials in s and an operator in the Laplace transform domain. Sometimes, the time domain and the Laplace transform domain will be mixed, and parentheses will determine the sense to be made of an expression. For example, $\hat{H}(u)$ or $\hat{H}\,\hat{u}$ is the output of the LTI system $\hat{H}$ with input u. $\hat{H}(u)v$ is $\hat{H}(u)$ multiplied by v in the time domain, while $\hat{H}(uv)$ is $\hat{H}$ operating on the product $u(t)v(t)$.

1.2 L_p SPACES, NORMS

We denote by $|x|$ the absolute value of x if x is a scalar and the euclidean norm of x if x is a vector. The notation $\| \ \|$ will be used to denote the induced norm of an operator, in particular the induced matrix norm

$$\|A\| = \sup_{|x| = 1} |Ax| \tag{1.2.1}$$

and for functions of time, the notation is used for the L_p norm

$$\|u\|_p = \left(\int_0^\infty |u(\tau)|^p d\tau\right)^{1/p} \tag{1.2.2}$$

for $p \in [1, \infty)$, while

$$\|u\|_\infty = \sup_{t \geq 0} |u(t)| \tag{1.2.3}$$

and we say that $u \in L_p$ when $\|u\|_p$ exists. When p is omitted, $\|u\|$ denotes the L_2 norm. Truncated functions are defined as

$$\begin{aligned} f_s(t) &= f(t) & t \leq s \\ &= 0 & t > s \end{aligned} \tag{1.2.4}$$

and the extended L_p spaces are defined by

$$L_{pe} = \{ f \mid \text{ for all } s < \infty, f_s \in L_p \} \tag{1.2.5}$$

For example, e^t does not belong to L_∞, but $e^t \in L_{\infty e}$. When $u \in L_{\infty e}$, we have

$$\|u_t\|_\infty = \sup_{\tau \leq t} |u(\tau)| \tag{1.2.6}$$

A function f may belong to L_1 and not be bounded. Conversely, a bounded function need not belong to L_1. However, if $f \in L_1 \cap L_\infty$, then $f \in L_p$ for all $p \in [1, \infty]$ (cf. Desoer & Vidyasagar [1975], p. 17).

Also, $f \in L_p$ does not imply that $f \to 0$ as $t \to \infty$. This is not even guaranteed if f is bounded. However, note the following results.

Lemma 1.2.1 Barbalat's Lemma

If $f(t)$ is a uniformly continuous function, such that $\lim_{t \to \infty} \int_0^t f(\tau)\, d\tau$ exists and is finite,

Then $f(t) \to 0$ as $t \to \infty$.

Proof of Lemma 1.2.1 cf. Popov [1973] p. 211.

Corollary 1.2.2

If $g, \dot{g} \in L_\infty$, and $g \in L_p$, for some $p \in [1, \infty)$,

Then $g(t) \to 0$ as $t \to \infty$.

Proof of Corollary 1.2.2

Direct from lemma 1.2.1, with $f = |g|^p$, since $g, \dot{g}$ bounded implies that f is uniformly continuous. □

1.3 POSITIVE DEFINITE MATRICES

Positive definite matrices are frequently found in work on adaptive systems. We summarize here several facts that will be useful. We consider real matrices. Recall that a scalar u, or a function of time $u(t)$, is said to be *positive* if $u \geq 0$, or $u(t) \geq 0$ for all t. It is *strictly positive* if $u > 0$, or, for some $\alpha > 0$, $u(t) \geq \alpha$ for all t. A square matrix $A \in R^{n \times n}$ is *positive semidefinite* if $x^T A x \geq 0$ for all x. It is *positive definite* if, for some $\alpha > 0$, $x^T A x \geq \alpha x^T x = \alpha |x|^2$ for all x. Equivalently, we can require $x^T A x \geq \alpha$ for all x such that $|x| = 1$. The matrix A is negative semidefinite if $-A$ is positive semidefinite and for symmetric matrices, we write $A \geq B$ if $A - B \geq 0$. Note that a matrix can be neither positive semidefinite nor negative semidefinite, so that this only establishes a partial order on symmetric matrices.

The eigenvalues of a positive semidefinite matrix lie in the closed right-half plane (RHP), while those of a positive definite matrix lie in the open RHP. If $A \geq 0$ and $A = A^T$, then A is *symmetric positive semidefinite.* In particular, if $A \geq 0$, then $A + A^T$ is symmetric positive semidefinite. The eigenvalues of a symmetric matrix are all real. Such a matrix also has n orthogonal eigenvectors, so that we can decompose A as

$$A = U^T \Lambda U \tag{1.3.1}$$

where U is the matrix of eigenvectors satisfying $U^T U = I$ (that is, U is a *unitary* matrix), and Λ is a diagonal matrix composed of the eigenvalues of A. When $A \geq 0$, the square root matrix $\Lambda^{1/2}$ is a diagonal matrix composed of the square roots of the eigenvalues of A, and

$$A^{1/2} = U^T \Lambda^{1/2} U \tag{1.3.2}$$

is the square root matrix of A, with $A = A^{1/2} \cdot A^{1/2}$ and $(A^{1/2})^T = A^{1/2}$.

If $A \geq 0$ and $B \geq 0$, then $A + B \geq 0$ but it is not true in general that $A \cdot B \geq 0$. However, if A, B are symmetric, positive semidefinite matrices, then AB—although not necessarily symmetric, or positive semidefinite—has all eigenvalues real positive.

Another property of symmetric, positive semidefinite matrices, following from (1.3.1), is

$$\lambda_{\min}(A) |x|^2 \leq x^T A x \leq \lambda_{\max}(A) |x|^2 \tag{1.3.3}$$

This simply follows from the fact that $x^T A x = x^T U^T \Lambda U x = z^T \Lambda z$ and $|z|^2 = z^T z = |x|^2$. We also have that

$$\| A \| = \lambda_{\max}(A) \tag{1.3.4}$$

and, when A is positive definite

$$\| A^{-1} \| = 1/\lambda_{\min}(A) \tag{1.3.5}$$

1.4 STABILITY OF DYNAMIC SYSTEMS

1.4.1 Differential Equations

This section is concerned with differential equations of the form

$$\dot{x} = f(t, x) \qquad x(t_0) = x_0 \tag{1.4.1}$$

where $x \in \mathbb{R}^n, t \geq 0$.

The system defined by (1.4.1) is said to be *autonomous*, or *time-invariant*, if f does not depend on t, and *non autonomous*, or *time-varying*, otherwise. It is said to be *linear* if $f(t,x) = A(t)x$ for some $A(.) : \mathbb{R}_+ \to \mathbb{R}^{n \times n}$ and *nonlinear* otherwise.

We will always assume that $f(t,x)$ is *piecewise continuous* with respect to t. By this, we mean that there are only a finite number of discontinuity points in any compact set.

We define by B_h the closed ball of radius h centered at 0 in $\mathbb{R}^n$. Properties will be said to be true:

- *locally*, if true for all x_0 in some ball B_h.

- *globally*, if true for all $x_0 \in \mathbb{R}^n$.
- *in any closed ball*, if true for all $x_0 \in B_h$, with h arbitrary.
- *uniformly*, if true for all $t_0 \geq 0$.

By default, properties will be true locally.

Lipschitz Condition and Consequences

The function f is said to be *Lipschitz* in x if, for some $h > 0$, there exists $l \geq 0$ such that

$$| f(t,x_1) - f(t,x_2) | \leq l | x_1 - x_2 | \tag{1.4.2}$$

for all $x_1, x_2 \in B_h$, $t \geq 0$. The constant l is called the *Lipschitz constant.* This defines locally Lipschitz functions. Globally Lipschitz functions satisfy (1.4.2) for all $x_1, x_2 \in \mathbb{R}^n$, while functions that are Lipschitz in any closed ball satisfy (1.4.2) for all $x_1, x_2 \in B_h$, with l possibly depending on h. The Lipschitz property is by default assumed to be satisfied uniformly, that is, l does not depend on t.

If f is Lipschitz in x, then it is continuous in x. On the other hand, if f has continuous and bounded partial derivatives in x, then it is Lipschitz. More formally, we denote

$$D_2 f := \left[\frac{\partial f_i}{\partial x_j} \right] \tag{1.4.3}$$

so that if $\| D_2 f \| \leq l$, then f is Lipschitz with constant l.

From the theory of ordinary differential equations (cf. Coddington & Levinson [1955]), it is known that f locally bounded, and f locally Lipschitz in x imply the existence and uniqueness of the solutions of (1.4.1) on some time interval (for as long as $x \in B_h$).

Definition Equilibrium Point

x is called an *equilibrium point* of (1.4.1), if $f(t,x) = 0$ for all $t \geq 0$.

By translating the origin to an equilibrium point x_0, we can make the origin 0 an equilibrium point. This is of great notational help, and we will assume henceforth that 0 is an equilibrium point of (1.4.1).

Proposition 1.4.1

If $x = 0$ is an equilibrium point of (1.4.1), f is Lipschitz in x with constant l and is piecewise continuous with respect to t

Then the solution $x(t)$ of (1.4.1) satisfies

$$|x_0|\, e^{l(t-t_0)} \ \geq\ |x(t)| \ \geq\ |x_0|\, e^{-l(t-t_0)} \tag{1.4.4}$$

as long as $x(t)$ remains in B_h.

Proof of Proposition 1.4.1

Note that $|x|^2 = x^T x$ implies that

$$\left|\frac{d}{dt}|x|^2\right| = 2|x| \left|\frac{d}{dt}|x|\right|$$

$$= 2\left|x^T \frac{d}{dt} x\right| \leq 2|x| \left|\frac{d}{dt} x\right| \tag{1.4.5}$$

so that

$$\left|\frac{d}{dt}|x|\right| \leq \left|\frac{d}{dt} x\right| \tag{1.4.6}$$

Since f is Lipschitz

$$-l|x| \leq \frac{d}{dt}|x| \leq l|x| \tag{1.4.7}$$

and there exists a positive function $s(t)$ such that

$$\frac{d}{dt}|x| = -l|x| + s \tag{1.4.8}$$

Solving (1.4.8)

$$|x(t)| = |x_0|\, e^{-l(t-t_0)} + \int_0^t e^{-l(t-\tau)} s(\tau)\, d\tau$$

$$\geq |x_0|\, e^{-l(t-t_0)} \tag{1.4.9}$$

The other inequality follows similarly from (1.3.7). □

Proposition 1.4.1 implies that solutions starting inside B_h will remain inside B_h for at least a finite time interval. Or, conversely, given a time interval, the solutions will remain in B_h provided that the initial conditions are sufficiently small. Also, f globally Lipschitz implies that $x \in L_{\infty e}$. Proposition 1.4.1 also says that x cannot tend to zero faster than exponentially.

The following lemma is an important result generalizing the well-known Bellman-Gronwall lemma (Bellman [1943]). The proof is similar to the proof of proposition 1.4.1, and is left to the appendix.

Lemma 1.4.2 Bellman-Gronwall Lemma

Let $x(.), a(.), u(.) : \mathbb{R}_+ \to \mathbb{R}_+$. Let $T \geq 0$.

If

$$x(t) \leq \int_0^t a(\tau)x(\tau)\,d\tau + u(t) \tag{1.4.10}$$

$$\text{for all } t \in [0, T]$$

Then

$$x(t) \leq \int_0^t a(\tau)\,u(\tau)\,e^{\int_\tau^t a(\sigma)\,d\sigma}\,d\tau + u(t) \tag{1.4.11}$$

$$\text{for all } t \in [0, T]$$

When $u(.)$ is differentiable

$$x(t) \leq u(0)\,e^{\int_0^t a(\sigma)\,d\sigma} + \int_0^t \dot{u}(\tau)\,e^{\int_\tau^t a(\sigma)\,d\sigma}\,d\tau \tag{1.4.12}$$

$$\text{for all } t \in [0, T]$$

Proof of Lemma 1.4.2 in Appendix.

1.4.2 Stability Definitions

Informally, $x = 0$ is a *stable* equilibrium point, if the trajectory $x(t)$ remains close to 0 if the initial condition x_0 is close to 0. More precisely, we say

Definition Stability in the Sense of Lyapunov

$x = 0$ is called a *stable* equilibrium point of (1.4.1), if, for all $t_0 \geq 0$ and $\epsilon > 0$, there exists $\delta(t_0, \epsilon)$ such that

$$|x_0| < \delta(t_0, \epsilon) \implies |x(t)| < \epsilon \qquad \text{for all } t \geq t_0$$

where $x(t)$ is the solution of (1.4.1) starting from x_0 at t_0.

Definition Uniform Stability

$x = 0$ is called a *uniformly stable* equilibrium point of (1.4.1) if, in the preceding definition, δ can be chosen independent of t_0.

Intuitively, this definition captures the notion that the equilibrium point is not getting progressively less stable with time. Stability is a very mild requirement for an equilibrium point. In particular, it does not require that trajectories starting close to the origin tend to the origin asymptotically. That property is made precise in the following definition.

Definition Asymptotic Stability

$x = 0$ is called an *asymptotically stable* equilibrium point of (1.4.1), if

(a) $x = 0$ is a stable equilibrium point of (1.4.1),

(b) $x = 0$ is *attractive*, that is, for all $t_0 \geq 0$, there exists $\delta(t_0)$, such that

$$|x_0| < \delta \Rightarrow \lim_{t \to \infty} |x(t)| = 0$$

Definition Uniform Asymptotic Stability (u.a.s.)

$x = 0$ is called a *uniformly asymptotically stable (u.a.s.)* equilibrium point of (1.4.1), if

(a) $x = 0$ is a uniformly stable equilibrium point of (1.4.1),

(b) the trajectory $x(t)$ converges to 0 uniformly in t_0. More precisely, there exists $\delta > 0$ and a function $\gamma(\tau, x_0)$: $\mathbb{R}_+ \times \mathbb{R}^n \to \mathbb{R}_+$, such that $\lim_{\tau \to \infty} \gamma(\tau, x_0) = 0$ for all x_0 and

$$|x_0| < \delta \Rightarrow |x(t)| \leq \gamma(t - t_0, x_0) \qquad \text{for all } t \geq 0$$

The previous definitions are *local*, since they concern neighborhoods of the equilibrium point. *Global* asymptotic stability is defined as follows.

Definition Global Asymptotic Stability

$x = 0$ is called a *globally asymptotically stable* equilibrium point of (1.4.1), if it is asymptotically stable and $\lim_{t \to \infty} |x(t)| = 0$, for all $x_0 \in \mathbb{R}^n$.

Global u.a.s. is defined likewise. Note that the speed of convergence is not quantified in the definitions of asymptotic stability. In the following definition, the convergence to zero is required to be at least exponential.

Definition Exponential Stability, Rate of Convergence

$x = 0$ is called an *exponentially stable* equilibrium point of (1.4.1) if there exist $m, \alpha > 0$ such that the solution $x(t)$ satisfies

$$|x(t)| \leq m e^{-\alpha(t - t_0)} |x_0| \qquad (1.4.13)$$

for all $x_0 \in B_h$, $t \geq t_0 \geq 0$. The constant α is called the *rate of convergence.*

Global exponential stability means that (1.4.13) is satisfied for any $x_0 \in \mathbb{R}^n$. Exponential stability in any closed ball is similar except that m and α may be functions of h. Exponential stability is assumed to be uniform with respect to t_0. It will be shown that uniform asymptotic stability is equivalent to exponential stability for linear systems (see Section 1.5.2), but it is not true in general.

1.4.3 Lyapunov Stability Theory

We now review some of the key concepts and results of Lyapunov stability theory for ordinary differential equations of the form (1.4.1). A more complete development is available, for instance, in the texts by Hahn [1967] and Vidyasagar [1978].

The so-called Lyapunov second method enables one to determine the nature of stability of an equilibrium point of (1.4.1) without explicitly integrating the differential equation. The method is basically a generalization of the idea that if some "measure of the energy" associated with a system is decreasing, then the system will tend to its equilibrium. To make this notion precise, we need to define exactly what we mean by a "measure of energy," that is, energy functions. For this, we first define class K functions (Hahn [1967], p. 7).

Definition Class K Functions

A function $\alpha(\epsilon) : \mathbb{R}_+ \to \mathbb{R}_+$ belongs to class K (denoted $\alpha(.) \in K$), if it is continuous, strictly increasing, and $\alpha(0) = 0$.

Definition Locally Positive Definite Functions

A continuous function $v(t,x) : \mathbb{R}_+ \times \mathbb{R}^n \to \mathbb{R}_+$ is called a *locally positive definite function (l.p.d.f.)* if, for some $h > 0$, and some $\alpha(.) \in K$

$$v(t,0) = 0 \quad \text{and} \quad v(t,x) \geq \alpha(|x|) \qquad \text{for all } x \in B_h, t \geq 0$$

An l.p.d.f. is locally like an "energy function." Functions which are globally like "energy functions" are called positive definite functions (p.d.f.) and are defined as follows.

Definition Positive Definite Functions

A continuous function $v(t,x) : \mathbb{R}_+ \times \mathbb{R}^n \to \mathbb{R}_+$ is called a *positive definite function (p.d.f.)*, if for some $\alpha(.) \in K$

$$v(t,0) = 0 \quad \text{and} \quad v(t,x) \geq \alpha(|x|) \qquad \text{for all } x \in \mathbb{R}^n, t \geq 0$$

and the function $\alpha(p) \to \infty$ as $p \to \infty$.

In the definitions of l.p.d.f. and p.d.f. functions, the energy like functions are not bounded from above as t varies. This follows in the next definition.

Definition Decrescent Function

The function $v(t,x)$ is called *decrescent*, if there exists a function $\beta(.) \in K$, such that

$$v(t,x) \le \beta(\|x\|) \qquad \text{for all } x \in B_h, t \ge 0$$

Examples

Following are several examples of functions, and their membership in the various classes:

$v(t,x) = |x|^2$: p.d.f., decrescent

$v(t,x) = x^T P x$, with $P > 0$: p.d.f., decrescent

$v(t,x) = (t+1)|x|^2$: p.d.f.

$v(t,x) = e^{-t}|x|^2$: decrescent

$v(t,x) = \sin^2(|x|^2)$: l.p.d.f., decrescent

Lyapunov Stability Theorems

Generally speaking, the theorems state that when $v(t,x)$ is a p.d.f., or an l.p.d.f., and $dv/dt(t,x) \le 0$, then we can conclude the stability of the equilibrium point. The derivative of v is taken along the trajectories of (1.4.1); that is,

$$\left.\frac{dv(t,x)}{dt}\right|_{(1.4.1)} = \frac{\partial v(t,x)}{\partial t} + \frac{\partial v(t,x)}{\partial x} f(t,x) \qquad (1.4.14)$$

Theorem 1.4.3 Basic Theorems of Lyapunov

Let $v(t,x)$ be continuously differentiable.

Then

Conditions on $v(t, x)$	Conditions on $-\dot{v}(t, x)$	Conclusions
l.p.d.f.	≥ 0 locally	stable
l.p.d.f., decrescent	≥ 0 locally	uniformly stable
l.p.d.f.	l.p.d.f.	asymptotically stable
l.p.d.f., decrescent	l.p.d.f.	uniformly asymptotically stable
p.d.f., decrescent	p.d.f.	globally u.a.s

Proof of Theorem 1.4.3 cf. Vidyasagar [1978], p. 148 and after.

Example

We refer to Vidyasagar [1978] for examples of application of the theorems. An interesting example is the second order system

$$\dot{x}_1 = x_1(x_1^2 + x_2^2 - 1) - x_2$$

$$\dot{x}_2 = x_1 + x_2(x_1^2 + x_2^2 - 1) \tag{1.4.15}$$

with the p.d.f.

$$v(x_1, x_2) = x_1^2 + x_2^2 \tag{1.4.16}$$

The derivative of v is given by

$$\dot{v}(x_1, x_2) = 2(x_1^2 + x_2^2)(x_1^2 + x_2^2 - 1) \tag{1.4.17}$$

so that $-\dot{v}$ is an l.p.d.f. (in B_h, where $h < 1$), establishing the local asymptotic stability of the origin. The origin is, in fact, not globally asymptotically stable: the system can be shown to have a circular limit cycle of radius 1 (by changing to polar coordinates).

From (1.4.17), we can also show that for all $x \in B_h$

$$\dot{v} \leq -2(1 - h^2)v \tag{1.4.18}$$

so that

$$v(t) \leq v(0)e^{-2(1-h^2)t}$$

$$|x(t)| \leq |x(0)|\, e^{-(1-h^2)t} \quad \text{for all } t \geq 0 \tag{1.4.19}$$

and, in fact, $x = 0$ is a locally exponentially stable equilibrium point. Note that the Lyapunov function is in fact the squared norm of the state, a situation that will often occur in the sequel.

Comments

The theorems of Lyapunov give sufficient conditions guaranteeing the stability of the system (1.4.1). It is a remarkable fact that the *converse* of theorem 1.4.3 is also true: for example, if an equilibrium point is stable, there exists an l.p.d.f. $v(t,x)$ with $\dot{v}(t,x) \le 0$. The usefulness of theorem 1.4.3 and its converse is limited by the fact that there is no general (and computationally non-intensive) prescription for generating the Lyapunov functions. A significant exception to this concerns exponentially stable systems, which are the topic of the following section.

1.5 EXPONENTIAL STABILITY THEOREMS

We will pay special attention to exponential stability for two reasons. When considering the convergence of adaptive algorithms, exponential stability means convergence, and the rate of convergence is a useful measure of how fast estimates converge to their nominal values. In Chapter 5, we will also observe that exponentially stable systems possess at least some tolerance to perturbations, and are therefore desirable in engineering applications.

1.5.1 Exponential Stability of Nonlinear Systems

The following theorem will be useful in proving several results and relates exponential stability to the existence of a specific Lyapunov function.

Theorem 1.5.1 Converse Theorem of Lyapunov

Assume that $f(t,x) : \mathbb{R}_+ \times \mathbb{R}^n \to \mathbb{R}^n$ has continuous and bounded first partial derivatives in x and is piecewise continuous in t for all $x \in B_h$, $t \ge 0$. Then, the following statements are equivalent:

(a) $x = 0$ is an *exponentially stable* equilibrium point of

$$\dot{x} = f(t,x) \quad x(t_0) = x_0 \tag{1.5.1}$$

(b) There exists a function $v(t,x)$, and some strictly positive constants h', $\alpha_1, \alpha_2, \alpha_3, \alpha_4$, such that, for all $x \in B_{h'}$, $t \ge 0$

$$\alpha_1 |x|^2 \le v(t,x) \le \alpha_2 |x|^2 \tag{1.5.2}$$

$$\left. \frac{dv(t,x)}{dt} \right|_{(1.5.1)} \le -\alpha_3 |x|^2 \tag{1.5.3}$$

$$\left| \frac{\partial v(t,x)}{\partial x} \right| \le \alpha_4 |x| \tag{1.5.4}$$

Comments

Again, the derivative in (1.5.3) is a derivative taken along the trajectories of (1.5.1), that is

$$\left.\frac{dv(t,x)}{dt}\right|_{(1.5.1)} = \frac{\partial v(t,x)}{\partial t} + \frac{\partial v(t,x)}{\partial x} f(t,x) \tag{1.5.5}$$

This means that we consider x to be a function of t to calculate the derivative along the trajectories of (1.5.1) passing through x at t. It does *not* require of x to be the solution $x(t)$ of (1.5.1) starting at $x(t_0)$.

Theorem 1.5.1 can be found in Krasovskii [1963] p. 60, and Hahn [1967] p. 273. It is known as one of the converse theorems. The proof of the theorem is constructive: it provides an explicit Lyapunov function $v(t,x)$. This is a rather unusual circumstance, and makes the theorem particularly valuable. In the proof, we derive explicit values of the constants involved in (1.5.2)–(1.5.4).

Proof of Theorem 1.5.1

(a) implies (b).

(i) Denote by $p(\tau, x, t)$ the solution at time τ of (1.5.1) starting at $x(t)$, t, and define

$$v(t, x) = \int_t^{t+T} |p(\tau,x,t)|^2 d\tau \tag{1.5.6}$$

where $T > 0$ will be defined in (ii). From the exponential stability and the Lipschitz condition

$$m\,|x|\, e^{-\alpha(\tau - t)} \geq |p(\tau,x,t)| \geq |x|\, e^{-l(\tau - t)} \tag{1.5.7}$$

and inequality (1.5.2) follows with

$$\alpha_1 := (1 - e^{-2lT})/2l \qquad \alpha_2 := m^2(1 - e^{-2\alpha T})/2\alpha \tag{1.5.8}$$

(ii) Differentiating (1.5.6) with respect to t, we obtain

$$\begin{aligned}\frac{dv(t,x)}{dt} &= |p(t+T,x,t)|^2 - |p(t,x,t)|^2 \\ &\quad + \int_t^{t+T} \frac{d}{dt}(|p(\tau,x,t)|^2)\,d\tau \end{aligned}\tag{1.5.9}$$

Note that d/dt is a derivative with respect to the *initial* time t and is taken along the trajectories of (1.5.1) By definition of the solution p,

$$p(\tau,x(t+\Delta t), t+\Delta t) = p(\tau,x(t),t) \tag{1.5.10}$$

for all Δt, so that the term in the integral is identically zero over $[t, t+T]$. The second term in the right-hand side of (1.5.9) is simply $|x|^2$, while the first is related to $|x|^2$ by the assumption of exponential stability. It follows that

$$\frac{dv(t,x)}{dt} \leq -(1-m^2 e^{-2\alpha T})|x|^2 \tag{1.5.11}$$

Inequality (1.5.3) follows, provided that $T > (1/\alpha)\ln m$ and

$$\alpha_3 := 1 - m^2 e^{-2\alpha T} \tag{1.5.12}$$

(iii) Differentiating (1.5.6) with respect to x_i, we have

$$\frac{\partial v(t,x)}{\partial x_i} = 2\int_t^{t+T} \sum_{j=1}^{n} p_j(\tau,x,t)\frac{\partial p_j(\tau,x,t)}{\partial x_i}\, d\tau \tag{1.5.13}$$

Under the assumptions, the partial derivative of the solution with respect to the initial conditions satisfies

$$\begin{aligned}
\frac{d}{d\tau}\left[\frac{\partial p_j(\tau,x,t)}{\partial x_i}\right] &= \frac{\partial}{\partial x_i}\left[\frac{d}{d\tau}\, p_j(\tau,x,t)\right] \\
&= \frac{\partial}{\partial x_i}\left[f_j(\tau, p(\tau,x,t))\right] \\
&= \sum_{k=1}^{n} \frac{\partial f_j}{\partial x_k}\bigg|_{\tau,p(\tau,x,t)} \cdot \frac{\partial p_k(\tau,x,t)}{\partial x_i}
\end{aligned} \tag{1.5.14}$$

(except possibly at points of discontinuity of $f(\tau,x)$). Denote

$$\begin{aligned}
Q_{ij}(\tau,x,t) &:= \partial p_i(\tau,x,t)/\partial x_j \\
A_{ij}(x,t) &:= \partial f_i(t,x)/\partial x_j
\end{aligned} \tag{1.5.15}$$

so that (1.5.14) becomes

$$\frac{d}{d\tau} Q(\tau,x,t) = A(p(\tau,x,t),\tau) \cdot Q(\tau,x,t) \tag{1.5.16}$$

Equation (1.5.16) defines $Q(\tau,x,t)$, when integrated from $\tau = t$ to $\tau = t+T$, with initial conditions $Q(t,x,t) = I$. Thus, $Q(\tau,x,t)$ is the *state transition matrix* (cf. Section 1.5.2) associated with the time varying matrix $A(p(\tau,x,t),\tau)$. By assumption, $\| A(.\,,.)\| \leq k$ for some k, so that

$$\| Q(\tau,x,t)\| \leq e^{k(\tau - t)} \tag{1.5.17}$$

and, using the exponential stability again, (1.5.14) becomes

$$\left| \frac{\partial v(t,x)}{\partial x} \right| \leq 2 \int_t^{t+T} m\,|x|\; e^{(k-\alpha)(\tau-t)} d\tau \tag{1.5.18}$$

which is (1.5.4) if we define

$$\alpha_4 := 2m\,(e^{(k-\alpha)T} - 1)\,/\,(k-\alpha) \tag{1.5.19}$$

Note that the function $v(t,x)$ is only defined for $x \in B_{h'}$ with $h' = h/m$, if we wish to guarantee that $p(\tau, x, t) \in B_h$ for all $\tau \geq t$.

(b) implies (a)

This direction is straightforward, using only (1.5.2)–(1.5.3), and we find

$$m := \left[\frac{\alpha_2}{\alpha_1} \right]^{\frac{1}{2}} \qquad \alpha := \frac{1}{2} \frac{\alpha_3}{\alpha_2} \tag{1.5.20}$$

□

Comments

The Lyapunov function $v(t, x)$ can be interpreted as an average of the squared norm of the state along the solutions of (1.5.1). This approach is actually the basis of exact proofs of exponential convergence presented in Sections 2.5 and 2.6 for identification algorithms. On the other hand, the approximate proofs presented in Chapter 4 rely on methods for averaging *the differential system* itself. Then the norm squared of the state itself becomes a Lyapunov function, from which the exponential convergence can be deduced.

Theorem 1.5.1 is mostly useful to establish the existence of the Lyapunov function corresponding to exponentially stable systems. To establish exponential stability from a Lyapunov function, the following theorem will be more appropriate. Again, the derivative is to be taken along the trajectories of (1.5.1).

Theorem 1.5.2 Exponential Stability Theorem

If there exists a function $v(t,x)$, and strictly positive constants α_1, α_2, α_3, and δ, such that for all $x \in B_h, t \geq 0$

$$\alpha_1 |x|^2 \leq v(t,x) \leq \alpha_2 |x|^2 \tag{1.5.21}$$

$$\frac{d}{dt} v(t, x(t)) \Big|_{(1.5.1)} \leq 0 \tag{1.5.22}$$

$$\int_t^{t+\delta} \frac{d}{d\tau} v(\tau, x(\tau)) \Big|_{(1.5.1)} d\tau \le -\alpha_3 |x(t)|^2 \tag{1.5.23}$$

Then $x(t)$ converges exponentially to 0.

Proof of Theorem 1.5.2
From (1.5.23)

$$v(t, x(t)) - v(t+\delta, x(t+\delta)) \ge (\alpha_3/\alpha_2)\, v(t, x(t)) \tag{1.5.24}$$

for all $t \ge 0$, so that

$$v(t+\delta, x(t+\delta)) \le (1-\alpha_3/\alpha_2)\, v(t, x(t)) \tag{1.5.25}$$

for all $t \ge 0$. From (1.5.22)

$$v(t_1, x(t_1)) \le v(t, x(t)) \qquad \text{for all } t_1 \in [t, t+\delta] \tag{1.5.26}$$

Choose for t the sequence t_0, $t_0+\delta$, $t_0+2\delta, \ldots$ so that $v(t, x(t))$ is bounded by a staircase $v(t_0, x(t_0))$, $v(t_0+\delta, x(t_0+\delta)), \ldots$ where the steps are related in geometric progression through (1.5.24). It follows that

$$v(t, x(t)) \le m_v e^{-\alpha_v (t-t_0)} v(t_0, x(t_0)) \tag{1.5.27}$$

for all $t \ge t_0 \ge 0$, where

$$m_v = \frac{1}{(1-\alpha_3/\alpha_2)} \qquad \alpha_v = \frac{1}{\delta} \ln \left[\frac{1}{(1-\alpha_3/\alpha_2)} \right] \tag{1.5.28}$$

Similarly,

$$|x(t)| \le m\, e^{-\alpha(t-t_0)} |x(t_0)| \tag{1.5.29}$$

where

$$m = \left[\frac{\alpha_2}{\alpha_1} \frac{1}{1-\alpha_3/\alpha_2} \right]^{\frac{1}{2}} \qquad \alpha = \frac{1}{2\delta} \ln \left[\frac{1}{1-\alpha_3/\alpha_2} \right] \tag{1.5.30}$$

□

1.5.2 Exponential Stability of Linear Time-Varying Systems

We now restrict our attention to linear time-varying (LTV) systems of the form

$$\dot{x} = A(t)x \qquad x(t_0) = x_0 \tag{1.5.31}$$

where $A(t) \in \mathbb{R}^{n \times n}$ is a piecewise continuous function belonging to $L_{\infty e}$.

Definition State-Transition Matrix

The *state-transition matrix* $\Phi(t, t_0) \in \mathbb{R}^{n \times n}$ associated with $A(t)$ is, by definition, the unique solution of the matrix differential equation

$$\frac{d}{dt}(\Phi(t, t_0)) = A(t)\Phi(t, t_0) \qquad \Phi(t_0, t_0) = I \tag{1.5.32}$$

Note that linear systems with $A(.) \in L_{\infty e}$ automatically satisfy the Lipschitz condition over any finite interval, so that the solutions of (1.5.31) and (1.5.32) are unique on any time interval. It is easy to verify that the solution of (1.5.31) is related to that of (1.5.32) through

$$x(t) = \Phi(t, t_0)x(t_0) \tag{1.5.33}$$

In particular, this expression shows that trajectories are "proportional" to the size of the initial conditions, so that local and global properties of LTV systems are identical.

The state-transition matrix satisfies the so-called *semigroup property* (cf. Kailath [1980], p. 599):

$$\Phi(t, t_0) = \Phi(t, \tau)\Phi(\tau, t_0) \qquad \text{for all } t \geq \tau \geq t_0 \tag{1.5.34}$$

and its inverse is given by

$$\Phi(t, t_0)^{-1} = \Phi(t_0, t) \tag{1.5.35}$$

A consequence is that

$$\begin{aligned} \frac{d}{dt_0}\Phi(t, t_0) &= \frac{d}{dt_0}(\Phi(t_0, t))^{-1} \\ &= -\Phi(t_0, t)^{-1}A(t_0)\Phi(t_0, t)\Phi(t_0, t)^{-1} \\ &= -\Phi(t, t_0)A(t_0) \end{aligned} \tag{1.5.36}$$

The following propositions relate the stability properties of (1.5.31) to properties of the state-transition matrix.

Proposition 1.5.3 Uniform Asymptotic Stability of LTV Systems

$x = 0$ is a uniformly asymptotically stable equilibrium point of (1.5.31) *if and only if* $x = 0$ is stable, which is guaranteed by

$$\sup_{t_0 \geq 0} \left(\sup_{t \geq t_0} \| \Phi(t, t_0) \| \right) < \infty \tag{1.5.37}$$

and $x = 0$ is attractive, which is guaranteed by

$$\| \Phi(t, t_0) \| \to 0 \text{ as } t \to \infty \qquad \text{uniformly in } t_0. \tag{1.5.38}$$

Proof of Proposition 1.5.3 direct from the expression of the solution (1.5.33).

Proposition 1.5.4 Exponential Stability of LTV Systems
$x = 0$ is an exponentially stable equilibrium point of (1.5.31)
if and only if for some $m, \alpha > 0$

$$\| \Phi(t, t_0) \| = m e^{-\alpha(t - t_0)} \tag{1.5.39}$$

for all $t \geq t_0 \geq 0$.
Proof of Proposition 1.5.4 direct from the expression of the solution (1.5.33).

A unique property of linear systems is the equivalence between uniform asymptotic stability and exponential stability, as stated in the following theorem.

Proposition 1.5.5 Exponential and Uniform Asymptotic Stability
$x = 0$ is a uniformly asymptotically stable equilibrium point of (1.5.31) *if and only if* $x = 0$ is an exponentially stable equilibrium point of (1.5.31).

Proof of Proposition 1.5.5
That exponential stability implies uniform asymptotic stability is obvious from their definitions and in particular from proposition 1.5.3 and proposition 1.5.4. We now show the converse. Proposition 1.5.3 implies that there exists $M > 0$ and $T > 0$, such that

$$\| \Phi(t, t_0) \| < M \qquad \text{for all } t \geq t_0 \geq 0 \tag{1.5.40}$$

and

$$\| \Phi(t_0 + T, t_0) \| < \frac{1}{2} \qquad \text{for all } t_0 \geq 0 \tag{1.5.41}$$

For all $t \geq t_0$, there exists an integer n such that $t \in [t_0 + nT, t_0 + (n+1)T]$. Using the semigroup property recursively, together with (1.5.40), (1.5.41)

$$\begin{aligned} \| \Phi(t, t_0) \| &\leq \| \Phi(t, t_0 + nT) \| \; \| \Phi(t_0 + nT, t_0) \| \\ &< M \left[\frac{1}{2} \right]^n \leq M \left[\frac{1}{2} \right]^{(t - t_0)/T} \end{aligned} \tag{1.5.42}$$

which can easily be expressed in the form of (1.5.39). □

Uniform Complete Observability—Definition and Results

Through the definition of uniform complete observability, some additional results on the stability of linear time-varying systems will now be established. We consider the linear time-varying system $[C(t), A(t)]$ defined by

$$\dot{x}(t) = A(t)x(t)$$
$$y(t) = C(t)x(t) \tag{1.5.43}$$

where $x(t) \in \mathbb{R}^n$, $y(t) \in \mathbb{R}^m$, while $A(t) \in \mathbb{R}^{n \times n}$, $C(t) \in \mathbb{R}^{m \times n}$, are piecewise continuous functions (therefore belonging to $L_{\infty e}$).

Definition Uniform Complete Observability (UCO)

The system $[C(t), A(t)]$ is called *uniformly completely observable* (UCO) if there exist strictly positive constants β_1, β_2, δ, such that, for all $t_0 \geq 0$

$$\beta_2 I \geq N(t_0, t_0+\delta) \geq \beta_1 I \tag{1.5.44}$$

where $N(t_0, t_0+\delta) \in \mathbb{R}^{n \times n}$ is the so-called *observability grammian*

$$N(t_0, t_0+\delta) = \int_{t_0}^{t_0+\delta} \Phi^T(\tau, t_0) C^T(\tau) C(\tau) \Phi(\tau, t_0) d\tau \tag{1.5.45}$$

Comments

Note that, using (1.5.33), condition (1.5.44) can be rewritten as

$$\beta_2 |x(t_0)|^2 \geq \int_{t_0}^{t_0+\delta} |C(\tau)x(\tau)|^2 d\tau \geq \beta_1 |x(t_0)|^2 \tag{1.5.46}$$

for all $x(t_0) \in \mathbb{R}^n$, $t_0 \geq 0$, where $x(t)$ is the solution of (1.5.43) starting at $x(t_0)$.

The observability is called uniform because (1.5.43) is satisfied uniformly for all t_0 and complete because (1.5.46) is satisfied for all $x(t_0)$. A specific $x(t_0)$ is observable on a specific time interval $[t_0, t_0+\delta]$ if condition (1.5.46) is satisfied on that interval. Then, $x(t_0)$ can be reconstructed from the knowledge of $y(.)$ using the expression

$$x(t_0) = N(t_0, t_0+\delta)^{-1} \int_{t_0}^{t_0+\delta} \Phi^T(\tau, t_0) C^T(\tau) y(\tau) d\tau \tag{1.5.47}$$

On the other hand, if condition (1.5.44) fails to be satisfied, (1.5.46) shows that there exists an $x(t_0) \neq 0$ such that the corresponding output satisfies

$$\int_{t_0}^{t_0+\delta} |y(\tau)|^2 d\tau = 0 \tag{1.5.48}$$

so that $x(t_0)$ is not distinguishable from 0.

Theorem 1.5.6 Exponential Stability of LTV Systems
The following statements are equivalent:

(a) $x = 0$ is an exponentially stable equilibrium point of (1.5.31).

(b) For all $C(t) \in \mathbb{R}^{m \times n}$ (with m arbitrary) such that the pair $[C(t), A(t)]$ is UCO, there exists a symmetric $P(t) \in \mathbb{R}^{n \times n}$, and some $\gamma_1, \gamma_2 > 0$, such that

$$\gamma_2 I \geq P(t) \geq \gamma_1 I \tag{1.5.49}$$

$$-\dot{P}(t) = A^T(t)P(t) + P(t)A(t) + C^T(t)C(t) \tag{1.5.50}$$

for all $t \geq 0$.

(c) For some $C(t) \in \mathbb{R}^{m \times n}$ (with m arbitrary) such that the pair $[C(t), A(t)]$ is UCO, there exists a symmetric $P(t) \in \mathbb{R}^{n \times n}$ and some $\gamma_1, \gamma_2 > 0$, such that (1.5.49) and (1.5.50) are satisfied.

Proof of Theorem 1.5.6
(a) implies (b)

Define

$$P(t) = \int_t^{\infty} \Phi^T(\tau, t) C^T(\tau) C(\tau) \Phi(\tau, t) d\tau \tag{1.5.51}$$

We will show that (1.5.49) is satisfied so that $P(t)$ is well defined, but first note that by differentiating (1.5.51), and using (1.5.36), it follows that $P(t)$ satisfies the linear differential equation (1.5.50). This equation is a linear differential equation, so that the solution is unique for given initial conditions. However, we did not impose initial conditions in (1.5.50), and $P(0)$ is in fact given by (1.5.51).

By the UCO assumption

$$\beta_2 I \geq \int_t^{t+\delta} \Phi^T(\tau, t) C^T(\tau) C(\tau) \Phi(\tau, t) d\tau \geq \beta_1 I \tag{1.5.52}$$

for all $t \geq 0$. Since the integral from t to ∞ is not less than from t to $t + \delta$, the lower inequality in (1.5.50) follows directly from (1.5.52). To show the upper inequality, we divide the interval of integration in intervals of size δ and sum the individual integrals. On the interval $[t+\delta, t+2\delta]$

$$\int_{t+\delta}^{t+2\delta} \Phi^T(\tau,t)C^T(\tau)C(\tau)\Phi(\tau,t)d\tau = \Phi^T(t+\delta,t)$$

$$\cdot \left[\int_{t+\delta}^{t+2\delta} \Phi^T(\tau,t+\delta)C^T(\tau)C(\tau)\Phi(\tau,t+\delta)d\tau \right] \phi(t+\delta,t)$$

$$\leq \beta_2 m^2 e^{-2\alpha\delta} I \tag{1.5.53}$$

where we used the UCO and exponential stability assumptions. Therefore

$$\int_t^{\infty} \Phi^T(\tau,t)C^T(\tau)C(\tau)\Phi(\tau,t)d\tau$$

$$\leq \beta_2(1+m^2 e^{-2\alpha\delta} + m^2 e^{-4\alpha\delta} + \cdots) I := \gamma_2 I \tag{1.5.54}$$

(b) implies (c) trivial.

(c) implies (a).

Consider the Lyapunov function

$$v(t,x) = x^T(t)P(t)x(t) \tag{1.5.55}$$

so that

$$\begin{aligned} -\dot{v} &= -x^T(t)P(t)A(t)x(t) - x^T(t)A^T(t)P(t)x(t) \\ &\quad - x^T(t)\dot{P}(t)x(t) \\ &= x^T(t)C^T(t)C(t)x(t) \leq 0 \end{aligned} \tag{1.5.56}$$

Using the UCO property

$$\int_t^{t+\delta} \dot{v}\,d\tau = -x^T(t)\left[\int_t^{t+\delta} \Phi^T(\tau,t)C^T(\tau)C(\tau)\Phi(\tau,t)d\tau \right] x(t)$$

$$\leq -\beta_1 |x(t)|^2 \tag{1.5.57}$$

Exponential convergence follows from theorem 1.5.2.

It is interesting to note that the Lyapunov function is identical to the function used in theorem 1.5.1, when $C = I$, and the upper bound of integration tends to infinity. □

1.5.3 Exponential Stability of Linear Time Invariant Systems

When the system (1.5.31) is time-invariant, even further simplification of the stability criterion results. Specifically, we consider the system

$$\dot{x} = A x \qquad x(t_0) = x_0 \tag{1.5.58}$$

Note that since the system is linear, local and global stability properties are identical, and since it is time-invariant, all properties are uniform (in t_0).

Theorem 1.5.7 Lyapunov Lemma

The following statements are equivalent:

(a) All eigenvalues of A lie in the open left-half plane

(b) $x = 0$ is an exponentially stable equilibrium point of (1.5.58)

(c) For all $C \in \mathbb{R}^{m \times n}$ (with m arbitrary), such that the pair $[C, A]$ is observable, there exists a symmetric positive definite $P \in \mathbb{R}^{n \times n}$ satisfying

$$A^T P + PA = -C^T C \tag{1.5.59}$$

(d) For *some* $C \in \mathbb{R}^{m \times n}$ (with m arbitrary), such that the pair $[C, A]$ is observable, there exists a symmetric positive definite $P \in \mathbb{R}^{n \times n}$ satisfying (1.5.59).

Proof of Theorem 1.5.7

(a) implies (b).

This follows from the fact that the transition matrix is given by the exponential matrix (cf. Vidyasagar [1978], pp. 171)

$$\Phi(t, t_0) = e^{A(t - t_0)} \tag{1.5.60}$$

(b) implies (c).

Let

$$S(t) := \int_0^t e^{A^T \tau} C^T C\, e^{A\tau}\, d\tau$$

$$= \int_0^t e^{A^T(t-\sigma)} C^T C\, e^{A(t-\sigma)} d\tau \qquad (1.5.61)$$

Clearly $S(t) = S^T(t)$, and since $[C, A]$ is observable, $S(t) > 0$ for all $t > 0$. Using both expressions in (1.5.61)

$$\dot{S}(t) = e^{A^T t} C^T C\, e^{At} = A^T S(t) + S(t) A + C^T C \qquad (1.5.62)$$

and using the exponential stability of A

$$\lim_{t \to \infty} \dot{S}(t) = 0 = A^T P + P A + C^T C \qquad (1.5.63)$$

where

$$P = \lim_{t \to \infty} S(t) \qquad (1.5.64)$$

(c) implies (d) trivial.

(d) implies (b) as in theorem 1.5.6. □

Remark

The original version of the Lyapunov lemma is stated with Q positive definite replacing $C^T C$ (which is only semi-definite), and leading to the so-called *Lyapunov equation*:

$$A^T P + P A = -Q \qquad (1.5.65)$$

Then, the observability hypothesis is trivial. We stated a more general version of the Lyapunov lemma partly to show the intimate connection between exponential stability and observability for linear systems.

1.6 GENERALIZED HARMONIC ANALYSIS

An appropriate tool for the study of stable linear time-invariant systems is Fourier transforms or harmonic analysis, since e^{jwt} is an eigenfunction of such systems. We will be concerned with the analysis of algorithms for identification and adaptive control of linear time-invariant systems. The overall adaptive systems are time-varying systems, but, under certain conditions, they become "asymptotically" time-invariant. For these systems, generalized harmonic analysis is a useful tool of analysis. The theory has been known since the early part of the century and was developed in Wiener's *Generalized Harmonic Analysis* (Wiener [1930]). Boyd & Sastry [1983] and [1986] illustrated its application to adaptive systems. Since the proofs of the various lemmas used in this book are neither difficult nor long, we provide them in this section.

Definition Stationarity, Autocovariance

A signal $u: \mathbb{R}_+ \to \mathbb{R}^n$ is said to be *stationary* if the following limit exists, uniformly in t_0

$$R_u(t) := \lim_{T \to \infty} \frac{1}{T} \int_{t_0}^{t_0+T} u(\tau) u^T(t+\tau)\, d\tau \quad \in \mathbb{R}^{n \times n} \tag{1.6.1}$$

in which instance, the limit $R_u(t)$ is called the *autocovariance* of u.

The concept of autocovariance is well known in the theory of stochastic systems. There is a strong analogy between (1.6.1) and $R_u^{stoch}(t) = \mathrm{E}\,[u(\tau) u^T(t+\tau)]$, when u is a wide sense stationary stochastic process. Indeed, for a wide sense stationary *ergodic* process $u(t, \omega)$ (ω here denotes a sample point of the underlying probability space), the autocovariance $R_u(t, \omega)$ exists, and is equal to $R_u^{stoch}(t)$ for almost all ω. But we emphasize that the autocovariance defined in (1.6.1) is completely deterministic.

Proposition 1.6.1

R_u is a *positive semi-definite matrix-valued function,* that is, for all $t_1, \ldots, t_k \in \mathbb{R}$, and $c_1, \ldots, c_k \in \mathbb{R}^n$

$$\sum_{i,j=1}^{k} c_i^T R_u(t_j - t_i) c_j \geq 0 \tag{1.6.2}$$

Proof of Proposition 1.6.1

Define the scalar valued function $v(t)$ by

$$v(t) := \sum_{i=1}^{k} c_i^T u(t + t_i) \tag{1.6.3}$$

Then, for all $T > 0$

$$0 \leq \frac{1}{T} \int_0^T |v(\tau)|^2 d\tau \tag{1.6.4}$$

$$= \sum_{i,j=1}^{k} c_i^T \Big(\frac{1}{T} \int_0^T u(\tau + t_i) u^T(\tau + t_j)\, d\tau \Big) c_j$$

$$= \sum_{i,j=1}^{k} c_i^T \Big(\frac{1}{T} \int_{t_i}^{t_i+T} u(\sigma) u^T(\sigma + t_j - t_i)\, d\sigma \Big) c_j \tag{1.6.5}$$

Since u has an autocovariance, the limit of (1.6.5) as $T \to \infty$ is

$$\sum_{i,j=1}^{k} c_i^T R_u(t_j - t_i) c_j \tag{1.6.6}$$

From (1.6.4), we find that (1.6.6) is positive, so that (1.6.2) follows. □

From proposition 1.6.1, it follows (see, for example, Widder [1971]) that R_u is the Fourier transform of a positive semi-definite matrix $S_u(d\omega)$ of bounded measures, that is,

$$R_u(t) = \frac{1}{2\pi} \int_{-\infty}^{\infty} e^{j\omega t} S_u(d\omega) \tag{1.6.7}$$

and

$$\int_{-\infty}^{\infty} S_u(d\omega) = 2\pi R_u(0) < \infty \tag{1.6.8}$$

The corresponding inverse transform is

$$S_u(d\omega) = \int_{-\infty}^{+\infty} e^{-j\omega\tau} R_u(\tau)\, d\tau \tag{1.6.9}$$

The matrix measure $S_u(d\omega)$ is referred to as the *spectral measure* of the signal u. In particular, if u has a sinusoidal component at frequency ω_0, then u is said to have a *spectral line* at frequency ω_0. $S_u(d\omega)$ has point mass (a delta function) at ω_0 and $-\omega_0$. If $u(t)$ is periodic, with period $2\pi/\omega_0$, then $S_u(d\omega)$ has point mass at ω_0, $2\omega_0, \ldots$ and $-\omega_0$, $-2\omega_0, \ldots$. Further, if $u(t)$ is almost periodic, its spectral measure has point masses concentrated at countably many points.

Since R_u is real, (1.6.9) shows that $\mathrm{Re}(S_u)$ is an even function of ω, and $\mathrm{Im}(S_u)$ is an odd function of ω. On the other hand, (1.6.1) shows that

$$R_u(t) = R_u^T(-t)$$

Therefore, (1.6.9) also shows that $\mathrm{Re}(S_u)$ is a symmetric matrix, while $\mathrm{Im}(S_u)$ is an antisymmetric matrix. In other words,

$$S_u^T(d\,\omega) = S_u^*(d\,\omega)$$

Equations (1.6.1) and (1.6.9) give a technique for obtaining the spectral content of stationary deterministic signals. For example, consider the scalar signal

$$u(t) = \sin(\omega_1 t) + \sin(\omega_2 t) + f(t) \tag{1.6.10}$$

where $f(t)$ is some continuous function that tends to zero as $t \to \infty$. The signal u has autocovariance

$$R_u(t) = \frac{1}{2}\cos(\omega_1 t) + \frac{1}{2}\cos(\omega_2 t) \tag{1.6.11}$$

showing spectral content at ω_1 and ω_2.

The autocovariance of the input and of the output signals of a stable linear time-invariant system can be related as in the following proposition.

Proposition 1.6.2 Linear Filter Lemma

Let $y = \hat{H}(u)$, where $\hat{H}$ is a proper, stable $m \times n$ matrix transfer function, with real impulse response $H(t)$.

If u is stationary, with autocovariance $R_u(t)$

Then y is stationary, with autocovariance

$$R_y(t) = \int_{-\infty}^{\infty}\int_{-\infty}^{\infty} H(\tau_1)R_u(t+\tau_1-\tau_2)H^T(\tau_2)\,d\tau_1\,d\tau_2 \tag{1.6.12}$$

and spectral measure

$$S_y(d\omega) = \hat{H}^*(j\omega)S_u(d\omega)\hat{H}^T(j\omega) \tag{1.6.13}$$

Proof of Proposition 1.6.2

We first establish that y has an autocovariance, by considering

$$\frac{1}{T}\int_{t_0}^{t_0+T} y(\tau)y^T(t+\tau)\,d\tau = \frac{1}{T}\int_{t_0}^{t_0+T}\left(\int H(\tau_1)u(\tau-\tau_1)\,d\tau_1\right)$$

$$\left(\int u^T(t+\tau-\tau_2)H^T(\tau_2)\,d\tau_2\right)d\tau \tag{1.6.14}$$

For all T, the integral in (1.6.14) exists absolutely, since $\hat{H}$ is stable, and therefore $H \in L_1$. Therefore, we may change the order of integration, to obtain

$$\int\int H(\tau_1)\,d\tau_1\left[\frac{1}{T}\int_{t_0-\tau_1}^{t_0-\tau_1+T} u(\sigma)u^T(t+\sigma+\tau_1-\tau_2)\,d\sigma\right]$$

$$(H^T(\tau_2)\,d\tau_2) \tag{1.6.15}$$

The expression in parenthesis converges to $R_u(t+\tau_1-\tau_2)$ as $T\to\infty$, uniformly in t_0. Further, it is bounded as a function of T, t_0, τ_1 and τ_2, since, by the Schwarz inequality

$$\left| \frac{1}{T} \int_{t_0-\tau_1}^{t_0-\tau_1+T} u(\sigma)\, u^T(t+\sigma+\tau_1-\tau_2)\, d\sigma \right|$$

$$\leq \sup_{t_0,T} \frac{1}{T} \int_{t_0}^{t_0+T} |u(\tau)|^2 d\tau \tag{1.6.16}$$

Hence, by dominated convergence, (1.6.15) converges uniformly in t_0, as $T\to\infty$, to (1.6.12), so that y has an autocovariance given by (1.6.12).

To complete the proof, we substitute (1.6.7) in (1.6.12) to get

$$\begin{aligned} R_y(t) &= \frac{1}{2\pi} \int\int H(\tau_1)\, d\tau_1 \int e^{j\omega(t+\tau_1-\tau_2)}\, S_u(d\omega)\, H^T(\tau_2)\, d\tau_2 \\ &= \frac{1}{2\pi} \int e^{j\omega t} \left(\int e^{j\omega\tau_1} H(\tau_1)\, d\tau_1\right) S_u(d\omega) \left(\int e^{-j\omega\tau_2} H(\tau_2)\, d\tau_2\right)^T \\ &= \frac{1}{2\pi} \int e^{j\omega t}\, \hat{H}(-j\omega)\, S_u(d\omega)\, \hat{H}^T(j\omega) \end{aligned} \tag{1.6.17}$$

Note that (1.6.17) is the Fourier representation of R_y, so that, using the fact that $H(t)$ is real

$$\begin{aligned} S_y(d\omega) &= \hat{H}(-j\omega)\, S_u(d\omega)\, \hat{H}^T(j\omega) \\ &= \hat{H}^*(j\omega)\, S_u(d\omega)\, \hat{H}^T(j\omega) \end{aligned} \tag{1.6.18}$$

□

Remark

It is easy to see that if u has a spectral line at frequency ω_0, so does y, and the intensity of the spectral line of y at ω_0 is given by (1.6.18). Note, however, that if $\hat{H}(s)$ has a zero of transmission at ω_0, then the amplitude of the spectral line at the output is zero.

We will also need to define the cross correlation and cross spectral density between two stationary signals.

Definition Cross Correlation

Let $u: \mathbb{R}_+ \to \mathbb{R}^n$ and $y: \mathbb{R}_+ \to \mathbb{R}^m$ be stationary. The *cross correlation* between u and y is defined to be the following limit, uniform in t_0

$$R_{yu}(t) := \lim_{T \to \infty} \frac{1}{T} \int_{t_0}^{t_0+T} y(\tau)\, u^T(t+\tau)\, d\tau \quad \in \mathbb{R}^{m \times n} \tag{1.6.19}$$

It may be verified that the relationship between R_{uy} and R_{yu} is

$$R_{yu}(t) = R_{uy}^T(-t) \tag{1.6.20}$$

and the Fourier transform of the cross correlation is referred to as the *cross spectral measure* of u and y.

The cross correlation between the input and the output of a stable LTI system can be related in much the same way as in proposition 1.6.2.

Proposition 1.6.3 Linear Filter Lemma—Cross Correlation

Let $y = \hat{H}(u)$, where $\hat{H}$ is a proper stable $m \times n$ matrix transfer function, with impulse response $H(t)$.

If u is stationary

Then the cross correlation between u and y is given by

$$R_{yu}(t) = \int_{-\infty}^{\infty} H(\tau_1) R_u(t+\tau_1)\, d\tau_1 \tag{1.6.21}$$

and the cross spectral measure is

$$S_{yu}(d\omega) = \hat{H}^*(j\omega) S_u(d\omega) \tag{1.6.22}$$

Proof of Proposition 1.6.3

The proof is analogous to the proof of proposition 1.6.2 and is omitted here.

CHAPTER 2
IDENTIFICATION

2.0 INTRODUCTION

In this chapter, we review some identification methods for single-input single-output (SISO), linear time invariant (LTI) systems. To introduce the subject, we first informally discuss a simple example. We consider the identification problem for a first order SISO LTI system described by a transfer function

$$\frac{\hat{y}_p(s)}{\hat{r}(s)} = \hat{P}(s) = \frac{k_p}{s + a_p} \tag{2.0.1}$$

The parameters k_p and a_p are unknown and are to be determined by the identification scheme on the basis of measurements of the input and output of the plant. The plant is assumed to be stable, i.e. $a_p > 0$.

Frequency Domain Approach

A standard approach to identification is the *frequency response* approach. Let the input r be a sinusoid

$$r(t) = \sin(\omega_0 t) \tag{2.0.2}$$

The steady-state response is then given by

$$y(t) = m \sin(\omega_0 t + \phi) \tag{2.0.3}$$

where

$$m = |\hat{P}(j\omega_0)| = \frac{k_p}{\sqrt{\omega_0^2 + a_p^2}}$$

$$\phi = \arg \hat{P}(j\omega_0) = -\ \text{arc tan} \left[\frac{\omega_0}{a_p} \right] \tag{2.0.4}$$

Measurements of the gain m and phase ϕ at a single frequency $\omega_0 \neq 0$ uniquely determine k_p and a_p by inversion of the above relationships. At $\omega_0 = 0$, that is when the input signal is constant, phase information is lost. Only one equation is left, giving the DC gain. Then, only the ratio of the parameters k_p and a_p is determined. Conversely, if several frequencies are used, each contributes two equations and the parameters are overdetermined.

Frequency response methods will not be further discussed in this book, because our goal is adaptive control. We will therefore concentrate on *recursive* approaches, where parameter estimates are updated in real-time. However, we will still *analyze* these algorithms in the frequency domain and obtain similar results as above for the recursive schemes.

Time Domain Approach

We now discuss schemes based on a time-domain expression of the plant (2.0.1), that is

$$\dot{y}_p(t) = -a_p y_p(t) + k_p r(t) \tag{2.0.5}$$

Measurements of y_p, $\dot{y}_p$ and r at *one* time instant t give us one equation with two unknown a_p and k_p. As few as two time instants may be sufficient to determine the unknown parameters from

$$\begin{bmatrix} -a_p \\ k_p \end{bmatrix} = \begin{bmatrix} y_p(t_1) & r(t_1) \\ y_p(t_2) & r(t_2) \end{bmatrix}^{-1} \begin{bmatrix} \dot{y}_p(t_1) \\ \dot{y}_p(t_2) \end{bmatrix} \tag{2.0.6}$$

assuming that the inverse exists. Note that, as in the frequency-domain approach, a constant input $r(t_1) = r(t_2)$ with constant output $y_p(t_1) = y_p(t_2)$ will prevent us from uniquely determining a_p and k_p.

We may refine this approach to avoid the measurement of $\dot{y}_p(t)$. Consider (2.0.1) and divide both sides by $s + \lambda$ for some $\lambda > 0$

$$\frac{s + a_p}{s + \lambda} \hat{y}_p = \frac{k_p}{s + \lambda} \hat{r} \tag{2.0.7}$$

This is equivalent to

$$\hat{y}_p = \frac{\lambda - a_p}{s + \lambda}\hat{y}_p + \frac{k_p}{s + \lambda}\hat{r} \tag{2.0.8}$$

Define the signals

$$\hat{w}^{(1)} = \frac{1}{s + \lambda}\hat{r} \qquad \hat{w}^{(2)} = \frac{1}{s + \lambda}\hat{y}_p \tag{2.0.9}$$

or, in the time domain

$$\dot{w}^{(1)} = -\lambda w^{(1)} + r \qquad \dot{w}^{(2)} = -\lambda w^{(2)} + y_p \tag{2.0.10}$$

Then, in the time domain, (2.0.8) reads

$$y_p(t) = k_p w^{(1)}(t) + (\lambda - a_p) w^{(2)}(t) \tag{2.0.11}$$

The signals $w^{(1)}$, $w^{(2)}$ may be obtained by stable filtering of the input and of the output of the plant. We have assumed zero initial conditions on $w^{(1)}$ and $w^{(2)}$. Nonzero initial conditions would only contribute exponential decaying terms with rate λ (arbitrary), but are not considered in this simplified derivation. Equation (2.0.11) is to be compared with (2.0.5). Again, measurements at one time instant give us one equation with two unknowns. However, we do not require differentiation, but instead stable filtering of available signals.

In the sequel, we will assume that measurements of r and y_p are made continuously between 0 and t. We will therefore look for algorithms that use the complete information and preferably update estimates only on the basis of new data, without storing the entire signals. But first, we transform (2.0.11) into the standard framework used later in this chapter. Define the vector of *nominal* identifier parameters

$$\theta^* := \begin{bmatrix} k_p \\ \lambda - a_p \end{bmatrix} \tag{2.0.12}$$

Knowledge of θ^* is clearly equivalent to the knowledge of the unknown parameters k_p and a_p. Similarly, define $\theta(t)$ to be a vector of identical dimension, called the *adaptive* identifier parameter. $\theta(t)$ is the estimate of θ^* based on input-output data up to time t. Letting

$$w(t) := \begin{bmatrix} w^{(1)}(t) \\ w^{(2)}(t) \end{bmatrix} \tag{2.0.13}$$

equation (2.0.11) may be written

$$y_p(t) = \theta^{*T} w(t) = w^T(t)\,\theta^* \tag{2.0.14}$$

Based on measurements of $r(t)$ and $y_p(t)$ up to time t, $w(t)$ may be calculated, and an estimate $\theta(t)$ derived. Since each time instant gives us one equation with two unknowns, it makes sense to consider the estimate that minimizes the *identification error*

$$e_1(t) = \theta^T(t)\, w(t) - y_p(t) = \left[\theta^T(t) - \theta^{*T} \right] w(t) \qquad (2.0.15)$$

Note that the identification error is linear in the parameter error $\theta - \theta^*$. We will therefore call (2.0.15) a *linear error equation.* The purpose of the identification scheme will be to calculate $\theta(t)$, on the basis of measurements of $e_1(t)$ and $w(t)$ up to time t.

Gradient and Least-Squares Algorithms

The *gradient algorithm* is a steepest descent approach to minimize $e_1^2(t)$. Since

$$\frac{\partial e_1^2}{\partial \theta} = 2\, e_1 \frac{\partial e_1}{\partial \theta} = 2\, e_1\, w \qquad (2.0.16)$$

we let the parameter update law

$$\dot{\theta} = -\, g\, e_1\, w \qquad g > 0 \qquad (2.0.17)$$

where g is an arbitrary gain, called the *adaptation gain.*

Another approach is the *least-squares algorithm* which minimizes the integral-squared-error (ISE)

$$ISE = \int_0^t e_1^2(\tau)\, d\tau \qquad (2.0.18)$$

Owing to the linearity of the error equation, the estimate may be obtained directly from the condition

$$\frac{\partial}{\partial \theta} \left[\int_0^t e_1^2(\tau)\, d\tau \right] = 2 \int_0^t w(\tau) \left[w^T(\tau)\, \theta(\tau) - y_p(\tau) \right] d\tau = 0 \qquad (2.0.19)$$

so that the *least-squares estimate* is given by

$$\theta_{LS}(t) = \left[\int_0^t w(\tau)\, w^T(\tau)\, d\tau \right]^{-1} \left[\int_0^t w(\tau)\, y_p(\tau)\, d\tau \right] \qquad (2.0.20)$$

Plugging (2.0.14) into (2.0.20) shows that $\theta_{LS}(t) = \theta^*$, *assuming that the inverse in (2.0.20) exists.*

For adaptive control applications, we are interested in *recursive formulations* such as (2.0.17), where parameters are updated continuously on the basis of input-output data. Such an expression may be obtained for the least-squares algorithms by defining

$$P(t) = \left[\int_0^t w(\tau)\, w^T(\tau)\, d\tau \right]^{-1} \tag{2.0.21}$$

so that

$$\frac{d}{dt} \left[P^{-1}(t) \right] = w(t)\, w^T(t) \tag{2.0.22}$$

Since

$$\begin{aligned} 0 = \frac{d}{dt} (I) &= \frac{d}{dt} \left[P(t)\, P^{-1}(t) \right] \\ &= \frac{d}{dt} \left[P(t) \right] P^{-1}(t) + P(t) \frac{d}{dt} \left[P^{-1}(t) \right] \end{aligned} \tag{2.0.23}$$

it follows that

$$\begin{aligned} \frac{d}{dt} \left[P(t) \right] &= - P(t) \frac{d}{dt} \left[P^{-1}(t) \right] P(t) \\ &= - P(t)\, w(t)\, w^T(t)\, P(t) \end{aligned} \tag{2.0.24}$$

On the other hand, (2.0.20) may be written

$$\theta_{LS}(t) = P(t) \int_0^t w(\tau)\, y_p(\tau)\, d\tau \tag{2.0.25}$$

so that, using (2.0.24)

$$\begin{aligned} \frac{d}{dt} \left[\theta_{LS}(t) \right] &= - P(t)\, w(t)\, w^T(t)\, \theta_{LS}(t) + P(t)\, w(t)\, y_p(t) \\ &= - P(t)\, w(t) \left[w^T(t)\, \theta_{LS}(t) - y_p(t) \right] \\ &= - P(t)\, w(t)\, e_1(t) \end{aligned} \tag{2.0.26}$$

Note that the recursive algorithm (2.0.24), (2.0.26) should be started with the correct initial conditions at some $t_0 > 0$ such that

$$P(t_0) = \left[\int_0^{t_0} w(\tau)\, w^T(\tau)\, d\tau \right]^{-1} \tag{2.0.27}$$

exists. In practice, the *recursive least-squares* algorithm is started with arbitrary initial conditions at $t_0 = 0$ so that

$$\dot{\theta}(t) = -P(t)\, w(t) \left[\theta^T(t)\, w(t) - y_p(t) \right] \qquad \theta(0) = \theta_0$$

$$\dot{P}(t) = -P(t)\, w(t)\, w^T(t)\, P(t) \qquad P(0) = P_0 > 0 \tag{2.0.28}$$

It may be verified that the solution of (2.0.28) is

$$\theta(t) = \left[P_0 + \int_0^t w(t)\, w^T(\tau)\, d\tau \right]^{-1} \left[P_0 \theta_0 + \int_0^t y_p(\tau)\, w(\tau)\, d\tau \right] \tag{2.0.29}$$

instead of (2.0.20). Since $y_p = \theta^{*T} w$, the *parameter error* is given by

$$\theta(t) - \theta^* = \left[P_0 + \int_0^t w(\tau)\, w^T(\tau) dt \right]^{-1} P_0(\theta_0 - \theta^*) \tag{2.0.30}$$

It follows that $\theta(t)$ *converges asymptotically* to θ^* if $\int_0^t w(\tau)\, w^T(\tau)\, d\tau$ is unbounded as $t \to \infty$. In this chapter, we will study conditions that guarantee *exponential convergence* of the parameter estimates to the nominal parameter. These are called *persistency of excitation* conditions and are closely related to the above condition. It is not obvious at this point how to relate the time domain condition on the vector w to frequency domain conditions on the input. This will be a goal of this chapter.

Model Reference Identification

We now discuss another family of identification algorithms, based on the so-called *model reference approach.* The algorithms have similarities to the previous ones, and are useful to introduce adaptive control techniques.

We first define a *reference model* transfer function

$$\frac{\hat{y}_m}{\hat{u}} = \hat{M}(s) = \frac{k_m}{s + a_m} \tag{2.0.31}$$

where $a_m,\ k_m > 0$. In the time domain

$$\dot{y}_m(t) \;=\; -a_m\, y_m(t) + k_m\, u(t) \tag{2.0.32}$$

Let the input u to the model be given by

$$u(t) \;=\; a_0(t)\, r(t) + b_0(t)\, y_m(t) \tag{2.0.33}$$

where $a_0(t)$, $b_0(t)$ are adaptive parameters and r is the input to the plant. The motivation is that there exist *nominal values* of the parameters, denoted a_0^*, b_0^*, such that the closed-loop transfer function matches any first order transfer function. Specifically, (2.0.32) and (2.0.33) give the closed-loop system

$$\dot{y}_m(t) \;=\; -(a_m - k_m\, b_0(t))\, y_m(t) + k_m\, a_0(t)\, r(t) \tag{2.0.34}$$

so that the nominal values of a_0, b_0 are

$$a_0^* \;=\; \frac{k_p}{k_m} \qquad b_0^* \;=\; \frac{a_m - a_p}{k_m} \tag{2.0.35}$$

Clearly, knowledge of a_0^*, b_0^* is equivalent to knowledge of a_p, k_p. We define the following vectors

$$\theta(t) \;=\; \begin{bmatrix} a_0(t) \\ b_0(t) \end{bmatrix} \qquad \theta^* \;=\; \begin{bmatrix} a_0^* \\ b_0^* \end{bmatrix} \qquad w(t) \;=\; \begin{bmatrix} r(t) \\ y_m(t) \end{bmatrix} \tag{2.0.36}$$

and the *identification error*

$$e_1(t) \;=\; y_m(t) - y_p(t) \tag{2.0.37}$$

so that

$$\begin{aligned} \dot{e}_1(t) \;&=\; -(a_m + k_m\, b_0(t))\, y_m(t) + k_m\, a_0(t)\, r(t) + a_p\, y_p(t) - k_p\, r(t) \\ &=\; -a_m\, e_1(t) + k_m\, ((a_0(t) - a_0^*)\, r(t) + (b_0(t) - b_0^*)\, y_p(t)) \\ &=\; -a_m\, e_1(t) + k_m\, (\theta^T(t) - \theta^{*T})\, w(t) \end{aligned} \tag{2.0.38}$$

In short

$$e_1(t) \;=\; \hat{M}\left[(\theta^T(t) - \theta^{*T})\, w(t)\right] \tag{2.0.39}$$

which may be compared to the linear error equation (2.0.15). This equation is still linear, but it involves the dynamics of the transfer function $\hat{M}$. Can we still find update algorithms for $\theta(t)$, based on the new error equation? The answer is yes, but the approach is now based on the Lyapunov function

$$v(e_1, \theta) = \frac{e_1^2}{2} + \frac{k_m}{2}(\theta^T - \theta^{*T})(\theta - \theta^*) \tag{2.0.40}$$

so that

$$\dot{v} = -a_m e_1 + k_m e_1 (\theta^T - \theta^{*T}) w + k_m (\theta^T - \theta^{*T}) \dot{\theta} \tag{2.0.41}$$

By letting

$$\dot{\theta} = -e_1 w \tag{2.0.42}$$

it follows that

$$\dot{v} = -a_m e_1^2 \leq 0 \tag{2.0.43}$$

Therefore, e_1 and θ are bounded. It may also be shown that $e_1 \to 0$ as $t \to \infty$ and that $\theta \to \theta^*$ under further conditions on the reference input.

The resulting update law (2.0.42) is identical to the gradient algorithm (2.0.17) obtained for the linear error equation (2.0.15). In this case however, it is *not* the gradient algorithm for (2.0.39), due to the presence of the transfer function $\hat{M}$. The motivation for the algorithm lies only in the Lyapunov stability proof.

Note that the derivation requires $a_m > 0$ (in (2.0.43)) and $k_m > 0$ (in (2.0.40)). In general, the conditions to be satisfied by $\hat{M}$ are that

- $\hat{M}$ is stable
- $\mathrm{Re}(\hat{M}(j\omega)) > 0$ for all $\omega \geq 0$.

These are very important conditions in adaptive control, defining *strictly positive real* transfer functions. They will be discussed in greater detail later in this chapter.

2.1 IDENTIFICATION PROBLEM

We now consider the general identification problem for single-input single-output (SISO) linear time invariant (LTI) systems. But first, a few definitions. A polynomial in s is called *monic* if the coefficient of the highest power in s is 1 and *Hurwitz* if its roots lie in the open left-half plane. Rational transfer functions are called *stable* if their denominator polynomial is Hurwitz and *minimum phase* if their numerator polynomial is Hurwitz. The *relative degree* of a transfer function is by definition the difference between the degrees of the denominator and numerator polynomials. A rational transfer function is called *proper* if its relative degree is at least 0 and *strictly proper* if its relative degree is at least 1.

In this chapter, we consider the identification problem of SISO LTI systems, given the following assumptions.

Assumptions

(A1) Plant Assumptions

The plant is a SISO LTI system, described by a transfer function

$$\frac{\hat{y}_p(s)}{\hat{r}(s)} = \hat{P}(s) = k_p \frac{\hat{n}_p(s)}{\hat{d}_p(s)} \tag{2.1.1}$$

where $\hat{r}(s)$ and $\hat{y}_p(s)$ are the Laplace transforms of the input and output of the plant, respectively, and $\hat{n}_p(s)$ and $\hat{d}_p(s)$ are monic, coprime polynomials of degrees m and n respectively. m is unknown, but the plant is strictly proper ($m \le n - 1$).

(A2) Reference Input Assumptions

The input $r(.)$ is piecewise continuous and bounded on $\mathbb{R}_+$.

The objective of the identifier is to obtain estimates of k_p and of the coefficients of the polynomials $\hat{n}_p(s)$ and $\hat{d}_p(s)$ from measurements of the input $r(t)$ and output $y_p(t)$ only. Note that we do not assume that $\hat{P}$ is stable.

2.2 IDENTIFIER STRUCTURE

The identifier structure presented in this section is generally known as an *equation error* identifier. (cf. Ljung & Soderstrom [1983]). The transfer function $\hat{P}(s)$ can be explicitly written as

$$\frac{\hat{y}_p(s)}{\hat{r}(s)} = \hat{P}(s) = \frac{\alpha_n s^{n-1} + \cdots + \alpha_1}{s^n + \beta_n s^{n-1} + \cdots + \beta_1} \tag{2.2.1}$$

where the $2n$ coefficients $\alpha_1 \ldots \alpha_n$ and $\beta_1 \ldots \beta_n$ are unknown. This expression is a *parameterization* of the unknown plant, that is a model in which only a finite number of parameters are to be determined. For identification purposes, it is convenient to find an expression which depends linearly on the unknown parameters. For example, the expression

$$s^n \hat{y}_p(s) = (\alpha_n s^{n-1} + \cdots + \alpha_1)\, \hat{r}(s) - (\beta_n s^{n-1} + \cdots + \beta_1)\, \hat{y}_p(s) \tag{2.2.2}$$

is linear in the parameters α_i and β_i. However, it would require explicit differentiations to be implemented. To avoid this problem, we introduce a monic nth order polynomial denoted $\hat{\lambda}(s) = s^n + \lambda_n s^{n-1} + \cdots + \lambda_1$. This polynomial is assumed to be Hurwitz but is otherwise arbitrary. Then, using (2.1.1)

$$\hat{\lambda}(s)\hat{y}_p(s) = k_p \hat{n}_p(s)\hat{r}(s) + (\hat{\lambda}(s) - \hat{d}_p(s))\,\hat{y}_p(s) \tag{2.2.3}$$

or, with (2.2.1)

$$\hat{y}_p(s) = \frac{\alpha_n s^{n-1} + \cdots + \alpha_1}{\hat{\lambda}(s)}\,\hat{r}(s) + \frac{(\lambda_n - \beta_n)s^{n-1} + \cdots + (\lambda_1 - \beta_1)}{\hat{\lambda}(s)}\,\hat{y}_p(s) \tag{2.2.4}$$

This expression is a new parameterization of the plant. Let

$$\begin{aligned} \hat{a}^*(s) &= \alpha_n s^{n-1} + \cdots + \alpha_1 = k_p \hat{n}_p(s) \\ \hat{b}^*(s) &= (\lambda_n - \beta_n)s^{n-1} + \cdots + (\lambda_1 - \beta_1) = \hat{\lambda}(s) - \hat{d}_p(s) \end{aligned} \tag{2.2.5}$$

so that the new representation of the plant can be written

$$\hat{y}_p(s) = \frac{\hat{a}^*(s)}{\hat{\lambda}(s)}\,\hat{r}(s) + \frac{\hat{b}^*(s)}{\hat{\lambda}(s)}\,\hat{y}_p(s) \tag{2.2.6}$$

The transfer function from $r \to y_p$ is given by

$$\frac{\hat{y}_p(s)}{\hat{r}(s)} = \frac{\hat{a}^*(s)}{\hat{\lambda}(s) - \hat{b}^*(s)} \tag{2.2.7}$$

and it is easy to verify that this transfer function is $\hat{P}(s)$ when $\hat{a}^*(s)$ and $\hat{b}^*(s)$ are given by (2.2.5). Further, this choice is unique when $\hat{n}_p(s)$ and $\hat{d}_p(s)$ are coprime: indeed, suppose that there exist $\hat{a}^*(s) + \delta\hat{a}(s)$, $\hat{b}^*(s) + \delta\hat{b}(s)$, such that the transfer function was still $k_p \hat{n}_p(s)/\hat{d}_p(s)$. The following equation would then have to be satisfied

$$\frac{\delta\hat{a}(s)}{\delta\hat{b}(s)} = -k_p \frac{\hat{n}_p(s)}{\hat{d}_p(s)} = -\hat{P}(s) \tag{2.2.8}$$

However, equation (2.2.8) has no solution since the degree of $\hat{d}_p$ is n, and $\hat{n}_p$, $\hat{d}_p$ are coprime, while the degree of $\delta\hat{b}$ is at most $n-1$.

State-Space Realization

A state-space realization of the foregoing representation can be found by choosing $\Lambda \in \mathbb{R}^{n \times n}$, $b_\lambda \in \mathbb{R}^n$ in controllable canonical form, such that

$$\Lambda = \begin{bmatrix} 0 & 1 & 0 & \cdot & \cdot & 0 \\ 0 & 0 & 1 & \cdot & \cdot & 0 \\ \cdot & \cdot & \cdot & \cdot & \cdot & \cdot \\ \cdot & \cdot & \cdot & \cdot & \cdot & 0 \\ 0 & 0 & \cdot & \cdot & \cdot & 1 \\ -\lambda_1 & \cdot & \cdot & \cdot & \cdot & -\lambda_n \end{bmatrix} \qquad b_\lambda = \begin{bmatrix} 0 \\ \cdot \\ \cdot \\ \cdot \\ 0 \\ 1 \end{bmatrix}$$

$$(sI - \Lambda)^{-1} b_\lambda = \frac{1}{\hat{\lambda}(s)} \begin{bmatrix} 1 \\ s \\ \cdot \\ \cdot \\ \cdot \\ s^{n-1} \end{bmatrix} \tag{2.2.9}$$

In analogy with (2.2.5), define

$$a^{*^T} := (\alpha_1, \ldots, \alpha_n) \qquad b^{*^T} := (\lambda_1 - \beta_1, \ldots, \lambda_n - \beta_n) \tag{2.2.10}$$

and the vectors $w_p^{(1)}(t)$, $w_p^{(2)}(t) \in \mathbb{R}^n$

$$\dot{w}_p^{(1)} = \Lambda w_p^{(1)} + b_\lambda r$$

$$\dot{w}_p^{(2)} = \Lambda w_p^{(2)} + b_\lambda y_p \tag{2.2.11}$$

with initial conditions $w_p^{(1)}(0), w_p^{(2)}(0)$. In Laplace transforms

$$\hat{w}_p^{(1)}(s) = (sI - \Lambda)^{-1} b_\lambda \hat{r}(s) + (sI - \Lambda)^{-1} w_p^{(1)}(0)$$

$$\hat{w}_p^{(2)}(s) = (sI - \Lambda)^{-1} b_\lambda \hat{y}_p(s) + (sI - \Lambda)^{-1} w_p^{(2)}(0) \tag{2.2.12}$$

With this notation, the description of the plant (2.2.6) becomes

$$\hat{y}_p(s) = a^{*^T} \hat{w}_p^{(1)}(s) + b^{*^T} \hat{w}_p^{(2)}(s) \tag{2.2.13}$$

and, since the plant parameters a^*, b^* are constant, the same expression is valid in the time domain

$$y_p(t) = a^{*^T} w_p^{(1)}(t) + b^{*^T} w_p^{(2)}(t) := \theta^{*^T} w_p(t) \tag{2.2.14}$$

where

$$\theta^{*^T} := (a^{*^T}, b^{*^T}) \in \mathbb{R}^{2n}$$

$$w_p(t)^T := (w_p^{(1)^T}(t), w_p^{(2)^T}(t)) \in \mathbb{R}^{2n} \tag{2.2.15}$$

Equations (2.2.10)–(2.2.14) define a realization of the new parameterization. The vector w_p is the *generalized state* of the plant and has dimension $2n$. Therefore, the realization of $\hat{P}(s)$ is not minimal, but the unobservable modes are those of $\hat{\lambda}(s)$ and are all stable.

The vector θ^* is a vector of unknown parameters related linearly to the original plant parameters α_i, β_i by (2.2.10)–(2.2.15). Knowledge of a set of parameters is equivalent to the knowledge of the other and each corresponds to one of the (equivalent) parameterizations. In the last form, however, the plant output depends linearly on the unknown parameters, so that standard identification algorithms can be used. This plant parameterization is represented in Figure 2.1.

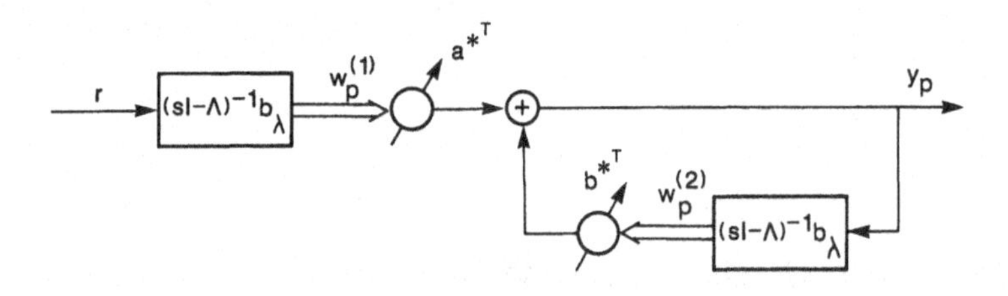

Figure 2.1: Plant Parameterization

Identifier Structure

The purpose of the identifier is to produce a recursive estimate $\theta(t)$ of the *nominal* parameter θ^*. Since r and y_p are available, we define the *observer*

$$\begin{aligned} \dot{w}^{(1)} &= \Lambda w^{(1)} + b_\lambda r \\ \dot{w}^{(2)} &= \Lambda w^{(2)} + b_\lambda y_p \end{aligned} \tag{2.2.16}$$

to reconstruct the states of the plant. The initial conditions in (2.2.16) are arbitrary. We also define the identifier signals

$$\begin{aligned} \theta^T(t) &:= (a^T(t), b^T(t)) \in \mathbb{R}^{2n} \\ w^T(t) &:= (w^{(1)^T}(t), w^{(2)^T}(t)) \in \mathbb{R}^{2n} \end{aligned} \tag{2.2.17}$$

By (2.2.11) and (2.2.16), the *observer error* $w(t) - w_p(t)$ decays exponentially to zero, *even when the plant is unstable.* We note that the generalized state of the plant $w_p(t)$ is such that it can be reconstructed from available signals, *without knowledge of the plant parameters.*

The plant output can be written

$$y_p(t) = \theta^{*^T} w(t) + \epsilon(t) \tag{2.2.18}$$

where the signal $\epsilon(t)$ is to remind one of the presence of an additive exponentially decaying term, given here by

$$\epsilon(t) = \theta^{*^T} (w_p(t) - w(t)) \tag{2.2.19}$$

This term is due to the initial conditions in the observer. We will first neglect the presence of the $\epsilon(t)$ term but later show that it does not affect the properties of the identifier.

In analogy with the expression of the plant output, the output of the identifier is defined to be

$$y_i(t) = \theta^T(t) w(t) \quad \in \mathbb{R} \tag{2.2.20}$$

We also define the *parameter error*

$$\phi(t) := \theta(t) - \theta^* \quad \in \mathbb{R}^{2n} \tag{2.2.21}$$

and the *identifier error*

$$e_1(t) := y_i(t) - y_p(t) = \phi^T(t) w(t) + \epsilon(t) \tag{2.2.22}$$

These signals will be used by the identification algorithm, and are represented in Figure 2.2.

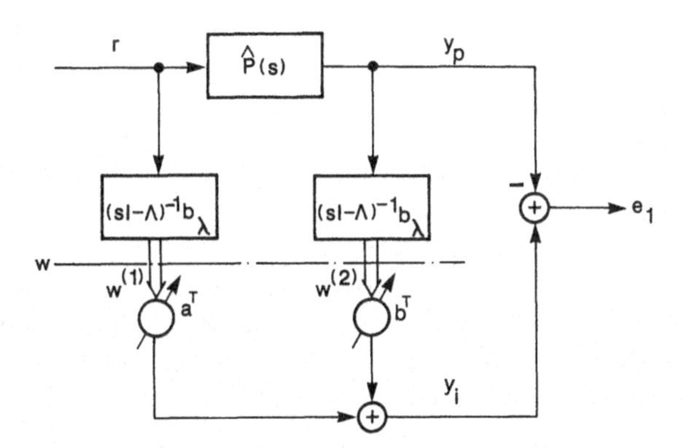

Figure 2.2: Identifier Structure

2.3 LINEAR ERROR EQUATION AND IDENTIFICATION ALGORITHMS

Many identification algorithms (cf. Eykhoff [1974], Ljung & Soderstrom [1983]) rely on a linear expression of the form obtained above, that is

$$y_p(t) = \theta^{*T} w(t) \tag{2.3.1}$$

where $y_p(t)$, $w(t)$ are known signals and θ^* is unknown. The vector $w(t)$ is usually called the *regressor* vector. With the expression of $y_p(t)$ is associated the standard *linear error equation*

$$e_1(t) = \phi^T(t) w(t) \tag{2.3.2}$$

We arbitrarily separated the identifier into an *identifier structure* and an *identification algorithm*. The identifier structure constructs the regressor w and other signals, related by the identifier error equation. The identification algorithm is defined by a differential equation, called the *update law*, of the form

$$\dot{\theta} = \dot{\phi} = F(y_p, e_1, \theta, w) \tag{2.3.3}$$

where F is a causal operator explicitly independent of θ^*, which defines the evolution of the identifier parameter θ.

2.3.1 Gradient Algorithms

The update law

$$\dot{\theta} = -g\, e_1 w \qquad g > 0 \tag{2.3.4}$$

defines the standard *gradient algorithm*. The right-hand side is proportional to the gradient of the output error squared, viewed as a function of θ, that is

$$\frac{\partial}{\partial \theta}\left(e_1^2(\theta) \right) = 2\, e_1 w \tag{2.3.5}$$

This update law can thus be seen as a *steepest descent* method. The parameter g is a *fixed*, strictly positive gain called the *adaptation gain*, and it allows us to vary the rate of adaptation of the parameters. The initial condition $\theta(0)$ is arbitrary, but it can be chosen to take any a priori knowledge of the plant parameters into account.

An alternative to this algorithm is the *normalized gradient algorithm*

$$\dot{\theta} = -g \frac{e_1 w}{1 + \gamma w^T w} \qquad g, \gamma > 0 \tag{2.3.6}$$

where g and γ are constants. This update law is equivalent to the previous update law, with w replaced by $w / \sqrt{1 + \gamma w^T w}$ in $\dot{\theta} = -g\, w w^T \phi$. The new regressor is thus a normalized form of w. The right-hand side of the differential equation (2.3.6) is globally Lipschitz in ϕ (using

(2.3.2)), even when w is unbounded.

When the nominal parameter θ^* is known *a priori* to lie in a set $\Theta \in \mathbb{R}^{2n}$ (which we will assume to be closed, convex and delimited by a smooth boundary), it is useful to modify the update law to take this information into account. For example, the *normalized gradient algorithm with projection* is defined by

$$\begin{aligned} \dot{\theta} &= -g \frac{e_1 w}{1+\gamma w^T w} & \theta \in \operatorname{int}(\Theta) \\ &= \Pr\left[-g \frac{e_1 w}{1+\gamma w^T w} \right] & \text{if } \theta \in \partial\Theta \text{ and } e_1 w^T \theta_{perp} < 0 \end{aligned} \tag{2.3.7}$$

where $\operatorname{int}\Theta$ and $\partial\Theta$ denote the interior and boundary of Θ, $\Pr(z)$ denotes the projection of the vector z onto the hyperplane tangent to $\partial\Theta$ at θ and θ_{perp} denotes the unit vector perpendicular to the hyperplane, pointing outward.

A frequent example of projection occurs when *a priori* bounds p_i^-, p_i^+ are known, that is

$$\theta_i^* \in [p_i^-, p_i^+] \tag{2.3.8}$$

The update law is then modified to

$$\begin{aligned} \dot{\theta}_i &= 0 \qquad \text{if } \theta_i = p_i^- \text{ and } \dot{\theta}_i < 0 \\ &\qquad\quad \text{or } \theta_i = p_i^+ \text{ and } \dot{\theta}_i > 0 \end{aligned} \tag{2.3.9}$$

The gradient algorithms can be used to identify the plant parameters with the identifier structure described in Section 2.2. Using the normalized gradient algorithm, for example, the implementation is as follows.

Identifier with Normalized Gradient Algorithm—Implementation

Assumptions

(A1)–(A2)

Data

n

Input

$r(t), y_p(t) \in \mathbb{R}$

Output

$\theta(t),\ y_i(t) \in \mathbb{R}$

Internal Signals

$w(t) \in \mathbb{R}^{2n}\ (w^{(1)}(t),\ w^{(2)}(t) \in \mathbb{R}^n)$

$\theta(t) \in \mathbb{R}^{2n}\ (a(t),\ b(t) \in \mathbb{R}^n)$

$y_i(t),\ e_1(t) \in \mathbb{R}$

Initial conditions are arbitrary.

Design Parameters

Choose

- $\Lambda \in \mathbb{R}^{n \times n}$, $b_\lambda \in \mathbb{R}^n$ in controllable canonical form such that $\det(sI - \Lambda) = \hat{\lambda}(s)$ is Hurwitz.
- $g, \gamma > 0$.

Identifier Structure

$\dot{w}^{(1)} = \Lambda w^{(1)} + b_\lambda r$

$\dot{w}^{(2)} = \Lambda w^{(2)} + b_\lambda y_p$

$\theta^T = (a^T, b^T)$ estimates of $(\alpha_1, \ldots, \alpha_n, \lambda_1 - \beta_1, \ldots, \lambda_n - \beta_n)$

$w^T = (w^{(1)^T}, w^{(2)^T})$

$y_i = \theta^T w$

$e_1 = y_i - y_p$

Normalized Gradient Algorithm

$$\dot{\theta} = -g \frac{e_1 w}{1 + \gamma w^T w}$$

□

Comment

We leave it to the reader to check that other choices of (Λ, b_λ) are possible, with minor adjustments. Indeed, all that is required for identification is that there exist unique a^*, b^* in the corresponding parameterization. Alternate choices of $(sI - \Lambda)^{-1} b_\lambda$ include

$$(sI - \Lambda)^{-1} b_\lambda = \begin{bmatrix} \frac{1}{s+a} \\ \cdot \\ \cdot \\ \cdot \\ \frac{1}{(s+a)^n} \end{bmatrix} \qquad a > 0$$

and

$$(sI - \Lambda)^{-1} b_\lambda = \begin{bmatrix} \frac{b_1}{s+a_1} \\ \cdot \\ \cdot \\ \cdot \\ \frac{b_n}{(s+a_n)} \end{bmatrix} \qquad a_i \neq a_j > 0 \,,\, b_i \neq 0$$

2.3.2 Least-Squares Algorithms

Least-squares (LS) algorithms can be derived by several methods. One approach was presented in the introduction. Another interesting approach is to connect the parameter identification problem to the *stochastic state estimation* problem of a linear time varying system. The parameter θ^* can be considered to be the unknown state of the system

$$\dot{\theta}^*(t) = 0 \tag{2.3.10}$$

with output

$$y_p(t) = w^T(t)\theta^*(t) \tag{2.3.11}$$

Assuming that the right-hand sides of (2.3.10)–(2.3.11) are perturbed by zero mean white gaussian noises of spectral intensities $Q \in \mathbb{R}^{2n \times 2n}$ and $1/g \in \mathbb{R}$, respectively, the least-squares estimator is the so-called *Kalman filter* (Kalman & Bucy [1961])

$$\begin{aligned} \dot{\theta} &= -g P w e_1 \\ \dot{P} &= Q - g P w w^T P \qquad Q, g > 0 \end{aligned} \tag{2.3.12}$$

Q and g are fixed design parameters of the algorithm. The update law for θ is very similar to the gradient update law, with the presence of the so-called *correlation* term $w e_1$. The matrix P is called the *covariance matrix* and acts in the θ update law as a time-varying, *directional* adaptation gain. The covariance update law in (2.3.12) is called the *covariance propagation equation.* The initial conditions are arbitrary, except that $P(0) > 0$. $P(0)$ is usually chosen to reflect the confidence in the initial estimate $\theta(0)$.

In the identification literature, the least-squares algorithm referred to is usually the algorithm with $Q = 0$, since the parameter θ^* is assumed to be constant. The covariance propagation equation is then replaced by

$$\frac{dP}{dt} = -gPww^TP \quad \text{or} \quad \frac{d(P^{-1})}{dt} = gww^T \quad g > 0 \tag{2.3.13}$$

where g is a constant.

The new expression for P^{-1} shows that $dP^{-1}/dt \geq 0$, so that P^{-1} may grow without bound. Then P will become arbitrarily small in some directions and the adaptation of the parameters in those directions becomes very slow. This so-called *covariance wind-up* problem can be prevented using the *least-squares with forgetting factor* algorithm, defined by

$$\frac{dP}{dt} = -g(-\lambda P + Pww^TP)$$

$$\text{or} \quad \frac{d(P^{-1})}{dt} = g(-\lambda P^{-1} + ww^T) \quad \lambda, g > 0 \tag{2.3.14}$$

Another possible remedy is the *covariance resetting,* where P is reset to a predetermined positive definite value, whenever $\lambda_{\min}(P)$ falls under some threshold.

The *normalized least-squares* algorithm is defined (cf. Goodwin & Mayne [1987]) by

$$\dot{\theta} = -g\frac{Pwe_1}{1 + \gamma w^TPw} \qquad g, \gamma > 0$$

$$\frac{dP}{dt} = -g\frac{Pww^TP}{1 + \gamma w^TPw}$$

$$\text{or} \quad \frac{d(P^{-1})}{dt} = g\frac{ww^T}{1 + \gamma w^T(P^{-1})^{-1}w} \tag{2.3.15}$$

Again g, γ are fixed parameters and $P(0) > 0$. The same modifications can also be made to avoid covariance wind-up.

The least-squares algorithms are somewhat more complicated to implement but are found in practice to have faster convergence properties.

Identifier with Normalized LS Algorithm and Covariance Resetting—Implementation

The algorithm is the same as for the normalized gradient algorithm, except

Internal Signals

In addition

$$P(t) \in \mathbb{R}^{2n \times 2n}$$

Design Parameters

Choose, in addition

- $k_0 > k_1 > 0$

Normalized LS Algorithm with Covariance Resetting

$$\dot{\theta} = -g \frac{P\, w\, e_1}{1 + \gamma\, w^T P\, w} \qquad \dot{P} = -g \frac{P\, w\, w^T P}{1 + \gamma\, w^T P\, w}$$

$$P(0) = P(t_r^+) = k_0 I > 0 \text{ where } t_r = \{\, t \mid \lambda_{\min}(P(t)) \le k_1 \,\}$$

□

2.4 PROPERTIES OF THE IDENTIFICATION ALGORITHMS—IDENTIFIER STABILITY

2.4.1 Gradient Algorithms

In this section, we establish properties of the gradient algorithm

$$\dot{\phi} = \dot{\theta} = -g\, e_1\, w \qquad g > 0 \tag{2.4.1}$$

and the normalized gradient algorithm

$$\dot{\phi} = \dot{\theta} = -g \frac{e_1\, w}{1 + \gamma\, w^T w} \qquad g, \gamma > 0 \tag{2.4.2}$$

assuming the linear error equation

$$e_1 = \phi^T w \tag{2.4.3}$$

Theorems 2.4.1–2.4.4 establish general properties of the gradient algorithms and concern solutions of the differential equations (2.4.1)–(2.4.2), with e_1 defined by (2.4.3). The properties do not require that the vector w originates from the identifier described in Section 2.2, but only require that w be a piecewise continuous function of time, to guarantee the existence of the solutions. The theorems are also valid for vectors w of any dimension, not necessarily even.

Theorem 2.4.1 Linear Error Equation with Gradient Algorithm
Consider the linear error equation (2.4.3), together with the gradient algorithm (2.4.1). Let $w : \mathbb{R}_+ \to \mathbb{R}^{2n}$ be piecewise continuous.

Then (a) $e_1 \in L_2$
(b) $\phi \in L_\infty$

Proof of Theorem 2.4.1
The differential equation describing ϕ is

$$\dot{\phi} = -g w w^T \phi$$

Let the Lyapunov function $v = \phi^T \phi$ so that the derivative along the trajectories is given by

$$\dot{v} = -2g(\phi^T w)^2 = -2g e_1^2 \leq 0$$

Hence, $0 \leq v(t) \leq v(0)$ for all $t \geq 0$, so that $v, \phi \in L_\infty$. Since v is a positive, monotonically decreasing function, the limit $v(\infty)$ is well-defined and

$$-\frac{1}{2} g \int_0^\infty \dot{v}\, dt = \int_0^\infty e_1^2\, dt < \infty$$

that is $e_1 \in L_2$. □

Theorem 2.4.2 Linear Error Equation with Normalized Gradient Algorithm
Consider the linear error equation (2.4.3) together with the normalized gradient algorithm (2.4.2). Let $w : \mathbb{R}_+ \to \mathbb{R}^{2n}$ be piecewise continuous.

Then (a) $\dfrac{e_1}{\sqrt{1+\gamma w^T w}} \in L_2 \cap L_\infty$

(b) $\phi \in L_\infty,\ \dot{\phi} \in L_2 \cap L_\infty$

(c) $\beta = \dfrac{\phi^T w}{1 + \| w_t \|_\infty} \in L_2 \cap L_\infty$

Proof of Theorem 2.4.2
Let $v = \phi^T \phi$, so that

$$\dot{v} = -\frac{2g\, e_1^2}{1 + \gamma w^T w} \leq 0$$

Hence, $0 \leq v(t) \leq v(0)$ for all $t \geq 0$, so that v, ϕ, $e_1 / \sqrt{1 + \gamma w^T w}$,

$\beta \in L_\infty$. Using the fact that $x/1+x \le 1$ for all $x \ge 0$, we get that $|\dot{\phi}| \le (g/\gamma)|\phi|$, and $\dot{\phi} \in L_\infty$. Since v is a positive, monotonically decreasing function, the limit $v(\infty)$ is well defined and

$$-\int_0^\infty \dot{v}\, dt < \infty$$

implies that $e_1/\sqrt{1+\gamma w^T w} \in L_2$. Note that

$$\beta = \frac{e_1}{\sqrt{1+\gamma w^T w}} \frac{\sqrt{1+\gamma w^T w}}{1+\| w_t \|_\infty}$$

where the first term is in L_2 and the second in L_∞, so that $\beta \in L_2$. Since

$$|\dot{\phi}|^2 \le \frac{g^2}{\gamma} \frac{e_1^2}{1+\gamma w^T w}$$

it follows that $\dot{\phi} \in L_2$. □

Effect of Initial Conditions and Projection

In the derivation of the linear error equation in Section 2.2, we found exponentially decaying terms, such that (2.4.3) is replaced by

$$e_1(t) = \phi^T(t)\, w(t) + \epsilon(t) \tag{2.4.4}$$

where $\epsilon(t)$ is an exponentially decaying term due to the initial conditions in the observer.

It may also be useful, or necessary, to replace the gradient algorithms by the algorithms with projection. The following theorem asserts that these modifications do not affect the previous results.

Theorem 2.4.3 Effect of Initial Conditions and Projection

If the linear error equation (2.4.3) is replaced by (2.4.4) and/or the gradient algorithms are replaced by the gradient algorithms with projection,

Then the conclusions of theorems 2.4.1 and 2.4.2 are valid.

Proof of Theorem 2.4.3

(a) Effect of initial conditions

Modify the Lyapunov function to

$$v = \phi^T \phi + \frac{g}{2} \int_t^{\infty} \epsilon^2(\tau)\, d\tau$$

Note that the additional term is bounded and tends to zero as t tends to infinity. Consider first the gradient algorithm (2.4.1), so that

$$\dot{v} = -2g(\phi^T w)^2 - 2g(\phi^T w)\epsilon - \frac{g}{2}\epsilon^2$$

$$= -2g(\phi^T w + \frac{\epsilon}{2})^2 \leq 0 \qquad (2.4.5)$$

The proof can be completed as in theorem 2.4.1, noting that $\epsilon \in L_2 \cap L_\infty$, and similarly for theorem 2.4.2.

(b) Effect of projection

Denote by z the right-hand side of the update law (2.4.1) or (2.4.2). When $\theta \in \partial\Theta$ and z is directed outside Θ, z is replaced by $\Pr(z)$ in the update law. Note that it is sufficient to prove that the derivative of the Lyapunov function on the boundary is less than or equal to its value with the original differential equation. Therefore, denote by z_{perp} the component of z perpendicular to the tangent plane at θ, so that $z = \Pr(z) + z_{perp}$. Since $\theta^* \in \Theta$ and Θ is convex, $(\theta^T - \theta^{*T}) \cdot z_{perp} = \phi^T z_{perp} \geq 0$. Using the Lyapunov function $v = \phi^T \phi$, we find that, for the original differential equation, $\dot{v} = 2\phi^T z$. For the differential equation with projection, $\dot{v}_{\Pr} = 2\phi^T \Pr(z) = \dot{v} - 2\phi^T \cdot z_{perp}$ so that $\dot{v}_{\Pr} \leq \dot{v}$, that is the projection can only improve the convergence of the algorithm. The proof can again be completed as before. □

2.4.2 Least-Squares Algorithms

We now turn to the *normalized LS algorithm with covariance resetting*, defined by the following update law

$$\dot{\phi} = \dot{\theta} = -g \frac{P w e_1}{1 + \gamma w^T P w} \qquad g, \gamma > 0 \qquad (2.4.6)$$

and a *discontinuous* covariance propagation

$$\frac{dP}{dt} = -g \frac{P w w^T P}{1 + \gamma w^T P w} \quad \text{or} \quad \frac{d(P^{-1})}{dt} = g \frac{w w^T}{1 + \gamma w^T (P^{-1})^{-1} w}$$

$$P(0) = P(t_r^+) = k_0 I > 0$$

$$\text{where } t_r = \{ t \mid \lambda_{\min}(P(t)) \leq k_1 < k_0 \}. \qquad (2.4.7)$$

This update law has similar properties as the normalized gradient update law, as stated in the following theorem.

Theorem 2.4.4 Linear Error Equation with Normalized LS Algorithm and Covariance Resetting

Consider the linear error equation (2.4.3), together with the normalized LS algorithm with covariance resetting (2.4.6)–(2.4.7).

Let $w : \mathbb{R}_+ \to \mathbb{R}^{2n}$ be piecewise continuous.

Then (a) $\dfrac{e_1}{\sqrt{1+\gamma w^T P w}} \in L_2 \cap L_\infty$

(b) $\phi \in L_\infty, \dot{\phi} \in L_2 \cap L_\infty$

(c) $\beta = \dfrac{\phi^T w}{1+\| w_t \|_\infty} \in L_2 \cap L_\infty$

Proof of Theorem 2.4.4

The covariance matrix P is a discontinuous function of time. Between discontinuities, the evolution is described by the differential equation in (2.4.7). We note that $d/dt\, P^{-1} \geq 0$, so that $P^{-1}(t_1) - P^{-1}(t_2) \geq 0$ for all $t_1 \geq t_2 \geq 0$ between covariance resettings. At the resettings, $P^{-1}(t_r^+) = k_0^{-1} I$, so that $P^{-1}(t) \geq P^{-1}(t_0) = k_0^{-1} I$, for all $t \geq 0$.

On the other hand, due to the resetting, $P(t) \geq k_1 I$ for all $t \geq 0$, so that

$$k_0 I \geq P(t) \geq k_1 I \qquad k_1^{-1} I \geq P^{-1}(t) \geq k_0^{-1} I \tag{2.4.8}$$

where we used results of Section 1.3.

Note that the interval between resettings is bounded below, since

$$\begin{aligned} \frac{d(P^{-1})}{dt} &\leq g \frac{|w|^2}{1+\gamma \lambda_{\min}(P)|w|^2} \\ &\leq \frac{g}{\gamma} \| P^{-1} \| \end{aligned} \tag{2.4.9}$$

where we used the fact that $x/1+x \leq 1$ for all $x \geq 0$. Thus, the differential equation governing P^{-1} is globally Lipschitz. It also follows that $\{ t_r \}$ is a set of measure zero.

Let now $v = \phi^T P^{-1} \phi$, so that

$$\dot{v} = -g \frac{e_1^2}{1+\gamma w^T P w} \leq 0$$

between resettings. At the points of discontinuity of P,

$$v(t_r^+) - v(t_r) = \phi^T (P^{-1}(t_r^+) - P^{-1}(t_r))\phi \le 0$$

It follows that $0 \le v(t) \le v(0)$, for all $t \ge 0$, and, from the bounds on P, we deduce that $\phi, \dot{\phi}, \beta \in L_\infty$. Also

$$-\int_0^\infty \dot{v}\, dt < \infty \quad \text{implies} \quad \frac{e_1}{\sqrt{1 + \gamma w^T P w}} \in L_2$$

Note that

$$\frac{\phi^T w}{1 + \| w_t \|_\infty} = \frac{\phi^T w}{\sqrt{1 + \gamma w^T P w}} \frac{\sqrt{1 + \gamma w^T P w}}{1 + \| w_t \|_\infty} \tag{2.4.10}$$

$$\dot{\phi} = -g \frac{e_1}{\sqrt{1 + \gamma w^T P w}} \frac{P w}{\sqrt{1 + \gamma w^T P w}} \tag{2.4.11}$$

where the first terms in the right-hand sides of (2.4.10)–(2.4.11) are in L_2 and the last terms are bounded. The conclusions follow from this observation. □

Comments

a) Theorems 2.4.1–2.4.4 state general properties of differential equations arising from the identification algorithms described in Section 2.3. The theorems can be directly applied to the identifier with the structure described in Section 2.2, and the results interpreted in terms of the parameter error ϕ and the identifier error e_1.

b) The conclusions of theorems 2.4.1–2.4.4 may appear somewhat weak, since none of the errors involved actually converge to zero. The reader should note however that the conclusions are valid under very general conditions regarding the input signal w. In particular, no assumption is made on the boundedness or on the differentiability of w.

c) The conclusions of theorem 2.4.2 can be interpreted in the following way. The function $\beta(t)$ is defined by

$$\beta(t) = \frac{\phi^T(t) w(t)}{1 + \| w_t \|_\infty} = \frac{e_1(t)}{1 + \| w_t \|_\infty} \tag{2.4.12}$$

so that

$$| e_1(t) | = | \phi^T(t) w(t) | = \beta(t) \| w_t \|_\infty + \beta(t) \tag{2.4.13}$$

The purpose of the identification algorithms is to reduce the parameter error ϕ to zero or at least the error e_1. In (2.4.12), β can be interpreted as a *normalized error*, that is e_1 normalized by $\| w_t \|_\infty$. In (2.4.13), β can be interpreted as the *gain* from w to $\phi^T w$. From theorem 2.4.2, this gain is guaranteed to become small as $t \to \infty$ in an L_2 sense.

2.4.3 Stability of the Identifier

We are not guaranteed the convergence of the *parameter error* ϕ to zero. Since only one output y_p is measured to determine a vector of unknown parameters, some additional condition on the signal w (see Section 2.5) must be satisfied in order to guarantee parameter convergence. In fact, we are not even guaranteed the convergence of the identifier error e_1 to zero. This can be obtained under the following additional assumption

(A3) Bounded Output Assumption
Assume that the plant is either stable or located in a control loop such that r and y_p are bounded.

Theorem 2.4.5 Stability of the Identifier
Consider the identification problem, with (A1)–(A3), the identifier structure of Section 2.2 and the gradient algorithms (2.4.1), (2.4.2) or the normalized LS algorithm with covariance resetting (2.4.6), (2.4.7).

Then the output error $e_1 \in L_2 \cap L_\infty$, $e_1 \to 0$ as $t \to \infty$ and the parameter error $\phi, \dot{\phi} \in L_\infty$.
The derivative of the parameter error $\dot{\phi} \in L_2 \cap L_\infty$ and $\dot{\phi} \to 0$ as $t \to \infty$.

Proof of Theorem 2.4.5
Since r and y_p are bounded, it follows from (2.2.16), (2.2.17), and the stability of Λ, that w and $\dot{w}$ are bounded. By theorems 2.4.1–2.4.4, ϕ and $\dot{\phi}$ are bounded so that e_1 and $\dot{e}_1$ are bounded. Also $e_1 \in L_2$, and by corollary 1.2.2, $e_1, \dot{e}_1 \in L_\infty$ and $e_1 \in L_2$ implies that $e_1 \to 0$ as $t \to \infty$. Similar conclusions follow directly for $\dot{\phi}$. □

Regular Signals

Theorem 2.4.5 relies on the boundedness of $w, \dot{w}$, guaranteed by (A3). It is of interest to relax this condition and to replace it by a weaker condition. We will present such a result using a *regularity* condition on the regressor w. This condition guarantees a certain degree of smoothness of the signal w and seems to be fundamental in excluding pathological signals in the course of the proofs presented in this book. In discrete time, such a condition is not necessary, because it is automatically verified. The definition presented here corresponds to a definition in Narendra, Lin, & Valavani [1980].

Definition Regular Signals

Let $z : \mathbb{R}_+ \to \mathbb{R}^n$, such that $z, \dot{z} \in L_{\infty e}$.
z is called *regular* if, for some $k_1, k_2 \geq 0$

$$| \dot{z}(t) | \leq k_1 \| z_t \|_\infty + k_2 \qquad \text{for all } t \geq 0 \qquad (2.4.14)$$

The class of regular signals includes bounded signals with bounded derivatives, but also unbounded signals (e.g., e^t). It typically excludes signals with "increasing frequency" such as $\sin(e^t)$. We will also derive some properties of regular signals in Chapter 3. Note that it will be sufficient for (2.4.14) to hold everywhere except on a set of measure zero. Therefore, piecewise differentiable signals can also be considered.

This definition allows us to state the following theorem, extending the properties derived in theorems 2.4.2–2.4.4 to the case when w is regular.

Proposition 2.4.6

Let $\phi, w : \mathbb{R}_+ \to \mathbb{R}^{2n}$ be such that $w, \dot{w} \in L_{\infty e}$ and $\phi, \dot{\phi} \in L_\infty$.

If (*a*) w is regular

(*b*) $\beta = \dfrac{\phi^T w}{1 + \| w_t \|_\infty} \in L_2$

Then $\beta, \dot{\beta} \in L_\infty$, and $\beta \to 0$ as $t \to \infty$.

Proof of Proposition 2.4.6

Clearly, $\beta \in L_\infty$ and since $\beta, \dot{\beta} \in L_\infty$, $\beta \in L_2$ implies that $\beta \to 0$ as $t \to \infty$ (corollary 1.2.2), we are left to show that $\dot{\beta} \in L_\infty$.

We have that

$$| \dot{\beta} | \leq \left| \dot{\phi}^T \frac{w}{1 + \| w_t \|_\infty} \right| + \left| \phi^T \frac{\dot{w}}{1 + \| w_t \|_\infty} \right|$$
$$+ \left| \frac{\phi^T w}{1 + \| w_t \|_\infty} \frac{(d / dt \| w_t \|_\infty)}{1 + \| w_t \|_\infty} \right| \qquad (2.4.15)$$

The first and second terms are bounded, since $\phi, \dot{\phi} \in L_\infty$ and w is regular. On the other hand

$$\left| \frac{d}{dt} \| w_t \|_\infty \right| = \left| \frac{d}{dt} \sup_{\tau \le t} | w(\tau) | \right|$$

$$\le \left| \frac{d}{dt} | w(t) | \right| \le \left| \frac{d}{dt} w(t) \right| \tag{2.4.16}$$

The regularity assumption then implies that the last term in (2.4.15) is bounded, and hence $\dot{\beta} \in L_\infty$. □

Stability of the Identifier with Unstable Plant

Proposition 2.4.6 shows that when w is possibly unbounded, but nevertheless satisfies the regularity condition, the relative error $e_1 / 1 + \| w_t \|_\infty$ or gain from $w \to \phi^T w$ tends to zero as $t \to \infty$.

The conclusions of proposition 2.4.6 are useful in proving stability in adaptive control, where the boundedness of the regressor w is not guaranteed a priori. In the identification problem, we are now allowed to consider the case of an unstable plant with bounded input, that is, to relax assumption (A3).

Theorem 2.4.7 Stability of the Identifier—Unstable Plant

Consider the identification problem with (A1) and (A2), the identifier structure of Section 2.2 and the normalized gradient algorithm (2.4.2), or the normalized LS with covariance resetting (2.4.6), (2.4.7).

Then The normalized error $\beta = \dfrac{\phi^T w}{1 + \| w_t \|_\infty} \in L_2 \cap L_\infty$, $\beta \to 0$ as $t \to \infty$ and $\phi, \dot{\phi} \in L_\infty$.

Proof of Theorem 2.4.7

It suffices to show that w is regular, to apply theorem 2.4.2 or 2.4.4 followed by proposition 2.4.6. Combining (2.2.16)–(2.2.18), it follows that

$$\dot{w}(t) = \begin{bmatrix} \Lambda & 0 \\ b_\lambda a^{*^T} & \Lambda + b_\lambda b^{*^T} \end{bmatrix} w(t) + \begin{bmatrix} b_\lambda \\ 0 \end{bmatrix} r(t) \tag{2.4.17}$$

Since r is bounded by (A2), (2.4.17) shows that w is regular. □

2.5 PERSISTENT EXCITATION AND EXPONENTIAL PARAMETER CONVERGENCE

In the previous section, we derived results on the stability of the identifiers and on the convergence of the output error $e_1 = \theta^T w - \theta^{*^T} w = \phi^T w$ to zero. We are now concerned with the convergence of the parameter θ to its nominal value θ^*, that is the convergence of the parameter error ϕ to zero.

The convergence of the identification algorithms is related to the asymptotic stability of the differential equation

$$\dot{\phi}(t) = -g\, w(t)\, w^T(t)\, \phi(t) \qquad g > 0 \tag{2.5.1}$$

which is of the form

$$\dot{\phi}(t) = -A(t)\, \phi(t) \tag{2.5.2}$$

where $A(t) \in \mathbb{R}^{2n \times 2n}$ is a positive semidefinite matrix for all t. Using the Lyapunov function $v = \phi^T \phi$

$$\dot{v} = -\phi^T (A + A^T)\, \phi$$

When $A(t)$ is uniformly positive definite, with $\lambda_{\min}(A + A^T) \geq 2\alpha$, then $\dot{v} \leq -2\alpha v$, which implies that system (2.5.2) is exponentially stable with rate α. For the original differential equation (2.5.1), such is never the case, however, since at any instant the matrix $w(t)\, w^T(t)$ is of rank 1. In fact, any vector ϕ perpendicular to w lies in the null space of $w\, w^T$ and results in $\dot{\phi} = 0$. However, since w varies with time, we can expect ϕ to still converge to 0 if w completely spans $\mathbb{R}^{2n}$ as t varies. This leads naturally to the following definition. For consistency, the dimension of w is assumed to be $2n$, but it is in fact arbitrary.

Definition Persistency of Excitation (PE)

A vector $w : \mathbb{R}_+ \to \mathbb{R}^{2n}$ is *persistently exciting (PE)* if there exist α_1, α_2, $\delta > 0$ such that

$$\alpha_2 I \geq \int_{t_0}^{t_0+\delta} w(\tau)\, w^T(\tau)\, d\tau \geq \alpha_1 I \qquad \text{for all } t_0 \geq 0 \tag{2.5.3}$$

Although the matrix $w(\tau)\, w^T(\tau)$ is singular for all τ, the PE condition requires that w rotates sufficiently in space that the integral of the matrix $w(\tau)\, w^T(\tau)$ is uniformly positive definite over any interval of some length δ.

The condition has another interpretation, by reexpressing the PE condition in scalar form

$$\alpha_2 \geq \int_{t_0}^{t_0+\delta} (w^T(\tau)\, x)^2\, d\tau \geq \alpha_1 \qquad \text{for all } t_0 \geq 0,\ |x| = 1 \tag{2.5.4}$$

which appears as a condition on the energy of w in all directions.

With this, we establish the following convergence theorem.

Theorem 2.5.1 PE and Exponential Stability

Let $w : \mathbb{R}_+ \to \mathbb{R}^{2n}$ be piecewise continuous.

If w is PE

Then (2.5.1) is globally exponentially stable

Comments

The proof of theorem 2.5.1 can be found in various places in the literature (Sondhi & Mitra [1976], Morgan & Narendra [1977a&b], Anderson [1977], Kreisselmeier [1977]). The proof by Anderson has the advantage of leading to interesting interpretations, while those by Sondhi & Mitra and Kreisselmeier give estimates of the convergence rates.

The idea of the proof of exponential stability by Anderson [1977] is to note that the PE condition is a UCO condition on the system

$$\dot{\theta}^*(t) = 0$$

$$y(t) = w^T(t)\theta^*(t) \tag{2.5.5}$$

which is the system described earlier in the context of the least-squares identification algorithms (cf. (2.3.10), (2.3.11)). We recall that the identification problem is equivalent to the state estimation problem for the system described by (2.5.5). We now find that the persistency of excitation condition, which turns out to be an *identifiability condition*, is equivalent to a uniform complete observability condition on system (2.5.5).

The proof of theorem 2.5.1 uses the following lemma by Anderson & Moore [1969], which states that the UCO of the system $[C, A]$ is equivalent to the UCO of the system with output injection $[C, A + KC]$. The proof of the lemma is given in the Appendix. It is an alternate proof to the original proof by Anderson & Moore and relates the eigenvalues of the associated observability grammians, thereby leading to estimates of the convergence rates in the proof of theorem 2.5.1 given afterward.

Lemma 2.5.2 Uniform Complete Observability Under Output Injection

Assume that, for all $\delta > 0$, there exists $k_\delta \geq 0$ such that, for all $t_0 \geq 0$

$$\int_{t_0}^{t_0+\delta} \| K(\tau) \|^2 d\tau \leq k_\delta \tag{2.5.6}$$

Then the system $[C, A]$ is uniformly completely observable *if and only if* the system $[C, A + KC]$ is uniformly completely observable.

Moreover, if the observability grammian of the system $[A, C]$ satisfies

$$\beta_2 I \geq N(t_0, t_0+\delta) \geq \beta_1 I$$

then the observability grammian of the system $[A + KC, C]$ satisfies these inequalities with identical δ and

$$\beta_1' = \beta_1 / (1 + \sqrt{k_\delta \beta_2})^2 \tag{2.5.7}$$

$$\beta_2' = \beta_2 \exp(k_\delta \beta_2) \tag{2.5.8}$$

Proof of Lemma 2.5.2 in Appendix.

Proof of Theorem 2.5.1

Let $v = \phi^T \phi$, so that $\dot{v} = -2g(w^T\phi)^2 \leq 0$ along the trajectories of (2.5.1). For all $t_0 \geq 0$

$$\int_{t_0}^{t_0+\delta} \dot{v}\, d\tau = -2g \int_{t_0}^{t_0+\delta} (w^T(\tau)\phi(\tau))^2 d\tau \tag{2.5.9}$$

By the PE assumption, the system $[0, w^T(t)]$ is UCO. Under output injection with $K(t) = -gw(t)$, the system becomes $[-gw(t)w^T(t), w^T(t)]$, with

$$k_\delta = \int_{t_0}^{t_0+\delta} |gw(\tau)|^2 d\tau$$

$$= g^2 \operatorname{tr}\left[\int_{t_0}^{t_0+\delta} w(\tau)w^T(\tau)\, d\tau\right] \leq 2ng^2\beta_2 \tag{2.5.10}$$

where $2n$ is the dimension of w. By lemma 2.5.2, the system with output injection is UCO. Therefore, for all $t_0 \geq 0$

$$\int_{t_0}^{t_0+\delta} \dot{v}\, d\tau \leq \frac{-2g\beta_1}{(1+\sqrt{2ng\beta_2})^2} |\phi(t_0)|^2 \tag{2.5.11}$$

Exponential convergence then follows from theorem 1.5.2. □

Exponential Convergence of the Identifier

Theorem 2.5.1 can be applied to the identification problem as follows.

Theorem 2.5.3 Exponential Convergence of the Identifier

Consider the identification problem with assumptions (A1)–(A3), the identifier structure of Section 2.2 and the gradient algorithms (2.4.1) or (2.4.2), or the normalized LS algorithm with covariance resetting (2.4.6), (2.4.7).

If w is PE

Then the identifier parameter θ converges to the nominal parameter θ^* exponentially fast.

Proof of Theorem 2.5.3

This theorem follows directly from theorem 2.5.1. Note that when w is bounded, w PE is equivalent to $w/\sqrt{1+\gamma w^T w}$ PE, so that the exponential convergence is guaranteed for both gradient update laws. The bounds on P obtained in the proof of theorem 2.4.4 allow us to extend the proof of exponential convergence to the LS algorithm. □

Exponential Convergence Rates

Estimates of the convergence rates can be found from the results in the proof of theorem 2.5.1. For the standard gradient algorithm (2.4.1), for example, the convergence rate is given by

$$\alpha = \frac{1}{2\delta} \ln \left(\frac{1}{1 - \dfrac{2g\alpha_1}{(1+\sqrt{2n}g\alpha_2)^2}} \right) \tag{2.5.12}$$

where g is the adaptation gain, α_1, α_2, and δ come from the PE definition (2.5.3) and n is the order of the plant. The influence of some design parameters can be studied with this relationship. The constants α_1, α_2 and δ depend in a complex manner on the input signal r and on the plant being identified. However, if r is multiplied by 2, then α_1, α_2 are multiplied by 4. In the limiting case when the adaptation gain g or the reference input r is made small, the rate of convergence $\alpha \to g\,\alpha_1/\delta$. In this case, the convergence rate is proportional to the adaptation gain g and to the lower bound in the PE condition. Through the PE condition, it is also proportional to the square of the amplitude of the reference input r. This result will be found again in Chapter 4, using averaging techniques.

When the adaptation gain and reference input get sufficiently large, this approximation is no longer valid and (2.5.12) shows that above some level, the convergence rate estimate saturates and even decreases (cf. Sondhi & Mitra [1976]).

It is also possible to show that the presence of the exponentially decaying terms due to initial conditions in the observer do not affect the exponential stability of the system. The rate of convergence will, however, be as found previously only if the rate of decay of the transients is faster than the rate of convergence of the algorithm (cf. Kreisselmeier [1977]).

2.6 MODEL REFERENCE IDENTIFIERS—SPR ERROR EQUATION

2.6.1 Model Reference Identifiers

In Section 2.2, we presented an identifier structure which led to a linear error equation. This structure was based on a convenient reparametrization of the plant. It is worth pointing out that there exist several ways to reparametrize the plant and many error equations that may be used for identification. The scheme discussed in Sections 2.2–2.3 is generally known as an *equation error* identifier. Landau [1979] discussed an interesting alternative called the *output error* approach. The resulting scheme has significant advantages in terms of noise bias, although its stability may only be guaranteed from prior knowledge about the plant. Another approach, which we will call the *model reference approach* (cf. Luders & Narendra [1973]) is discussed now. We start from an arbitrary reference model, with transfer function $\hat{M}$ satisfying the following conditions

(A3) Reference Model Assumptions

The reference model is a SISO LTI system, described by a transfer function

$$\hat{M}(s) = k_m \frac{\hat{n}_m(s)}{\hat{d}_m(s)} \tag{2.6.1}$$

where $\hat{n}_m(s)$, $\hat{d}_m(s)$ are monic, coprime polynomials of degrees l and $k \le n$, respectively. Assume that the reference model is strictly proper, but that its relative degree is no greater than the relative degree of the plant, that is $1 \le k - l \le n - m$. The reference model is stable, i.e. $\hat{d}_m$ is Hurwitz.

A simple choice of $\hat{M}$ is $1/s+1$. As previously, the plant is assumed to satisfy assumptions (A1) and (A2) (cf. Section 2.1).

Consider now the representation of the plant of Figure 2.3.

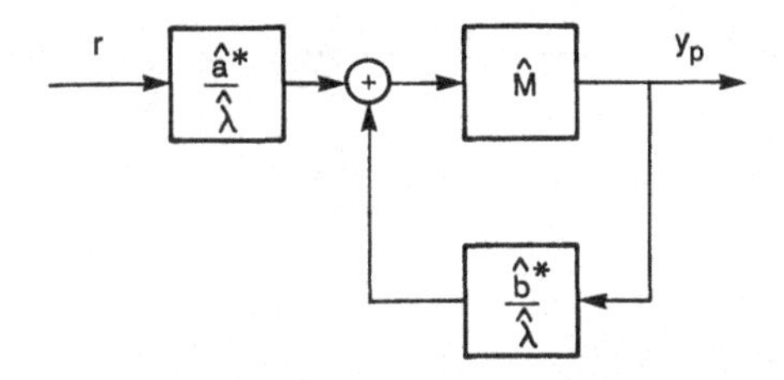

Figure 2.3: Model Reference Reparameterization

Although we will have to show (see proposition 2.6.1) that the plant can indeed be represented this way, we may already interpret it as being obtained by modifying the reference model through feedback and feed-forward action, so as to match the plant transfer function. Alternatively, we might interpret this structure as a more general parametrization of the plant than the one in Figure 2.1. In that case, the model transfer function was simply the identity.

The polynomial $\hat{\lambda}$ is a monic, Hurwitz polynomial of degree $n-1$. It serves a similar purpose in Section 2.2, and although the degree is now $n-1$, instead of n in Section 2.2, the following derivations can also be carried out with a degree equal to n, with only minor adjustments (the reference model may then be assumed to be proper and not strictly proper). We will have to require that the zeros of $\hat{\lambda}(s)$ contain those of $\hat{n}_m(s)$ and therefore we write

$$\hat{\lambda}(s) = \hat{n}_m(s)\hat{\lambda}_0(s) \tag{2.6.2}$$

where $\hat{\lambda}_0(s)$ is a monic, Hurwitz polynomial of degree $n-l-1$.

The polynomials $\hat{a}^*(s)$, $\hat{b}^*(s)$ in Figure 2.3 have degrees at most $n-1$ and serve similar purposes as before. Therefore, we start with the following proposition.

Proposition 2.6.1

There exist unique $\hat{a}^*$, $\hat{b}^*$ such that the transfer function from $r \to y_p$ in Figure 2.3 is $\hat{P}(s)$.

Proof of Proposition 2.6.1

Existence

The transfer function from $r \to y_p$ in Figure 2.3 is given by

$$\frac{\hat{y}_p}{\hat{r}} = \frac{\hat{a}^*}{\hat{\lambda}} \frac{\dfrac{k_m \hat{n}_m}{\hat{d}_m}}{1 - k_m \dfrac{\hat{n}_m}{\hat{d}_m} \dfrac{\hat{b}^*}{\hat{\lambda}}} = \frac{\hat{a}^* k_m \hat{n}_m}{\hat{\lambda} \hat{d}_m - k_m \hat{n}_m \hat{b}^*}$$

$$= \frac{k_m \hat{a}^*}{\hat{\lambda}_0 \hat{d}_m - k_m \hat{b}^*} \tag{2.6.3}$$

This transfer function will be equal to $\hat{P}(s)$ if and only if

$$\hat{\lambda}_0 \hat{d}_m - k_m \hat{b}^* = \frac{k_m}{k_p} \hat{d}_p \frac{\hat{a}^*}{\hat{n}_p} \tag{2.6.4}$$

The problem is therefore to find polynomials $\hat{a}^*$, $\hat{b}^*$ of degrees at most $n - 1$ that satisfy (2.6.4).

A solution can be found by inspection. Divide $\hat{\lambda}_0 \hat{d}_m$ by $\hat{d}_p$: denote by $\hat{q}$ the quotient of degree $k - l - 1$ and let $k_m \hat{b}^*$ be the remainder of degree $n - 1$. In other words, let

$$\hat{\lambda}_0 \hat{d}_m = \hat{q} \hat{d}_p + k_m \hat{b}^* \tag{2.6.5}$$

This defines $\hat{b}^*$ appropriately. Equality (2.6.4) is satisfied if $\hat{a}^*$ is defined by

$$\hat{a}^* = \frac{k_p}{k_m} \hat{q} \hat{n}_p \tag{2.6.6}$$

The degree of the polynomial in the right-hand side is $m + k - l - 1$, which is at most $n - 1$ by assumption (A3), so that the degree requirements are satisfied.

Uniqueness

Assume that there exist $\hat{a}^* + \delta \hat{a}$, $\hat{b}^* + \delta \hat{b}$ satisfying

$$\hat{\lambda}_0 \hat{d}_m - k_m (\hat{b}^* + \delta \hat{b}) = \frac{k_m}{k_p} \hat{d}_p \frac{(\hat{a} + \delta \hat{a})}{\hat{n}_p} \tag{2.6.7}$$

Subtracting (2.6.4) from (2.6.7), we find that

$$\frac{\delta \hat{a}}{\delta \hat{b}} = -k_p \frac{\hat{n}_p}{\hat{d}_p} = -\hat{P} \tag{2.6.8}$$

Recall that $\hat{n}_p$, $\hat{d}_p$ are assumed to be coprime, while the degree of $\hat{d}_p$ and $\delta\hat{b}$ are n and at most $n-1$, respectively. Therefore, equation (2.6.8) cannot have any solution. □

Identifier Structure

The identifier structure is obtained in a similar way as in Section 2.2. We let $\Lambda \in \mathbb{R}^{n-1 \times n-1}$, $b_\lambda \in \mathbb{R}^{n-1}$ in controllable canonical form such that $\det(sI - \Lambda) = \hat{\lambda}(s)$ and

$$(sI - \Lambda)^{-1} b_\lambda = \frac{1}{\hat{\lambda}(s)} \begin{bmatrix} 1 \\ s \\ \cdot \\ \cdot \\ \cdot \\ s^{n-2} \end{bmatrix} \tag{2.6.9}$$

Given any plant and model transfer functions satisfying (A1) and (A3), we showed that there exist unique polynomials $\hat{a}^*(s)$, $\hat{b}^*(s)$. We now let $a_0^*, b_0^* \in \mathbb{R}$ and $a^*, b^* \in \mathbb{R}^{n-1}$ such that

$$\begin{aligned} \frac{\hat{a}^*(s)}{\hat{\lambda}(s)} &= a_0^* + a^{*^T}(sI - \Lambda)^{-1} b_\lambda \\ \frac{\hat{b}^*(s)}{\hat{\lambda}(s)} &= b_0^* + b^{*^T}(sI - \Lambda)^{-1} b_\lambda \end{aligned} \tag{2.6.10}$$

We define the observer through

$$\begin{aligned} \dot{w}^{(1)} &= \Lambda w^{(1)} + b_\lambda r \\ \dot{w}^{(2)} &= \Lambda w^{(2)} + b_\lambda y_p \end{aligned} \tag{2.6.11}$$

where $w^{(1)}, w^{(2)} \in \mathbb{R}^{n-1}$ and the initial conditions are arbitrary. The regressor vector is now

$$w^T(t) := \left[r(t),\, w^{(1)^T}(t),\, y_p(t),\, w^{(2)^T}(t) \right] \in \mathbb{R}^{2n} \tag{2.6.12}$$

and the nominal parameter is

$$\theta^{*^T} := \left[a_0^*,\, a^{*^T},\, b_0^*,\, b^{*^T} \right] \in \mathbb{R}^{2n} \tag{2.6.13}$$

Using proposition 2.6.1, we find that any plant can be represented uniquely as

$$y_p(t) = \hat{M}\left[\theta^{*^T} w(t) \right] \tag{2.6.14}$$

Rigorously, one should add an exponentially decaying term due to the initial conditions in the observer and in the reference model transfer function, but we will neglect this term for simplicity.

In similarity with (2.6.14), the *identifier output* is defined to be

$$\begin{aligned} y_i(t) &:= \hat{M}\left[a_0(t)r(t) + a^T(t)w^{(1)}(t) + b_0(t)y_p(t) \right.\\ &\qquad\qquad \left. + b^T(t)w^{(2)}(t)\right] \\ &:= \hat{M}\left[\theta^T(t)w(t)\right] \end{aligned} \tag{2.6.15}$$

where

$$\theta^T(t) := \left[a_0(t), a^T(t), b_0(t), b^T(t)\right] \in \mathbb{R}^{2n} \tag{2.6.16}$$

is the vector of adaptive parameters. We define the parameter error

$$\phi(t) := \theta(t) - \theta^* \tag{2.6.17}$$

so that the *identifier error*

$$e_1(t) = y_i(t) - y_p(t) \tag{2.6.18}$$

is also given, for the purpose of the analysis, by

$$e_1(t) = \hat{M}\left[\phi^T(t)w(t)\right] \tag{2.6.19}$$

The error equation is linear in the parameter error, but it now involves the dynamics of the transfer function $\hat{M}$. We will show in the next section that the same gradient algorithm as used previously can still be used for identification, provided that the following additional assumption is satisfied.

(A4) Positive Real Model

$\hat{M}$ is strictly positive real.

We will define and discuss this assumption in the next section. The overall identifier structure is represented in Figure 2.4 and we summarize the implementation of the algorithm hereafter.

Model Reference Identifier—Implementation

Assumptions

(A1)–(A4)

Data

n, upper bound on $n - m$ (e.g., n)

Input

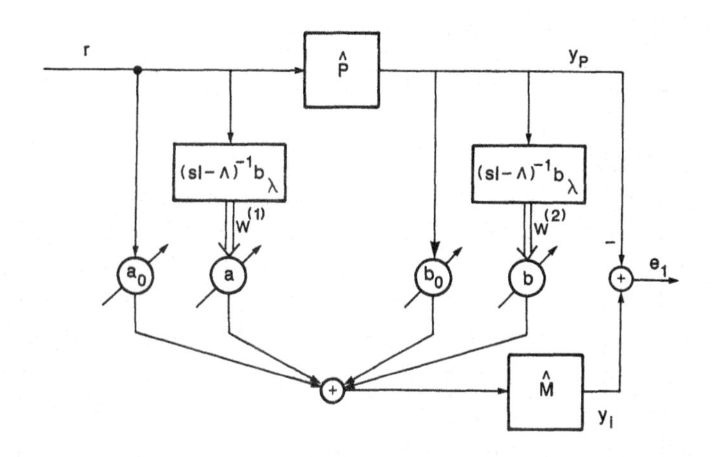

Figure 2.4: Model Reference Identifier Structure

$r(t)\,,\ y_p(t) \in \mathbb{R}$

Output

$\theta(t),\ y_i(t) \in \mathbb{R}$

Internal Signals

$w(t) \in \mathbb{R}^{2n}\ [\,w^{(1)}(t)\,,\ w^{(2)}(t) \in \mathbb{R}^{n-1}]$

$\theta(t) \in \mathbb{R}^{2n}\ [\,a_0(t) \in \mathbb{R}\,,\ a(t) \in \mathbb{R}^{n-1},\ b_0(t) \in \mathbb{R}\,,\ b(t) \in \mathbb{R}^{n-1}]$

$y_i(t)\,,\ e_1(t) \in \mathbb{R}$

Initial conditions are arbitrary.

Design Parameters

- $\hat{M}$ satisfying (A3)–(A4)
- $\Lambda \in \mathbb{R}^{n-1 \times n-1}$, $b_\lambda \in \mathbb{R}^{n-1}$ in controllable canonical form such that $\det(sI - \Lambda)$ is Hurwitz, and contains the zeros of $\hat{n}_m(s)$.
- $g > 0$

Identifier Structure

$$
\begin{aligned}
\dot{w}^{(1)} &= \Lambda\, w^{(1)} + b_\lambda\, r \\
\dot{w}^{(2)} &= \Lambda\, w^{(2)} + b_\lambda\, y_p \\
\theta^T &= [\,a_0\,,\ a^T\,,\ b_0\,,\ b^T\,] \\
w^T &= [\,r\,,\ w^{(1)^T}\,,\ y_p\,,\ w^{(2)^T}\,] \\
y_i &= \hat{M}(\theta^T w) \\
e_1 &= y_i - y_p
\end{aligned}
$$

Gradient Algorithm

$$\dot{\theta} = -g\, e_1 w$$

□

Comments

The model reference identifier is an alternative to the equation error scheme discussed previously. It may result in savings in computations. Indeed, the state variable filters are of order $n-1$, instead of n previously. If the transfer function $\hat{M}$ is of order 1, the realization requires one less state. The presence of $\hat{M}$, however, precludes the use of the least-squares algorithms, which are usually faster. We will also show in Chapter 4 how the transfer function $\hat{M}$ influences the convergence properties of the gradient algorithm.

2.6.2 Strictly Positive Real Error Equation and Identification Algorithms

In Section 2.6.1, we encountered a more general error equation, which we will call the *strictly positive real* (SPR) error equation

$$e_1(t) = \hat{M}\left[\phi^T(t) w(t)\right]$$

where $\hat{M}$ is a strictly positive real transfer function. This error equation is still linear, but it now involves additional dynamics contained in $\hat{M}$. In this section, we will establish general properties involving this error equation. For uniformity with previous discussions, we assume that $w : \mathbb{R}_+ \rightarrow \mathbb{R}^{2n}$, but the dimension of w is in fact completely arbitrary.

The definitions of *positive real* (PR) and *strictly positive real* (SPR) functions originate from network theory. A rational transfer function is the driving point impedance of a *passive* network if and only if it is PR. Similarly, it is the driving point impedance of a *dissipative* network if and only if it is SPR. The following definitions are deduced from these properties.

Definition Positive Real (PR) and Strictly Positive Real (SPR) Functions

A rational function $\hat{M}(s)$ of the complex variable $s = \sigma + j\omega$ is *positive real* (PR), if $\hat{M}(\sigma) \in \mathbb{R}$ for all $\sigma \in \mathbb{R}$ and $\mathrm{Re}(\hat{M}(\sigma + j\omega)) \geq 0$ for all $\sigma > 0$, $\omega \geq 0$. It is *strictly positive real* (SPR) if, for some $\epsilon > 0$, $\hat{M}(s-\epsilon)$ is PR.

It may be shown (cf. Ioannou & Tao [1987]) that a *strictly proper* transfer function $\hat{M}(s)$ *is SPR if and only if*

- $\hat{M}(s)$ is stable
- $\mathrm{Re}(\hat{M}(j\omega)) > 0$, for all $\omega \geq 0$
- $\lim\limits_{\omega \to \infty} \omega^2 \mathrm{Re}(\hat{M}(j\omega)) > 0$

For example, the transfer function

$$\hat{M}(s) = \frac{s+c}{(s+a)(s+b)}$$

is SPR if and only if $a > 0$, $b > 0$, $a + b > c > 0$.

SPR *transfer functions* form a rather restricted class. In particular, an SPR transfer function must be minimum phase and its phase may never exceed 90°. An important lemma concerning SPR transfer functions is the *Kalman-Yacubovitch-Popov lemma* given next.

Lemma 2.6.2 Minimal Realization of an SPR Transfer Function

Let $[A, b, c^T]$ be a minimal realization of a strictly proper, stable, rational transfer function $\hat{M}(s)$. Then, the following statements are equivalent

(a) $\hat{M}(s)$ is *SPR*

(b) There exist symmetric positive definite matrices P, Q, such that

$$PA + A^T P = -Q$$
$$Pb = c \qquad (2.6.20)$$

Proof of Lemma 2.6.2 cf. Anderson & Vongpanitlerd [1973].

SPR Error Equation with Gradient Algorithm

A remarkable fact about SPR transfer functions is that the gradient update law

$$\dot{\phi}(t) = \dot{\theta}(t) = -g e_1(t) w(t) \qquad g > 0 \qquad (2.6.21)$$

has similar properties when e_1 is defined by the SPR error equation (2.6.19), and when e_1 is defined by the linear error equation (2.4.3). Note that in the case of the SPR error equation, the algorithm is *not* the true gradient anymore, although we will keep using this terminology for the similarity.

Using lemma 2.6.2, a state-space realization of $\hat{M}(s)$ with state e_m can be obtained so that

$$\dot{e}_m(t) = A e_m(t) + b \phi^T(t) w(t)$$

$$
\begin{aligned}
e_1(t) &= c^T(t)\, e_m(t) \\
\dot{\phi}(t) &= -g c^T e_m(t)\, w(t) \qquad g>0
\end{aligned}
\tag{2.6.22}
$$

Theorem 2.6.3 SPR Error Equation with Gradient Algorithm

Let $w : \mathbb{R}_+ \to \mathbb{R}^{2n}$ be piecewise continuous. Consider the SPR error equation (2.6.19) with $\hat{M}(s)$ SPR, together with the gradient update law (2.6.21). Equivalently, consider the state-space realization (2.6.22) where $[A\,,\, b\,,\, c^T]$ satisfy the conditions of lemma 2.6.2.

Then

(a) $e_m, e_1 \in L_2$

(b) $e_m, e_1, \phi \in L_\infty$

Proof of Theorem 2.6.3

Let $P\,,\, Q$ be as in lemma 2.6.2 and $v = g e_m^T P e_m + \phi^T \phi$. Along the trajectories of (2.6.22)

$$
\begin{aligned}
\dot{v} &= g e_m^T P A e_m + g e_m^T P b \phi^T w + g e_m^T A^T P e_m \\
&\qquad + g\, \phi^T w b^T P e_m - 2 g c^T e_m \phi^T w \\
&= -g e_m^T Q e_m \le 0
\end{aligned}
\tag{2.6.23}
$$

where we used (2.6.20). The conclusions follow as in theorem 2.4.1, since P and Q are positive definite. □

Modified SPR Error Equation

The normalized gradient update law presented for the linear error equation is not usually applied to the SPR error equation. Instead, a *modified SPR error equation* is considered

$$
e_1(t) = \hat{M}\left[\phi^T(t)\, w(t) - \gamma\, w^T(t)\, w(t)\, e_1(t)\right] \qquad \gamma > 0 \tag{2.6.24}
$$

where γ is a constant. The same gradient algorithm may be applied with this error equation, so that in state-space form

$$
\begin{aligned}
\dot{e}_m(t) &= A\, e_m(t) + b\, \left[\phi^T(t) w(t) - \gamma\, w^T(t)\, w(t)\, c^T e_m(t)\right] \\
e_1(t) &= c^T e_m(t) \\
\dot{\phi}(t) &= -g c^T e_m(t)\, w(t) \qquad\qquad g\,,\gamma > 0
\end{aligned}
\tag{2.6.25}
$$

Theorem 2.6.4 Modified SPR Error Equation with Gradient Algorithm

Let $w : \mathbb{R}_+ \to \mathbb{R}^{2n}$ be piecewise continuous. Consider the modified SPR error equation (2.6.24) with $\hat{M}(s)$ SPR, together with the gradient update law (2.6.21). Equivalently, consider the state-space realization (2.6.25), where $[A, b, c^T]$ satisfy the conditions of lemma 2.6.2.

Then

(a) $e_m, e_1, \dot{\phi} \in L_2$

(b) $e_m, e_1, \phi \in L_\infty$

Proof of Theorem 2.6.4

Let P, Q be as in lemma 2.6.2 and $v = g e_m^T P e_m + \phi^T \phi$. Along the trajectories of (2.6.25)

$$\dot{v} = -g e_m^T Q e_m - 2g\gamma (e_1 w)^T (e_1 w) \leq 0 \tag{2.6.26}$$

Again, it follows that e_m, e_1, ϕ are bounded, and e_m, $e_1 \in L_2$. Moreover, it also follows now that $e_1 w \in L_2$, so that $\dot{\phi} \in L_2$. □

2.6.3 Exponential Convergence of the Gradient Algorithms with SPR Error Equations

As stated in the following theorem, the gradient algorithm is also exponentially convergent with the SPR error equations, under the PE condition.

Theorem 2.6.5 Exponential Convergence of the Gradient Algorithms with SPR Error Equations

Let $w : \mathbb{R}_+ \to \mathbb{R}^{2n}$. Let $[A, b, c^T]$ satisfy the conditions of lemma 2.6.2.

If w is PE and $w, \dot{w} \in L_\infty$

Then (2.6.22) and (2.6.25) are globally exponentially stable.

The proof given hereafter is similar to the proof by Anderson [1977] (with some significant differences however). The main condition for exponential convergence is the PE condition, as required previously, and again, the main idea is to interpret the condition as a UCO condition. The additional boundedness requirement on $\dot{w}$ guarantees that PE is not lost through the transfer function $\hat{M}$ (cf. lemma 2.6.7 hereafter). It is sufficient that the boundedness conditions hold almost everywhere, so that piecewise differentiable signals may be considered.

Auxiliary Lemmas on PE Signals

The following auxiliary lemmas will be useful in proving the theorem. Note that the sum of two PE signals is not necessarily PE. On the other hand, an L_2 signal is necessarily *not* PE. Lemma 2.6.6 asserts that PE is not altered by the addition of a signal belonging to L_2. In particular, this implies that terms due to initial conditions do not affect PE. Again, we assume the dimension of the vectors to be $2n$, for uniformity, but the dimension is in fact arbitrary.

Lemma 2.6.6 PE and L_2 Signals

Let $w, e : \mathbb{R}_+ \to \mathbb{R}^{2n}$ be piecewise continuous.

If w is PE
$e \in L_2$

Then $w + e$ is PE.

Proof of Lemma 2.6.6 in the Appendix.

Lemma 2.6.7 shows that PE is not lost if the signal is filtered by a stable, minimum phase transfer function, provided that the signal is sufficiently smooth.

Lemma 2.6.7 PE Through LTI Systems

Let $w : \mathbb{R}_+ \to \mathbb{R}^{2n}$.

If w is PE and $w, \dot{w} \in L_\infty$
$\hat{H}$ is a stable, minimum phase, rational transfer function

Then $\hat{H}(w)$ is PE.

Proof of Lemma 2.6.7 in the Appendix.

We now prove theorem 2.6.5.

Proof of Theorem 2.6.5

As previously, let $v = g e_m^T P e_m + \phi^T \phi$, so that for both SPR error equations

$$\int_{t_0}^{t_0+\delta} \dot{v}\, d\tau \le -g \int_{t_0}^{t_0+\delta} e_m^T Q e_m\, d\tau$$

$$\le -g \frac{\lambda_{\min}(Q)}{|c|^2} \int_{t_0}^{t_0+\delta} e_1^2\, d\tau \le 0 \tag{2.6.27}$$

By theorem 1.5.2, exponential convergence will be guaranteed if, for some $\alpha_3 > 0$

$$\int_{t_0}^{t_0+\delta} e_1^2(\tau)\,d\tau \;\geq\; \alpha_3\left(|\,e_m(t_0)|^2 + |\,\phi(t_0)|^2\right) \tag{2.6.28}$$

for all t_0, $e_m(t_0)$, $\phi(t_0)$.

Derivation of 2.6.28: This condition can be interpreted as a UCO condition on the system

$$\begin{aligned} \dot{e}_m &= Ae_m + b\phi^T w \\ \dot{\phi} &= -gc^T e_m\, w \\ e_1 &= c^T e_m \end{aligned} \tag{2.6.29}$$

An additional term $-\,b\gamma\, w^T w c^T e_m$ is added in the differential equation governing e_m in the case of the modified SPR error equation. Using lemma 2.5.2 about UCO under output injection, we find that inequality (2.6.28) will be satisfied if the following system

$$\begin{aligned} \dot{e}_m &= Ae_m + b\phi^T w \\ \dot{\phi} &= 0 \\ e_1 &= c^T e_m \end{aligned} \tag{2.6.30}$$

is UCO. For this, we let

$$K = \begin{bmatrix} 0 \\ gw \end{bmatrix} \qquad \text{or} \qquad K = \begin{bmatrix} b\gamma\, w^T w \\ gw \end{bmatrix} \tag{2.6.31}$$

for the basic SPR and modified SPR error equations, respectively. The condition on K in lemma 2.5.3 is satisfied, since w is bounded.

We are thus left to show that system (2.6.30) is UCO, i.e. that

$$\begin{aligned} e_1(t) &= c^T e^{A(t-t_0)} e_m(t_0) + \int_{t_0}^{t} c^T e^{A(t-\tau)} b\, w^T(\tau)\,d\tau\; \phi(t_0) \\ &:= x_1(t) + x_2(t) \end{aligned} \tag{2.6.32}$$

satisfies, for some $\beta_1, \beta_2, \delta > 0$

$$\beta_2\left[\,|\,e_m(t_0)|^2 + |\,\phi(t_0)|^2\right] \;\geq\; \int_{t_0}^{t_0+\delta} e_1^2(\tau)\,d\tau$$

$$\geq \beta_1 \left[| e_m(t_0)|^2 + | \phi(t_0)|^2 \right] \tag{2.6.33}$$

for all t_0, $e_m(t_0)$, $\phi(t_0)$.

Derivation of 2.6.33: By assumption, w is PE and $w, \dot{w} \in L_\infty$. Therefore, using lemma 2.6.7, we have that, for all $t_0 \geq 0$, the signal

$$w_f(t) = \int_{t_0}^{t} c^T e^{A(t-\tau)} b\, w(\tau)\, d\tau \tag{2.6.34}$$

is PE. This means that, for some $\alpha_1, \alpha_2, \sigma > 0$

$$\alpha_2 \, | \phi(t_0)|^2 \geq \int_{t_1}^{t_1+\sigma} x_2^2(\tau)\, d\tau \geq \alpha_1 | \phi(t_0)|^2 \tag{2.6.35}$$

for all $t_1 \geq t_0 \geq 0$ and $\phi(t_0)$.

On the other hand, since A is stable, there exist $\gamma_1, \gamma_2 > 0$, such that

$$\int_{t_0+m\sigma}^{\infty} x_1^2(\tau)\, d\tau \leq \gamma_1 | e_m(t_0)|^2 e^{-\gamma_2 m \sigma} \tag{2.6.36}$$

for all t_0, $e_m(t_0)$ and an arbitrary integer $m > 0$ to be defined later. Since $[A, c^T]$ is observable, there exists $\gamma_3(m\sigma) > 0$, with $\gamma_3(m\sigma)$ increasing with $m\sigma$ such that

$$\int_{t_0}^{t_0+m\sigma} x_1^2(\tau)\, d\tau \geq \gamma_3(m\sigma) | e_m(t_0)|^2 \tag{2.6.37}$$

for all t_0, $e_m(t_0)$ and $m > 0$.

Let $n > 0$ be another integer to be defined and let $\delta = (m+n)\sigma$. Using the triangle inequality

$$\int_{t_0}^{t_0+\delta} e_1^2(\tau)\, d\tau \geq \int_{t_0}^{t_0+m\sigma} x_1^2(\tau)\, d\tau - \int_{t_0}^{t_0+m\sigma} x_2^2(\tau)\, d\tau$$

$$+ \int_{t_0+m\sigma}^{t_0+\delta} x_2^2(\tau)\, d\tau - \int_{t_0+m\sigma}^{t_0+\delta} x_1^2(\tau)\, d\tau$$

$$\geq \gamma_3(m\sigma) | e_m(t_0)|^2 - m\alpha_2 | \phi(t_0)|^2$$

$$+ n\alpha_1 \, | \, \phi(t_0) |^2 - \gamma_1 e^{-\gamma_2 m \sigma} | \, e_m(t_0) |^2 \tag{2.6.38}$$

Let m be large enough to get

$$\gamma_3(m\,\sigma) - \gamma_1 e^{-\gamma_2 m \sigma} \geq \gamma_3(m\,\sigma)/2 \tag{2.6.39}$$

and n sufficiently large to obtain

$$n\alpha_1 - m\alpha_2 \geq \alpha_1 \tag{2.6.40}$$

Further, define

$$\beta_1 = \min\,(\alpha_1, \gamma_3(m\,\sigma)/2) \tag{2.6.41}$$

The lower inequality in (2.6.33) follows from (2.6.38), with β_1 as defined, while the upper inequality is easily found to be valid with

$$\beta_2 = \max\,(\gamma_1, (m+n)\alpha_2) \tag{2.6.42}$$

□

Comments

a) Although the proof of theorem 2.6.5 is somewhat long and tedious, it has some interesting features. First, it relies on the same basic idea as the proof of exponential convergence for the linear error equation (cf. theorem 2.5.1). It interprets the condition for exponential convergence as a uniform complete observability condition. Then, it uses lemma 2.5.2 concerning UCO under output injection to transform the UCO condition to a UCO condition on a similar system, but *where the vector* ϕ *is constant* (cf. (2.6.30)). The UCO condition leads then to a PE condition on a vector w_f, which is a filtered version of w, through the LTI system $\hat{M}(s)$.

b) The steps of the proof can be followed to obtain guaranteed rates of exponential convergence. Although such rates would be useful to the designer, the expression one obtains is quite complex and examination of the proof leaves little hope that the estimate would be tight. A more successful approach is found in Chapter 4, using averaging techniques.

c) The results presented in this section are very general. They do not rely on the structure of the identifier, but only on the SPR error equation (2.6.19). We leave it to the reader to specialize these results and obtain stability and convergence properties of the model reference identifier of Section 2.6.1.

2.7 Frequency Domain Conditions for Parameter Convergence

Theorems 2.5.3 and 2.6.5 give a condition on the regressor vector w, namely that w be PE, to guarantee exponential convergence of the parameter error. The difficulty with the PE condition is that it is not an *explicit* condition on the reference input r. In this section, we give frequency domain conditions on the input r to guarantee that w is PE.

The result of this section will make precise the following intuitive argument: assume that the parameter vector *does* converge (but not necessarily to the nominal value). Then, the plant loop is "asymptotically time-invariant." If the reference input contains frequencies $\omega_1, \ldots, \omega_k$, we expect that y_p and y_i will too. Since $y_i \to y_p$ as $t \to \infty$, the asymptotic identifier transfer function must match the plant transfer function at $s = j\omega_1, \ldots, j\omega_k$. If k is large enough, this will imply that the asymptotic identifier transfer function is *precisely* the plant transfer function and, therefore, that the parameter error converges to zero. Thus, we will show that the reference signal must be "rich enough," that is, "contains enough frequencies," for the parameter error to converge to zero. Roughly speaking, we will show

> A reference input r results in parameter error convergence to zero *unless* its spectrum is concentrated on $k < 2n$ points, where $2n$ is the number of unknown parameters in the adaptive scheme.

We will also discuss partial parameter convergence when the input is not sufficiently rich. We will use the results of generalized harmonic analysis developed in Section 1.6 and restrict our attention to stationary reference signals.

2.7.1 Parameter Convergence

From the definition of the regressor w in (2.2.16), we see that w is the output of a LTI system with input r and transfer function

$$\hat{H}_{wr}(s) = \begin{bmatrix} (sI - \Lambda)^{-1} b_\lambda \\ (sI - \Lambda)^{-1} b_\lambda \hat{P}(s) \end{bmatrix} \tag{2.7.1}$$

We will assume that the input r is stationary and that $\hat{P}$ is stable. Then, by the linear filter lemma (proposition 1.6.2), it follows that w is also stationary, that is has an autocovariance. For stationary signals, persistency of excitation is directly related to the positive definiteness of the autocovariance, as stated in the following proposition. Note that $R_w(0)$ is a symmetric positive semidefinite matrix.

Proposition 2.7.1 PE and Autocovariance

Let $w(t) \in \mathbb{R}^{2n}$ be stationary.

(a) w is PE *if and only if* (b) $R_w(0) > 0$.

Proof of Proposition 2.7.1

(a) implies (b)

From the PE condition on w, we have that for all $c \in \mathbb{R}^{2n}$ and for any positive integer k

$$\frac{1}{k\delta} \int_{t_0}^{t_0 + k\delta} (w^T(\tau) c)^2 \, d\tau \;\geq\; \frac{\alpha_1}{\delta} \, |c|^2 \tag{2.7.2}$$

Hence, for all $T \geq \delta$

$$\frac{1}{T} \int_{t_0}^{t_0 + T} (w^T(\tau) c)^2 \, d\tau \;\geq\; \frac{T - \delta}{T} \; \frac{\alpha_1}{\delta} \, |c|^2 \tag{2.7.3}$$

Since w has an autocovariance

$$\lim_{T \to \infty} \frac{1}{T} \int_{t_0}^{t_0 + T} (w^T(\tau) c)^2 \, d\tau \;=\; c^T R_w(0)\, c \tag{2.7.4}$$

Combining (2.7.3) and (2.7.4) yields that $R_w(0) \geq \dfrac{\alpha_1}{\delta} I > 0$.

(b) implies (a)

From the definition of autocovariance, for all $\epsilon > 0$, there exists δ such that

$$\left| \frac{1}{\delta} \int_{t_0}^{t_0 + \delta} (w^T(\tau) c)^2 \, d\tau - c^T R_w(0)\, c \right| \;<\; \epsilon \tag{2.7.5}$$

Therefore

$$\frac{1}{\delta} \int_{t_0}^{t_0 + \delta} (w^T(\tau) c)^2 \, d\tau \;\geq\; \lambda_{\min}(R_w(0)) |c|^2 - \epsilon \tag{2.7.6}$$

which implies PE for ϵ sufficiently small. □

We may relate the PE condition to the frequency content of w through the formula

$$R_w(0) \;=\; \int S_w(d\omega) \tag{2.7.7}$$

where $S_w(d\omega)$ is the spectral measure of w. In turn, using the linear filter lemma (proposition 1.6.2) and (2.7.7), we see that

$$R_w(0) = \frac{1}{2\pi}\int \hat{H}^*_{wr}(j\omega)\hat{H}^T_{wr}(j\omega)S_r(d\omega) \tag{2.7.8}$$

The expression (2.7.8) allows us to relate the frequency content of r to the persistency of excitation of w. This leads us to the following definition.

Definition Sufficient Richness of Order k

A stationary signal $r : \mathbb{R}_+ \to \mathbb{R}$ is called *sufficiently rich of order k*, if the support of the spectral density of r, namely, $S_r(d\omega)$ contains at least k points.

Comment

Note that a single sinusoid in the input contributes 2 points to the spectrum: at $+\omega_0$ and at $-\omega_0$. On the other hand, a DC signal contributes only one point in the support of the spectral density.

Theorem 2.7.2 PE and Sufficient Richness

Let $w(t) \in \mathbb{R}^{2n}$ be the output of a stable LTI system with transfer function $\hat{H}_{wr}(s)$ and stationary input $r(t)$. Assume that $\hat{H}_{wr}(j\omega_1), \ldots,$ $\hat{H}_{wr}(j\omega_{2n})$, are linearly independent on $\mathbb{C}^{2n}$ for all $\omega_1, \omega_2, \ldots,$ $\omega_{2n} \in \mathbb{R}$.

(a) w is PE *if and only if* (b) r is sufficiently rich of order $2n$.

Proof of Theorem 2.7.2

By proposition 2.7.1, (a) is equivalent to $R_w(0) > 0$.

(a) implies (b)

We prove this by contradiction. Assume that r is not sufficiently rich of order $2n$, that is the support of $S_r(d\omega)$ is $\omega_1, \ldots, \omega_k$ with $k < 2n$. Then, from (2.7.8), it follows that

$$R_w(0) = \frac{1}{2\pi}\sum_{i=1}^{k} \hat{H}^*_{wr}(j\omega_i)\hat{H}^T_{wr}(j\omega_i)S_r(\{\omega_i\}) \tag{2.7.9}$$

The right hand side of (2.7.9) is the sum of k dyads, so that the rank of $R_w(0)$ is at most $k < 2n$, contradicting the PE of w.

(b) implies (a)

We also prove this by contradiction. If w is not PE, there exists $c \in \mathbb{R}^{2n}$ such that

$$c^T R_w(0) c = 0 \tag{2.7.10}$$

Using (2.7.8), we see that (2.7.10) implies that

$$\int | \hat{H}_{wr}^T(j\omega) c |^2 S_r(d\omega) = 0 \tag{2.7.11}$$

Since $S_r(d\omega)$ is a continuous, positive measure, we may conclude that

$$\hat{H}_{wr}^T(j\omega) c = 0 \qquad \text{for all } \omega \in \text{support } S_r(d\omega) \tag{2.7.12}$$

Since the support of $S_r(d\omega)$ has at least $2n$ points, say $\omega_1, \ldots, \omega_{2n}$, we have that

$$\hat{H}_{wr}^T(j\omega_i) c = 0 \qquad i = 1, \ldots, 2n \tag{2.7.13}$$

contradicting the hypothesis of linear independence of $\hat{H}_{wr}(j\omega_i)$ for $i = 1, \ldots, 2n$. □

We saw that sufficient richness of order $2n$ in the reference input translates to w PE, provided that the $\hat{H}_{wr}(j\omega_i)$ are independent for every set of $2n$ ω_i's. It remains to be verified that this property holds for the $\hat{H}_{wr}(s)$ given in (2.7.1).

Theorem 2.7.3 Exponential Parameter Convergence and Sufficient Richness

Consider the identification problem of Section 2.1, with assumptions (A1) and (A2) and $\hat{P}$ stable, the identifier structure of Section 2.2 and the gradient algorithms (2.4.1) or (2.4.2), or the normalized LS algorithm with covariance resetting (2.4.6)–(2.4.7).

If r is stationary and sufficiently rich of order $2n$

Then the identifier parameter θ converges to the nominal parameter θ^* exponentially fast.

Proof of Theorem 2.7.3

Using theorems 2.5.3 and 2.7.2, we are left to show that, for every $\omega_1, \ldots, \omega_{2n} \in \mathbb{R}$, the vectors $\hat{H}_{wr}(j\omega_i)$ (with $\hat{H}_{wr}$ defined in (2.7.1)) are linearly independent in $\mathbb{C}^{2n}$.

We show the result by contradiction. Assume that there existed $\omega_1, \ldots, \omega_{2n}$ such that

$$\hat{H}_{wr}^T(j\omega_i) c = 0 \tag{2.7.14}$$

for some $c \in \mathbb{C}^{2n}$ and for all $i = 1, \ldots, 2n$. Using the fact that (Λ, b_λ) are in controllable canonical form and that $\hat{P}(s) = k_p \hat{n}_p(s) / \hat{d}_p(s)$,

(2.7.14) becomes

$$\left[1, j\omega_i, \ldots, (j\omega_i)^{n-1}, k_p \frac{\hat{n}_p(j\omega_i)}{\hat{d}_p(j\omega_i)}, \right.$$

$$\left. (j\omega_i) k_p \frac{\hat{n}_p(j\omega_i)}{\hat{d}_p(j\omega_i)}, \ldots, (j\omega_i)^{n-1} k_p \frac{\hat{n}_p(j\omega_i)}{\hat{d}_p(j\omega_i)} \right] c = 0 \qquad (2.7.15)$$

for $i = 1, \ldots, 2n$, that is

$$[\hat{d}_p(s), s\hat{d}_p(s), \ldots, s^{n-1}\hat{d}_p(s), k_p\hat{n}_p(s),$$

$$k_p s\hat{n}_p(s), \ldots, s^{n-1}k_p\hat{n}_p(s)]\, c = 0 \qquad (2.7.16)$$

for $s = j\omega_1, \ldots, j\omega_{2n}$. Equation (2.7.16) may be written more compactly as

$$\hat{c}^{(1)}(s)\hat{d}_p(s) + \hat{c}^{(2)}(s) k_p \hat{n}_p(s) = 0 \qquad (2.7.17)$$

for $s = j\omega_1, \ldots, j\omega_{2n}$, where

$$\hat{c}^{(1)}(s) = c_1 + c_2 s + \cdots + c_n s^{n-1}$$

and

$$\hat{c}^{(2)}(s) = c_{n+1} + c_{n+2}s + \cdots + c_{2n}s^{2n-1} \qquad (2.7.18)$$

The polynomial on the left-hand side of (2.7.17) is of degree at most $2n - 1$. Since it vanishes at $2n$ points, it must be identically zero. Consequently, we have

$$\frac{k_p\hat{n}_p(s)}{\hat{d}_p(s)} = -\frac{\hat{c}^{(1)}(s)}{\hat{c}^{(2)}(s)} \qquad \text{for all } s \qquad (2.7.19)$$

By assumption, $\hat{d}_p(s)$ is of degree n and $\hat{n}_p(s)$, $\hat{d}_p(s)$ are coprime. Therefore, (2.7.19) cannot be satisfied since the degree of $\hat{c}^{(2)}$ is at most $n - 1$. This establishes the contradiction. □

Comments

It would appear that the developments of this section would enable us to give an explicit expression for the rate of exponential convergence: it is in fact possible to carry out this program rigorously when the rate of adaptation is slow (see Chapter 4 for a further and more detailed description of this point).

If w is *not* PE, then the parameter error need not converge to zero. In this case, S_r is concentrated on $k < 2n$ frequencies $\omega_1, \ldots, \omega_k$. Intuition suggests that although θ need not converge to θ^*, it should converge to the set of θ's for which the *identifier transfer function matches the plant transfer function* at the frequencies $s = j\omega_1, \ldots, j\omega_k$. This is indeed the case.

2.7.2 Partial Parameter Convergence

We now consider the case when the spectrum $S_r(d\omega)$ is concentrated on $k < 2n$ points. Before stating the theorem, let us discuss the idea informally. From the structure of the identifier, we see that the plant output is given by

$$y_p = \theta^{*^T} \hat{H}_{wr}(r) \tag{2.7.20}$$

and the identifier output by

$$y_i = \theta^T(t) \hat{H}_{wr}(r) \tag{2.7.21}$$

Consequently, if the identifier output matches the plant output at $s = j\omega_1, \ldots, j\omega_k$, the asymptotic value of θ must satisfy

$$\begin{bmatrix} \hat{H}_{wr}(j\omega_1)^T \\ \cdot \\ \cdot \\ \cdot \\ \hat{H}_{wr}(j\omega_k)^T \end{bmatrix} \theta = \begin{bmatrix} \hat{H}_{wr}(j\omega_1)^T \\ \cdot \\ \cdot \\ \cdot \\ \hat{H}_{wr}(j\omega_k)^T \end{bmatrix} \theta^* \tag{2.7.22}$$

Let us then call Θ, the set of θ's, which satisfy (2.7.22). Clearly, $\theta^* \in \Theta$ and

$$\Theta = \theta^* + \text{null space} \begin{bmatrix} \hat{H}_{wr}(j\omega_1)^T \\ \cdot \\ \cdot \\ \cdot \\ \hat{H}_{wr}(j\omega_k)^T \end{bmatrix} \tag{2.7.23}$$

The k row vectors $\hat{H}_{wr}(j\omega_i)^T$ are linearly independent. Consequently, Θ has dimension $2n - k$.

In terms of the parameter error vector $\phi = \theta - \theta^*$, Θ has the simple description

$$\theta \in \Theta \iff R_w(0)\phi = 0 \tag{2.7.24}$$

We leave the verification of this fact to the reader, recalling that

$$R_w(0) = \sum_{i=1}^{k} \hat{H}_{wr}(j\omega_i)\hat{H}^*_{wr}(j\omega_i)S_r(\{\omega_i\}) \tag{2.7.25}$$

Proceeding more rigorously, we have the following theorem.

Theorem 2.7.4 Partial Convergence Theorem

Consider the identification problem of Section 2.1, with assumptions (A1)–(A2) and $\hat{P}$ stable, the identifier structure of Section 2.2 and the gradient algorithms (2.4.1) or (2.4.2), or the normalized LS algorithm with covariance resetting (2.4.6)–(2.4.7).

If r is stationary

Then $\lim_{t \to \infty} R_w(0)\phi(t) = 0$

Proof of Theorem 2.7.4

The proof relies on the fact, proved in theorem 2.4.5, that the output error $e_1 = \phi^T w \to 0$ as $t \to \infty$, and so does the derivative of the parameter error $\dot{\phi}$. Since ϕ and w are bounded, let k be such that $|\phi(t)|, |w(t)| \le k$, for all $t \ge 0$.

We will show that, for all $\epsilon > 0$, there exists $t_1 \ge 0$ such that, for all $t \ge t_1$, $\phi^T(t)R_w(0)\phi(t) \le \epsilon$. This means that $\phi^T(t)R_w(0)\phi(t)$ converges to zero as $t \to \infty$ and since $R_w(0)$ is symmetric positive definite, it also implies that $|R_w(0)\phi(t)|$ tends to zero as $t \to \infty$.

Since w has an autocovariance, for T large enough

$$\left| R_w(0) - \frac{1}{T}\int_{t_0}^{t_0+T} w(\tau)w^T(\tau)\,d\tau \right| \le \frac{\epsilon}{3k^2} \tag{2.7.26}$$

for all $t_0 \ge 0$ and therefore

$$\left| \phi(t)^T R_w(0)\phi(t) - \phi(t)^T \frac{1}{T}\int_{t_0}^{t_0+T} w(\tau)w(\tau)^T\,d\tau\,\phi(t) \right| \le \frac{\epsilon}{3} \tag{2.7.27}$$

From the fact that $\dot{\phi} \to 0$ as $t \to \infty$ (cf. theorem 2.4.5) and $\phi^T w \to 0$ as $t \to \infty$, we find that there exists t_1 such that, for all $t \ge t_1$

$$|\phi^T(t)w(t)| \le \frac{\epsilon}{3} \tag{2.7.28}$$

and

$$|\dot{\phi}(t)| \;\le\; \frac{\epsilon}{6k^3 T} \tag{2.7.29}$$

From (2.7.29), we have that $|\phi(\tau) - \phi(t)| \le \epsilon(\tau - t)/6k^3 T$ for all $\tau \ge t \ge t_1$. Together with the boundedness of ϕ and w, this implies that, for $t \ge t_1$

$$\left| \phi(t)^T \left[\frac{1}{T} \int_t^{t+T} w(\tau) w^T(\tau) \, d\tau \right] \phi(t) - \frac{1}{T} \int_t^{t+T} \left[\phi^T(\tau) w(\tau) w^T(\tau) \phi(\tau) \right] d\tau \right|$$

$$= \left| \frac{1}{T} \int_t^{t+T} w^T(\tau) \left(\phi(t) - \phi(\tau) \right) w^T(\tau) \left(\phi(t) + \phi(\tau) \right) d\tau \right|$$

$$\le \frac{\epsilon}{3} \tag{2.7.30}$$

Using (2.7.28), we have that, for $t \ge t_1$

$$\left| \frac{1}{T} \int_t^{t+T} \phi^T(\tau) w(\tau) w^T(\tau) \phi(\tau) \, d\tau \right| \;\le\; \frac{\epsilon}{3} \tag{2.7.31}$$

Now, combining (2.7.27), (2.7.30) and (2.7.31), we have that, for $t \ge t_1$

$$\phi^T(t) R_w(0) \phi(t) \;\le\; \epsilon \tag{2.7.32}$$

which completes the proof of the theorem. □

Comments

The proof relies on the fact that for both the identification schemes discussed *the parameter error eventually becomes orthogonal to the regressor* w, and the *updating slows down.* These are very common properties of identification schemes.

While the $2n - k$ dimensional set Θ to which $\theta(t)$ converges depends only on the frequencies $\omega_1, \ldots, \omega_k$ and not on the average powers $S_r(\{\omega_1\}), \ldots, S_r(\{\omega_k\})$ contained in the reference signal at those frequencies, the *rate of convergence* of θ to Θ depends on both.

As opposed to the proof of theorem 2.7.3, the proof of theorem 2.7.4 does not rely on theorem 2.5.3 relating PE and exponential convergence. If w is PE and $R_w(0) > 0$, the proof of theorem 2.7.4 is an alternate proof of the parameter convergence results of Section 2.5, with the additional assumption of stationarity of the reference input $r(t)$.

2.8 CONCLUSIONS

In this chapter, we derived a simple identification scheme for SISO LTI plants. The scheme involved a generic linear error equation, relating the identifier error, the regressor and the parameter error. Several gradient and least-squares algorithms were reviewed and common properties were established, that are valid under general conditions. It was shown that for any of these algorithms and provided that the regressor was a bounded function of time, the identifier error converged to zero as t approached infinity. The parameter error was also guaranteed to remain bounded. When the regressor was not bounded, but satisfied a regularity condition, then it was shown that a normalized error still converged to zero.

The exponential convergence of the parameter error to its nominal value followed from a persistency of excitation condition on the regressor. Guaranteed rates of exponential convergence were also obtained and showed the influence of various design parameters. In particular, the reference input was found to be a dominant factor influencing the parameter convergence.

The stability and convergence properties were further extended to strictly positive real error equations. Although more complex to analyze, the SPR error equation was found to have similar stability and convergence properties. In particular, PE appeared as a fundamental condition to guarantee exponential parameter convergence.

Finally, the PE conditions were transformed into conditions on the input. We assumed stationarity of the input, so that a frequency-domain analysis could be carried out. It was shown that parameter convergence was guaranteed, if the input contained the same number of spectral components as there were unknown parameters. If the input was a sum of sinusoids, for example, their number should be greater than or equal to the order of the plant.

CHAPTER 3
ADAPTIVE CONTROL

3.0 INTRODUCTION

In this chapter, we derive and analyze algorithms for adaptive control. Our attention is focused on model reference adaptive control. Then, the objective is to design an adaptive controller such that the behavior of the controlled plant remains close to the behavior of a desirable model, despite uncertainties or variations in the plant parameters. More formally, a *reference model* $\hat{M}$ is given, with input $r(t)$ and output $y_m(t)$. The unknown *plant* $\hat{P}$ has input $u(t)$ and output $y_p(t)$. The control objective is to design $u(t)$ such that $y_p(t)$ asymptotically tracks $y_m(t)$, with all generated signals remaining bounded.

We will consider linear time invariant systems of arbitrary order, and establish the stability and convergence properties of the adaptive algorithms. In this section however, we start with an informal discussion for a first order system with two unknown parameters. This will allow us to introduce the algorithms and the stability results in a simpler context.

Consider a first order single-input single-output (SISO) linear time invariant (LTI) plant with transfer function

$$\hat{P} = \frac{k_p}{s + a_p} \tag{3.0.1}$$

where k_p and a_p are unknown. The reference model is a stable SISO LTI system of identical order

$$\hat{M} = \frac{k_m}{s + a_m} \tag{3.0.2}$$

where k_m and $a_m > 0$ are arbitrarily chosen by the designer. In the time domain, the plant is described by

$$\dot{y}_p(t) = -a_p\, y_p(t) + k_p\, u(t) \tag{3.0.3}$$

and the reference model by

$$\dot{y}_m(t) = -a_m\, y_m(t) + k_m\, r(t) \tag{3.0.4}$$

The next steps are similar to those followed for model reference identification in Section 2.0. Let the control input be given by

$$u(t) = c_0(t)\, r(t) + d_0(t)\, y_p(t) \tag{3.0.5}$$

the motivation being that there exist *nominal* parameter values

$$c_0^* = \frac{k_m}{k_p} \qquad d_0^* = \frac{a_p - a_m}{k_p} \tag{3.0.6}$$

such that the closed-loop transfer function matches the reference model transfer function. Specifically, (3.0.3) and (3.0.5) yield

$$\begin{aligned} \dot{y}_p(t) &= -a_p\, y_p(t) + k_p\, (\, c_0(t)\, r(t) + d_0(t)\, y_p(t)\,) \\ &= -(\, a_p - k_p\, d_0(t)\,)\, y_p(t) + k_p\, c_0(t)\, r(t) \end{aligned} \tag{3.0.7}$$

which becomes

$$\dot{y}_p(t) = -a_m\, y_p(t) + k_m\, r(t) \tag{3.0.8}$$

when $c_0(t) = c_0^*$, $d_0(t) = d_0^*$.

For the analysis, it is convenient to introduce an error formulation. Define the *output error*

$$e_0 = y_p - y_m \tag{3.0.9}$$

and the *parameter error*

$$\phi = \begin{bmatrix} \phi_r(t) \\ \phi_y(t) \end{bmatrix} = \begin{bmatrix} c_0(t) - c_0^* \\ d_0(t) - d_0^* \end{bmatrix} \tag{3.0.10}$$

Subtracting (3.0.4) from (3.0.7)

$$\dot{e}_0 = -a_m\, (y_p - y_m) + (a_m - a_p + k_p\, d_0)\, y_p + k_p\, c_0\, r - k_m\, r$$

$$= -a_m e_0 + k_p \left[(c_0 - c_0^*) r + (d_0 - d_0^*) y_p \right]$$

$$= -a_m e_0 + k_p (\phi_r r + \phi_y y_p) \tag{3.0.11}$$

We may represent (3.0.11) in compact form as

$$e_0 = \frac{k_p}{s + a_m} \hat{M} (\phi_r r + \phi_y y_p)$$

$$= \frac{k_p}{k_m} \hat{M} (\phi_r r + \phi_y y_p) = \frac{1}{c_0^*} \hat{M} (\phi_r r + \phi_y y_p) \tag{3.0.12}$$

Although this notation will be very convenient in this book, we caution the reader that it mixes time domain operations $(\phi_r r + \phi_y y_p)$ and filtering by the LTI operator $\hat{M}$.

Equation (3.0.12) is of the form of the *SPR error equation* of Chapter 2. Therefore, we tentatively choose the update laws

$$\dot{c}_0 = -g e_0 r$$

$$\dot{d}_0 = -g e_0 y_p \qquad g > 0 \tag{3.0.13}$$

assuming that $k_p / k_m > 0$ and that $\hat{M}$ is SPR. The first condition requires prior knowledge about the plant (sign of the *high frequency gain* k_p), while the second condition restricts the class of models which may be chosen.

Note that there is a significant difference with the model reference identification case, namely that the signal y_p which appears in (3.0.12) is not exogeneous, but is itself a function of e_0. However, the stability proof proceeds along exactly the same lines. First assume that r is bounded, so that y_m is also bounded. The adaptive system is described by (3.0.3)–(3.0.5) and (3.0.13). Alternatively, the error formulation is

$$\dot{e}_0 = -a_m e_0 + k_p (\phi_r r + \phi_y e_0 + \phi_y y_m)$$

$$\dot{\phi}_r = -g e_0 r$$

$$\dot{\phi}_y = -g e_0^2 - g e_0 y_m \tag{3.0.14}$$

In this representation, the right-hand sides only contain states (e_0, ϕ_r, ϕ_y) and exogeneous signals (r, y_m). Consider then the Lyapunov function

$$v(e_0, \phi_r, \phi_y) = \frac{e_0^2}{2} + \frac{k_p}{2g} (\phi_r^2 + \phi_y^2) \tag{3.0.15}$$

so that, along the trajections of (3.0.14)

$$\begin{aligned}\dot{v} &= -a_m e_0^2 + k_p \phi_r e_0 r + k_p \phi_y e_0^2 + k_p \phi_y e_0 y_m \\ &\quad - k_p \phi_r e_0 r - k_p \phi_y e_0^2 - k_p \phi_y e_0 y_m \\ &= -a_m e_0^2 \qquad \le 0 \end{aligned} \tag{3.0.16}$$

It follows that the adaptive system is stable (in the sense of Lyapunov) and that, for all initial conditions, e_0, ϕ_r and ϕ_y are bounded. From (3.0.14), $\dot{e}_0$ is also bounded. Since v is monotonically decreasing and bounded below, $\lim v(t)$ as $t \to \infty$ exists, so that $e_0 \in L_2$. Since $e_0 \in L_2 \cap L_\infty$, and $\dot{e}_0 \in L_\infty$, it follows that $e_0 \to 0$ as $t \to \infty$.

The above approach is elegant, and relatively simple. However, it is not straightforward to extend it to the case when the relative degree of the plant is greater than 1. Further, it places restrictions on the reference model which are in fact not necessary. We now discuss an alternate approach which we will call the *input error approach*. The resulting scheme is slightly more complex for this first order example, but has significant advantages, which will be discussed later.

Instead of using e_0 in the adaptation procedure, we use the error

$$e_2 = c_0 y_p + d_0 \hat{M}(y_p) - \hat{M}(u) \tag{3.0.17}$$

The motivation behind this choice will be made clear from the analysis. Equation (3.0.17) determines how e_2 is calculated in the implementation (as (3.0.9) in the output error scheme). For the derivation of the adaptation scheme and for the analysis, we relate e_2 to the parameter error (as (3.0.12) in the output error scheme). First note that

$$\begin{aligned}\hat{M} &= \frac{k_m}{s + a_m} = \frac{k_m}{k_p} \frac{s + a_p}{s + a_m} \frac{k_p}{s + a_p} \\ &= \frac{k_m}{k_p} \frac{k_p}{s + a_p} + \frac{k_m}{k_p} \frac{a_p - a_m}{s + a_m} \frac{k_p}{s + a_p} \\ &= \frac{k_m}{k_p} \hat{P} + \frac{a_p - a_m}{k_p} \hat{M} \hat{P} \\ &= c_0^* \hat{P} + d_0^* \hat{M} \hat{P} \end{aligned} \tag{3.0.18}$$

where we used (3.0.6). Therefore, applying (3.0.18) to the signal u

$$\hat{M}(u) = c_0^* y_p + d_0^* \hat{M}(y_p) \tag{3.0.19}$$

and with (3.0.17)

$$\begin{aligned} e_2 &= (c_0 - c_0^*) y_p + (d_0 - d_0^*) \hat{M}(y_p) \\ &= \phi_r \, y_p + \phi_y \, \hat{M}(y_p) \end{aligned} \tag{3.0.20}$$

Equation (3.0.20) is of the form of the *linear error equation* studied in Chapter 2. Therefore, we may now use any of the identification algorithms, including the least-squares algorithms. No condition is posed on the reference model by the identification/adaptation algorithm. Proving stability is however more difficult, and is not addressed in this introduction.

The algorithms described above are both *direct algorithms*, for which update laws are designed to directly update the controller parameters c_0 and d_0. An alternate approach is the *indirect approach.* Using any of the procedures discussed in Chapter 2, we may design a recursive identifier to provide estimates of the plant parameters k_p and a_p. If k_p and a_p were known, the equations in (3.0.6) would determine the nominal controller parameters c_0^* and d_0^*. In an indirect approach, one replaces k_p and a_p in (3.0.6) by their estimates, thereby defining c_0 and d_0. This is a very intuitive approach to adaptive control which we will also discuss in this chapter.

After extending the above algorithms to more general schemes for plants of arbitrary order, we will study the stability and convergence properties of the adaptive systems. Unfortunately, the simple Lyapunov stability proof presented for the output error scheme does not extend to the general case, or to the other schemes. Instead, we will use tools from functional analysis, together with a set of standard lemmas which we will first derive. The global stability of the adaptive control schemes will be established. Conditions for parameter convergence will also follow, together with input signal conditions similar to those encountered in Chapter 2 for identification.

3.1 MODEL REFERENCE ADAPTIVE CONTROL PROBLEM

We now turn to the general model reference adaptive control problem considered in this chapter. The following assumptions will be in effect.

Assumptions

(A1) Plant Assumptions

The plant is a single-input, single-output (SISO), linear time-invariant (LTI) system, described by a transfer function

$$\frac{\hat{y}_p(s)}{\hat{u}(s)} = \hat{P}(s) = k_p \frac{\hat{n}_p(s)}{\hat{d}_p(s)} \tag{3.1.1}$$

where $\hat{n}_p(s)$, $\hat{d}_p(s)$ are monic, coprime polynomials of degree m and n, respectively. The plant is strictly proper and minimum phase. The sign of the so-called *high-frequency gain* k_p is known and, without loss of generality, we will assume $k_p > 0$.

(A2) Reference Model Assumptions

The reference model is described by

$$\frac{\hat{y}_m(s)}{\hat{r}(s)} = \hat{M}(s) = k_m \frac{\hat{n}_m(s)}{\hat{d}_m(s)} \tag{3.1.2}$$

where $\hat{n}_m(s)$, $\hat{d}_m(s)$ are monic, coprime polynomials of degree m and n respectively (that is, the same degrees as the corresponding plant polynomials). The reference model is stable, minimum phase and $k_m > 0$.

(A3) Reference Input Assumptions

The reference input $r(.)$ is piecewise continuous and bounded on $\mathbb{R}_+$.

Note that $\hat{P}(s)$ is assumed to be minimum phase, but is *not* assumed to be stable.

3.2 CONTROLLER STRUCTURE

To achieve the control objective, we consider the controller structure shown in Figure 3.1.

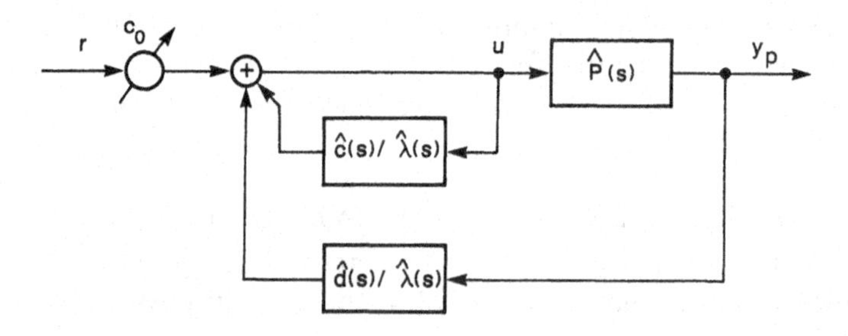

Figure 3.1: Controller Structure

By inspection of the figure, we see that

$$u = c_0 r + \frac{\hat{c}(s)}{\hat{\lambda}(s)}(u) + \frac{\hat{d}(s)}{\hat{\lambda}(s)}(y_p) \tag{3.2.1}$$

where c_0 is a scalar, $\hat{c}(s)$, $\hat{d}(s)$ and $\hat{\lambda}(s)$ are polynomials of degrees $n-2$, $n-1$ and $n-1$, respectively. From (3.2.1),

$$u = \frac{\hat{\lambda}}{\hat{\lambda}-\hat{c}}\left(c_0 r + \frac{\hat{d}}{\hat{\lambda}}(y_p)\right) \tag{3.2.2}$$

which is shown in Figure 3.2.

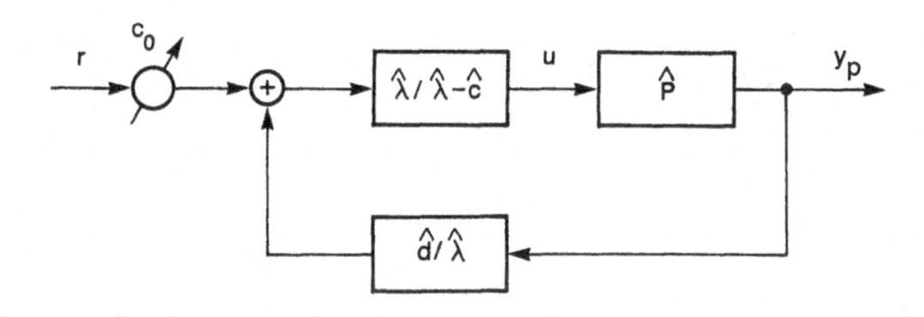

Figure 3.2: Controller Structure–Equivalent Form

Since

$$y_p = k_p \frac{\hat{n}_p}{\hat{d}_p}(u) \tag{3.2.3}$$

the transfer function from r to y_p is

$$\frac{\hat{y}_p}{\hat{r}} = \frac{c_0 k_p \hat{\lambda} \hat{n}_p}{(\hat{\lambda}-\hat{c})\hat{d}_p - k_p \hat{n}_p \hat{d}} \tag{3.2.4}$$

Note that the derivation of (3.2.4) relies on the cancellation of polynomials $\hat{\lambda}(s)$. Physically, this would correspond to the exact cancellation of modes of $\hat{c}(s)/\hat{\lambda}(s)$ and $\hat{d}(s)/\hat{\lambda}(s)$. For numerical considerations, we will therefore require that $\hat{\lambda}(s)$ is a Hurwitz polynomial.

The following proposition indicates that the controller structure is adequate to achieve the control objective, that is, that it is possible to make the transfer function from r to y_p equal to $\hat{M}(s)$. For this, it is clear from (3.2.4) that $\hat{\lambda}(s)$ must contain the zeros of $\hat{n}_m(s)$, so that we write

$$\hat{\lambda}(s) = \hat{\lambda}_0(s)\,\hat{n}_m(s) \tag{3.2.5}$$

where $\hat{\lambda}_0(s)$ is an arbitrary Hurwitz polynomial of degree $n-m-1$.

Proposition 3.2.1 Matching Equality

There exist unique c_0^*, $\hat{c}^*(s)$, $\hat{d}^*(s)$ such that the transfer function from $r \to y_p$ is $\hat{M}(s)$.

Proof of Proposition 3.2.1

1. Existence

The transfer function from r to y_p is $\hat{M}$ if and only if the following *matching equality* is satisfied

$$(\hat{\lambda} - \hat{c}^*)\hat{d}_p - k_p \hat{n}_p \hat{d}^* = c_0^* \frac{k_p}{k_m} \hat{\lambda}_0 \hat{n}_p \hat{d}_m \tag{3.2.6}$$

The solution can be found by inspection. Divide $\hat{\lambda}_0 \hat{d}_m$ by $\hat{d}_p$, let $\hat{q}$ be the quotient (of degree $n - m - 1$) and $-k_p \hat{d}^*$ the remainder (of degree $n - 1$). Thus $\hat{d}^*$ is given by

$$\hat{d}^* = \frac{1}{k_p}(\hat{q}\hat{d}_p - \hat{\lambda}_0 \hat{d}_m) \tag{3.2.7}$$

Let $\hat{c}^*$ (of degree $n - 2$), c_0^* be given by

$$\hat{c}^* = \hat{\lambda} - \hat{q}\hat{n}_p \tag{3.2.8}$$

$$c_0^* = \frac{k_m}{k_p} \tag{3.2.9}$$

Equations (3.2.7)–(3.2.9) define a solution to (3.2.6), as can easily be seen by substituting c_0^*, c^* and d^* in (3.2.6).

2. Uniqueness

Assume that there exist $c_0 = c_0^* + \delta c_0$, $\hat{c} = \hat{c}^* + \delta\hat{c}$, $\hat{d} = \hat{d}^* + \delta\hat{d}$ satisfying (3.2.6). The following equality must then be satisfied

$$\delta\hat{c}\,\hat{d}_p + k_p \hat{n}_p \delta\hat{d} = -\delta c_0 \frac{k_p}{k_m} \hat{\lambda}_0 \hat{n}_p \hat{d}_m \tag{3.2.10}$$

Recall that $\hat{d}_p$, $\hat{n}_p$, $\hat{\lambda}_0$ and $\hat{d}_m$ have degrees n, m, $n - m - 1$ and n, respectively, with $m \le n - 1$, and $\delta\hat{c}$ and $\delta\hat{d}$ have degrees at most $n - 2$ and $n - 1$. Consequently, the right-hand side is a polynomial of degree $2n - 1$ and the left-hand side is a polynomial of degree at most $2n - 2$. No solution exists unless $\delta c_0 = 0$, so that c_0^* is unique. Let, then, $\delta c_0 = 0$, so that (3.2.10) becomes

$$\frac{\delta \hat{c}}{\delta \hat{d}} = -k_p \frac{\hat{n}_p}{\hat{d}_p} = -\hat{P} \tag{3.2.11}$$

This equation has no solution since $\hat{n}_p$, $\hat{d}_p$ are coprime, so that $\hat{c}^*$ and $\hat{d}^*$ are also unique. □

Comments

a) The coprimeness of $\hat{n}_p$, $\hat{d}_p$ is only necessary to guarantee a *unique* solution. If this assumption is not satisfied, a solution can still be found using (3.2.7)–(3.2.9). Equation (3.2.11) characterizes the set of solutions in this case.

b) Using (3.2.2), the controller structure can be expressed as in Figure 3.2, with a forward block $\hat{\lambda}/\hat{\lambda} - \hat{c}$ and a feedback block $\hat{d}/\hat{\lambda}$. When matching with the model occurs, (3.2.7), (3.2.8) show that the compensator becomes

$$\frac{\hat{\lambda}}{\hat{\lambda} - \hat{c}^*} = \frac{\hat{\lambda}_0 \hat{n}_m}{\hat{q}\, \hat{n}_p} \tag{3.2.12}$$

and

$$\frac{\hat{d}^*}{\hat{\lambda}} = \frac{1}{k_p} \frac{\hat{q}\, \hat{d}_p - \hat{\lambda}_0 \hat{d}_m}{\hat{\lambda}_0 \hat{n}_m} \tag{3.2.13}$$

Thus the forward block actually cancels the zeros of $\hat{P}$ and replaces them by the zeros of $\hat{M}$.

c) The transfer function from r to y_p is of order n, while the plant and controller have $3n - 2$ states. It can be checked (see Section 3.5) that the $2n - 2$ extra modes are those of $\hat{\lambda}$, $\hat{\lambda}_0$ and $\hat{n}_p$. The modes corresponding to $\hat{\lambda}$, $\hat{\lambda}_0$ are stable by choice and those of $\hat{n}_p$ are stable by assumption (A1).

d) The structure of the controller is not unique. In particular, it is equivalent to the familiar structure found, for example, in Callier & Desoer [1982], p. 164, and represented in Figure 3.3. The polynomials found in this case are related to the previous ones through

$$\hat{n}_\pi = c_0 \hat{\lambda} \qquad \hat{d}_c = \hat{\lambda} - \hat{c} \qquad \hat{n}_f = -\hat{d} \tag{3.2.14}$$

The motivation in using the previous controller structure is to obtain an expression that is *linear* in the unknown parameters. These parameters are the coefficients of the polynomials $\hat{c}$, $\hat{d}$ and the gain c_0. The expression in (3.2.1) shows that the control signal is the sum of the parameters multiplied by known or reconstructible signals.

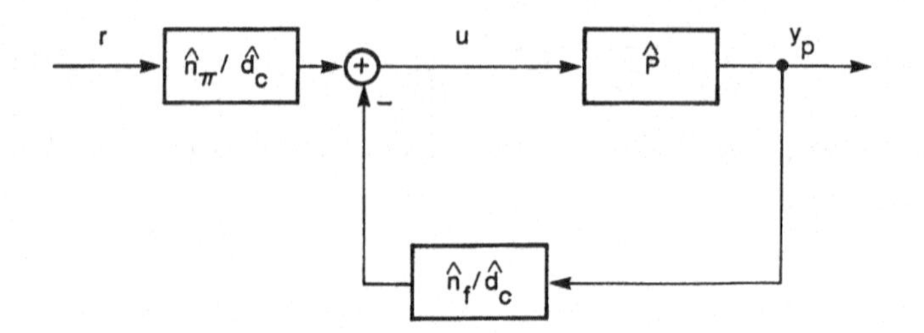

Figure 3.3: Alternate Controller Structure

State-Space Representation

To make this more precise, we consider a state-space representation of the controller. Choose $\Lambda \in \mathbb{R}^{n-1 \times n-1}$ and $b_\lambda \in \mathbb{R}^{n-1}$, such that (Λ, b_λ) is in controllable canonical form as in (2.2.9) and det $(sI - \Lambda) = \hat{\lambda}(s)$. It follows that

$$(sI - \Lambda)^{-1} b_\lambda = \frac{1}{\hat{\lambda}(s)} \begin{bmatrix} 1 \\ s \\ \cdot \\ \cdot \\ \cdot \\ s^{n-2} \end{bmatrix} \tag{3.2.15}$$

Let $c \in \mathbb{R}^{n-1}$ be the vector of coefficients of the polynomial $\hat{c}(s)$, so that

$$\frac{\hat{c}(s)}{\hat{\lambda}(s)} = c^T (sI - \Lambda)^{-1} b_\lambda \tag{3.2.16}$$

Consequently, this transfer function can be realized by

$$\begin{aligned} \dot{w}^{(1)} &= \Lambda w^{(1)} + b_\lambda u \\ \frac{\hat{c}}{\hat{\lambda}}(u) &= c^T w^{(1)} \end{aligned} \tag{3.2.17}$$

where the state $w^{(1)} \in \mathbb{R}^{n-1}$ and the initial condition $w^{(1)}(0)$ is arbitrary. Similarly, there exist $d_0 \in \mathbb{R}$ and $d \in \mathbb{R}^{n-1}$, such that

$$\frac{\hat{d}(s)}{\hat{\lambda}(s)} = d_0 + d^T (sI - \Lambda)^{-1} b_\lambda \tag{3.2.18}$$

and

$$\dot{w}^{(2)} = \Lambda w^{(2)} + b_\lambda y_p$$

$$\frac{\hat{d}}{\hat{\lambda}}(y_p) = d_0 y_p + d^T w^{(2)} \tag{3.2.19}$$

where the state $w^{(2)} \in \mathbb{R}^{n-1}$ and the initial condition $w^{(2)}(0)$ is arbitrary. The controller can be represented as in Figure 3.4, with

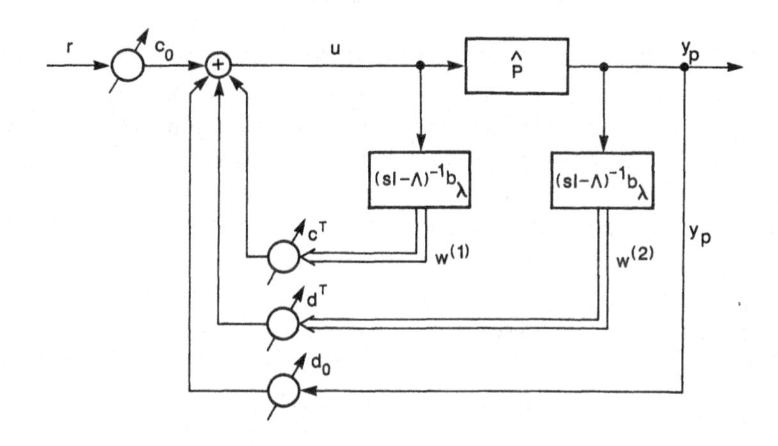

Figure 3.4: Controller Structure—Adaptive Form

$$\begin{aligned} u &= c_0 r + c^T w^{(1)} + d_0 y_p + d^T w^{(2)} \\ &:= \theta^T w \end{aligned} \tag{3.2.20}$$

where

$$\theta^T := (c_0, \bar{\theta}^T) := (c_0, c^T, d_0, d^T) \in \mathbb{R}^{2n} \tag{3.2.21}$$

is the vector of *controller parameters* and

$$w^T := (r, \bar{w}^T) := (r, w^{(1)^T}, y_p, w^{(2)^T}) \in \mathbb{R}^{2n} \tag{3.2.22}$$

is a vector of signals that can be obtained without knowledge of the plant parameters. Note the definitions of $\bar{\theta}$ and $\bar{w}$ which correspond to the vectors θ and w with their first components removed.

In analogy to the previous definitions, we let

$$\theta^{*^T} := (c_0^*, \bar{\theta}^{*^T}) := (c_0^*, c^{*^T}, d_0^*, d^{*^T}) \in \mathbb{R}^{2n} \tag{3.2.23}$$

be the vector of *nominal* controller parameters that achieves a matching of the transfer function $r \to y_p$ to the model transfer function $\hat{M}$. We also define the *parameter errors*

$$\phi := \theta - \theta^* \in \mathbb{R}^{2n} \qquad \bar{\phi} := \bar{\theta} - \bar{\theta}^* \in \mathbb{R}^{2n-1} \tag{3.2.24}$$

The linear dependence of u on the parameters is clear in (3.2.20). In the sequel, we will consider *adaptive* control algorithms and the parameter θ will be a function of time. Similarly, $\hat{c}(s)$, $\hat{d}(s)$ will be polynomials in s whose coefficients vary with time. Equations (3.2.17) and (3.2.19) give a meaning to (3.2.1) in that case.

3.3 ADAPTIVE CONTROL SCHEMES

In Section 3.2, we showed how a controller can be designed to achieve tracking of the reference output y_m by the plant output y_p, when the plant transfer function is known. We now consider the case when the plant is unknown and the control parameters are updated recursively using an identifier. Several approaches are possible. In an indirect adaptive control scheme, the plant parameters (i.e., k_p and the coefficients of $\hat{n}_p(s)$, $\hat{d}_p(s)$) are identified using a recursive identification scheme, such as those described in Chapter 2. The estimates are then used to compute the control parameters through (3.2.7)–(3.2.9).

In a direct adaptive control scheme, an identification scheme is designed that *directly* identifies the controller parameters c_0, c, d_0 and d. A typical procedure is to derive an identifier error signal which depends linearly on the parameter error ϕ. The output error $e_0(t) = y_p(t) - y_m(t)$ is the basis for output error adaptive control schemes such as those of Narendra & Valavani [1978], Narendra, Lin, & Valavani [1980], and Morse [1980]. An output error direct adaptive control scheme and an indirect adaptive control scheme will be described in Sections 3.3.2 and 3.3.3, but we will first turn to an input error direct adaptive control scheme in Section 3.3.1 (this scheme is discussed in Bodson [1986] and Bodson & Sastry [1987]).

Note that we made the distinction between controller and identifier, even in the case of direct adaptive control. The *controller* is by definition the system that determines the value of the control input, using some controller parameters as in a nonadaptive context. The *identifier* obtains estimates of these parameters—directly or indirectly.

As in Chapter 2, we also make the distinction, within the identifier, between the *identifier structure* and the *identification algorithm.* The identifier structure constructs signals which are related by some error equation and are to be used by the identification algorithm. The identification algorithm defines the evolution of the identifier parameters, from which the controller parameters depend. Given an identifier structure with linear error equation for example, several identification algorithms exist from which we can choose (cf. Section 2.3).

Although we make the distinction between controller and identifier, we will see that, for efficiency, some internal signals will be shared by both systems.

3.3.1 Input Error Direct Adaptive Control

Define

$$r_p = \hat{M}^{-1}(y_p) = \hat{M}^{-1}\hat{P}(u) \tag{3.3.1}$$

and let the *input error* e_i be defined by

$$\begin{aligned} e_i &:= r_p - r \\ &= \hat{M}^{-1}(y_p - y_m) = \hat{M}^{-1}(e_0) \end{aligned} \tag{3.3.2}$$

where $e_0 = y_p - y_m$ is the *output error.*

By definition, an input error adaptive control scheme is a scheme based on this error, or a modification of it.

Input Error and Linear Error Equation

The interest of the input error in (3.3.2) is to lead to a linear error equation such as studied in Section 2.3 in the context of identification. We first present an intuitive derivation of this linear error equation.

Consider the matching equality (3.2.6), and divide both sides by $\hat{\lambda}\,\hat{d}_p$. Using (3.2.5)

$$1 - \frac{\hat{c}^*}{\hat{\lambda}} - k_p \frac{\hat{n}_p}{\hat{d}_p}\frac{\hat{d}^*}{\hat{\lambda}} = c_0^* \frac{k_p \hat{n}_p}{\hat{d}_p}\frac{\hat{d}_m}{k_m \hat{n}_m} \tag{3.3.3}$$

With the definitions of $\hat{P}$ and $\hat{M}$, (3.3.3) becomes

$$1 - \frac{\hat{c}^*}{\hat{\lambda}} - \frac{\hat{d}^*}{\hat{\lambda}}\hat{P} = c_0^* \hat{M}^{-1}\hat{P} \tag{3.3.4}$$

We may interpret this last equality as an equality of two polynomial ratios, but also as an equality of two LTI system transfer functions. Applying both transfer functions to the signal u, we have

$$u - \frac{\hat{c}^*}{\hat{\lambda}}(u) - \frac{\hat{d}^*}{\hat{\lambda}}(y_p) = c_0^* \hat{M}^{-1}(y_p) \tag{3.3.5}$$

Now, we recall that $\overline{w} \in \mathbb{R}^{2n} - 1$ is given by

$$\overline{w} = \begin{bmatrix} (sI - \Lambda)^{-1} b_\lambda(u) \\ y_p \\ (sI - \Lambda)^{-1} b_\lambda(y_p) \end{bmatrix} \tag{3.3.6}$$

so that, with (3.3.5)

$$\bar{\theta}^{*T}\bar{w} = \frac{\hat{c}^*}{\hat{\lambda}}(u) + \frac{\hat{d}^*}{\hat{\lambda}}(y_p)$$

$$= u - c_0^* \hat{M}^{-1}(y_p) \tag{3.3.7}$$

The control input is given by (3.2.20)

$$u = \theta^T w = c_0 r + \bar{\theta}^T \bar{w} \tag{3.3.8}$$

so that

$$\bar{\theta}^{*T}\bar{w} = c_0 r + \bar{\theta}^T\bar{w} - c_0^* \hat{M}^{-1}(y_p)$$

$$= c_0(r - \hat{M}^{-1}(y_p)) + \bar{\theta}^T\bar{w} + (c_0 - c_0^*)\hat{M}^{-1}(y_p)$$

$$= -c_0 e_i + \bar{\theta}^T\bar{w} + (c_0 - c_0^*)\hat{M}^{-1}(y_p) \tag{3.3.9}$$

We now define the signal

$$z^T := (r_p, \bar{w}^T) := (\hat{M}^{-1}(y_p), \bar{w}^T) \in \mathbb{R}^{2n} \tag{3.3.10}$$

so that (3.3.9) becomes

$$e_i = \frac{1}{c_0}\phi^T z \tag{3.3.11}$$

This equation is of the form of the linear error equation studied in Section 2.3 (the gain $1/c_0$ being known may be merged either with e_i or z, as will be seen later). It could thus be used to derive an identification procedure to identify the parameter θ directly. As presented, however, the scheme and its derivation show several problems:

a) Since the relative degree of $\hat{M}$ is at least 1, its inverse is not proper. Although $\hat{M}^{-1}(.)$ is well defined, provided that the argument is sufficiently smooth, the gain of the operator $\hat{M}^{-1}$ is arbitrarily large at high frequencies. Therefore, due to the presence of measurement noise, the use of $\hat{M}^{-1}$ is not desirable in practice. Although we will use $\hat{M}^{-1}(.)$ in the analysis, we will consider it not implementable, so that r_p and e_i are not available.

b) The derivation of the error equation (3.3.11) relies on (3.3.8) being satisfied at all times. Although this is not a crucial problem, we will discuss the advantages of avoiding it.

c) We were somewhat careless with initial conditions, going from (3.3.4) to (3.3.5), since $\hat{P}$ may be unstable.

We now derive a modified input error, leading to a scheme that does not have the above disadvantages **a)** and **b)**, and resolve the technical question in **c)**.

Fundamental Identity

Since $\hat{M}$ is minimum phase with relative degree $n-m$, for any stable, minimum phase transfer function $\hat{L}^{-1}$ of relative degree $n-m$, the transfer function $\hat{M}\hat{L}$ has a proper and stable inverse. For example, we can let $\hat{L}$ be a Hurwitz polynomial of degree $n-m$. The signal $\hat{L}^{-1}(r_p)$ is available since

$$\hat{L}^{-1}(r_p) = (\hat{M}\hat{L})^{-1}(y_p) \tag{3.3.12}$$

where $(\hat{M}\hat{L})^{-1}$ is a proper, stable transfer function.

Divide both sides of (3.2.6) by $\hat{\lambda}\hat{d}_p\hat{L}$ so that it becomes, using (3.2.5) and the definitions of $\hat{P}$ and $\hat{M}$,

$$\left[\hat{L}^{-1}\frac{\hat{d}^*}{\hat{\lambda}} + c_0^*(\hat{M}\hat{L})^{-1}\right]\hat{P} = \hat{L}^{-1} - \hat{L}^{-1}\frac{\hat{c}^*}{\hat{\lambda}} \tag{3.3.13}$$

Consider (3.3.13) as an equality of two transfer functions. The right-hand side is a stable transfer function, while the left-hand side is possibly unstable (since $\hat{P}$ is not assumed to be stable).

To transform (3.3.13) into an equality in the time domain, care must be taken of the effect of the initial conditions related to the unstable modes of $\hat{P}$. These will be unobservable or uncontrollable, depending on the realization of the transfer function. If the left-hand side is realized by $\hat{P}$ followed by

$$L^{-1}\frac{\hat{d}^*}{\hat{\lambda}} + c_0^*(\hat{M}\hat{L})^{-1}$$

the unstable modes of $\hat{P}$ will be controllable and, therefore, unobservable.

The operator equality (3.3.13) can then be transformed to a signal equality by applying both operators to u, so that

$$\hat{L}^{-1}\frac{\hat{d}^*}{\hat{\lambda}}(y_p) + c_0^*(\hat{M}\hat{L})^{-1}(y_p) = \hat{L}^{-1}(u) - \hat{L}^{-1}\frac{\hat{c}^*}{\hat{\lambda}}(u) + \epsilon(t) \tag{3.3.14}$$

where $\epsilon(t)$ reminds us of the presence of exponentially *decaying* terms due to initial conditions. These are decaying because the transfer functions are stable and the unstable modes are unobservable. Therefore,

(3.3.14) is valid for arbitrary initial conditions in the realizations of $\hat{L}^{-1}, \hat{\lambda}$, and $(\hat{M}\hat{L})^{-1}$.

Since $\bar{\theta}^*$ is constant, $\bar{\theta}^{*T}\hat{L}^{-1}(\bar{w})$ is given by

$$\bar{\theta}^{*T}\hat{L}^{-1}(\bar{w}) = \hat{L}^{-1}(\bar{\theta}^{*T}\bar{w})$$

$$= \hat{L}^{-1}\left[\frac{\hat{c}^*}{\hat{\lambda}}(u) + \frac{\hat{d}^*}{\hat{\lambda}}(y_p)\right]$$

$$= \hat{L}^{-1}(u) - c_0^*(\hat{M}\hat{L})^{-1}(y_p) + \epsilon(t) \tag{3.3.15}$$

where we used (3.3.14). Define now

$$v^T := \left[\hat{L}^{-1}(r_p), \hat{L}^{-1}(\bar{w}^T)\right] = \left[(\hat{M}\hat{L})^{-1}(y_p), \hat{L}^{-1}(\bar{w}^T)\right] \in \mathbb{R}^{2n} \tag{3.3.16}$$

so that (3.3.15) can be written

$$\hat{L}^{-1}(u) = \theta^{*T}v + \epsilon(t) \tag{3.3.17}$$

where θ^* is defined in (3.2.7)–(3.2.9), with (3.2.23). Equation (3.3.17) is essential to subsequent derivations, so that we summarize the result in the following proposition.

Proposition 3.3.1 Fundamental Identity

Let $\hat{P}$ and $\hat{M}$ satisfy assumptions (A1) and (A2). Let $\hat{L}^{-1}$ be any stable, minimum phase transfer function of relative degree $n - m$. Let v and $\bar{w}$ be as defined by (3.3.16) and (3.3.6), with arbitrary initial conditions in the realizations of the transfer functions. Let θ^* be defined by (3.2.7)–(3.2.9), with (3.2.23).

Then for all piecewise continuous $u \in L_{\infty e}$, (3.3.17) is satisfied.

Input Error Identifier Structure

Equation (3.3.17) is of the form studied in Section 2.3 for recursive identification. Both the signal $\hat{L}^{-1}(u)$ and v are available from measurements, and the expression is linear in the unknown parameter θ^*.

Therefore, we define the *modified input error* to be

$$e_2 := \theta^T v - \hat{L}^{-1}(u) \tag{3.3.18}$$

so that, using (3.3.17)

$$e_2 = \phi^T v + \epsilon(t) \tag{3.3.19}$$

which is of the form of the *linear error equation* studied in Section 2.3. Although we considered the input error e_i not to be available, because it would require the realization of a nonproper transfer function, the approximate input error e_2, and the signal v are available, given these considerations.

We also observed in Chapter 2 that standard properties of the identification algorithms are not affected by the $\epsilon(t)$ term. For simplicity, we will omit this term in subsequent derivations. We now consider the practical implementation of the algorithm, with the required assumptions.

Assumptions

The algorithm relies on assumptions (A1)–(A3) and the following additional assumption.

(A4) Bound on the High-Frequency Gain

Assume that an upper bound on k_p is known, that is, that $k_p \le k_{\max}$ for some $k_{\max}$.

The structure of the controller and identifier is shown in Figure 3.5, while the complete algorithm is summarized hereafter. The need for assumption (A4), and for the projection of c_0, will be discussed later, in connection with alternate schemes. It will be more obvious from the proof of stability of the algorithm in Section 3.7.

Input Error Direct Adaptive Control Algorithm—Implementation

Assumptions

(A1)–(A4)

Data

$n\,,\, m\,,\, k_{\max}$

Input

$r(t),\, y_p(t) \in \mathbb{R}$

Output

$u(t) \in \mathbb{R}$

Internal Signals

$$w(t) \in \mathbb{R}^{2n}\ [w^{(1)}(t),\, w^{(2)}(t) \in \mathbb{R}^{n-1}]$$

$$\theta(t) \in \mathbb{R}^{2n}\ [c_0(t),\, d_0(t) \in \mathbb{R},\, c(t),\, d(t) \in \mathbb{R}^{n-1}]$$

$$v(t) \in \mathbb{R}^{2n},\, e_2(t) \in \mathbb{R}$$

Initial conditions are arbitrary, except $c_0(0) \ge c_{\min} = k_m/k_{\max} > 0$.

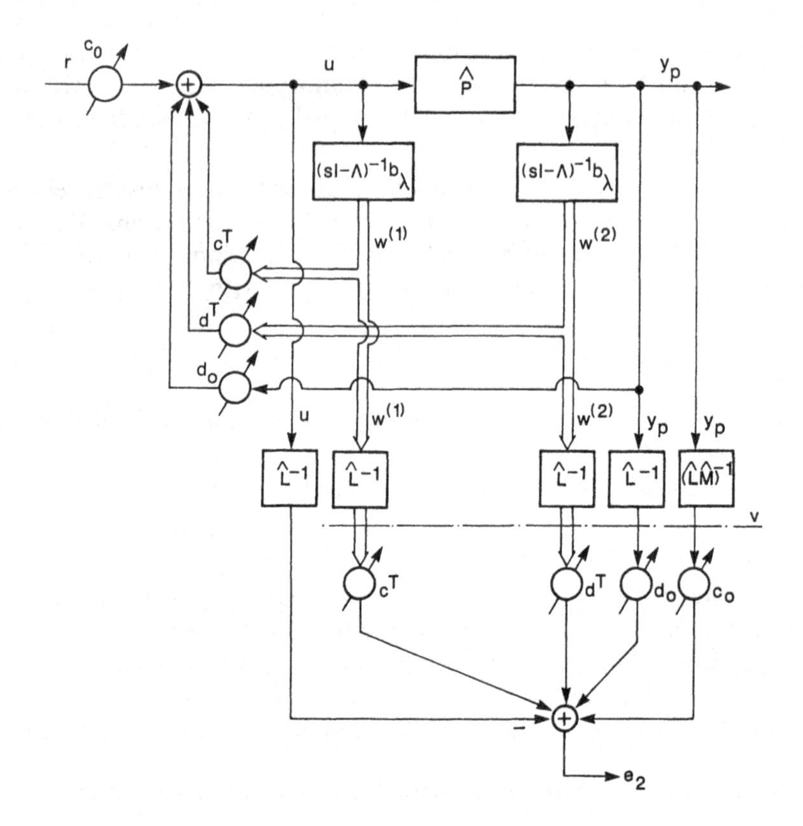

Figure 3.5: Controller and Input Error Identifier Structures

Design Parameters

Choose

- $\hat{M}$ (i.e. k_m, $\hat{n}_m$, $\hat{d}_m$) satisfying (A2).
- $\Lambda \in \mathbb{R}^{n-1 \times n-1}$, $b_\lambda \in \mathbb{R}^{n-1}$ in controllable canonical form, such that $\det(sI-\Lambda)$ is Hurwitz and contains the zeros of $\hat{n}_m(s)$.
- $\hat{L}^{-1}$ stable, minimum phase transfer function of relative degree $n-m$.
- $g, \gamma > 0$.

Controller Structure

$$\dot{w}^{(1)} = \Lambda w^{(1)} + b_\lambda u$$

$$\dot{w}^{(2)} = \Lambda w^{(2)} + b_\lambda y_p$$

$$\theta^T = (c_0, c^T, d_0, d^T)$$
$$w^T = (r, w^{(1)}, y_p, w^{(2)})$$
$$u = \theta^T w$$

Identifier Structure

$$v^T = [(\hat{M}\hat{L})^{-1}(y_p), \hat{L}^{-1}(w^{(1)^T}), \hat{L}^{-1}(y_p), \hat{L}^{-1}(w^{(2)^T})]$$
$$e_2 = \theta^T v - \hat{L}^{-1}(u)$$

Normalized Gradient Algorithm with Projection

$$\dot{\theta} = -g \frac{e_2 v}{1 + \gamma v^T v}$$

if $c_0 = c_{\min}$ and $\dot{c}_0 < 0$, then set $\dot{c}_0 = 0$.

□

Comments

Adaptive Observer

The signal generators for $w^{(1)}$ and $w^{(2)}$ ((3.2.17) and (3.2.19)) are almost identical to those used in Chapter 2 for identification of the plant parameters (their dimensions are now $n-1$ instead of n previously). They are shared by the controller and the identifier. The signal generators (sometimes called *state-variable filters) for* $w^{(1)}$ and $w^{(2)}$ form a *generalized observer,* reconstructing the states of the plant in a specific parameterization. This parameterization has the characteristic of allowing the reconstruction of the states without knowledge of the parameters. The states are used for the state feedback of the controller to the input in a certainty equivalence manner, meaning that the parameters used for feedback are the current estimates multiplying the states *as if* they were the true parameters. The identifier with the generalized observer is sometimes called an *adaptive observer* since it provides at the same time estimates of the states and of the parameters.

Separation of Identification and Control

Although we have derived a direct adaptive control scheme, the identifier and the controller can be distinguished. The gains c_0, c, d_0 and d serving to generate u are associated with the controller, while those used to compute e_2 are associated with the identifier. In fact, it is not necessary that these be identical for the identifier error to be as defined in (3.3.19). This is because (3.3.19) was derived using the fundamental identity (3.3.17), which is valid no matter how u is actually computed. In other words, the identifier can be used off-line, without actually updating the controller parameters if necessary. This is also useful, for example, in case of input saturation (cf. Goodwin & Mayne

[1987]). If the actual input to the LTI plant is different from the computed input $u = \theta^T w$ (due to actuator saturation, for example), the identifier will still have consistent input signals, provided that the signal u entering the identifier is the actual input entering the LTI plant.

3.3.2 Output Error Direct Adaptive Control

An output error scheme is based on the output error $e_0 = y_p - y_m$. Note that by applying $\hat{M}\hat{L}$ to both sides of (3.3.17), we find

$$\hat{M}(u) = c_0^* y_p + \hat{M}(\bar{\theta}^{*T}\bar{w}) \tag{3.3.20}$$

As before, the control input u is set equal to $u = \theta^T w$, but now, this equality is used to derive the identifier error equation

$$\begin{aligned} e_0 = y_p - y_m &= \frac{1}{c_0^*}\hat{M}(u - \bar{\theta}^{*T}\bar{w}) - \hat{M}(r) \\ &= \frac{1}{c_0^*}\hat{M}((c_0 - c_0^*)r + (\bar{\theta}^T - \bar{\theta}^{*T})\bar{w}) \\ &= \frac{1}{c_0^*}\hat{M}(\phi^T w) \end{aligned} \tag{3.3.21}$$

which has the form of the *basic* strictly positive real (SPR) error equation of Chapter 2. The gradient identification algorithms of Section 2.6 can therefore be used, provided that $\hat{M}$ is SPR. However, since this requires $\hat{M}$ to have relative degree at most 1, this scheme does not work for plants with relative degree greater than 1.

The approach can however be saved by modifying the scheme, as for example in Narendra, Lin, & Valavani [1980]. We now review their scheme for the case when the high-frequency gain k_p is known, and we let $c_0 = c_0^*$.

The controller structure of the output error scheme is identical to the controller structure of the input error scheme, while the identifier structure is different. It relies on the identifier error

$$e_1 = \frac{1}{c_0^*}\hat{M}\hat{L}(\bar{\phi}^T\bar{v} - \gamma\bar{v}^T\bar{v}e_1) \tag{3.3.22}$$

which is now of the form of the *modified* SPR error equation of Chapter 2. As previously, $\bar{v}$ is identical to v, but with the first component removed. Practically, (3.3.22) is not implemented as such. Instead, we use (3.3.17) to obtain

$$e_1 = \frac{1}{c_0^*} \hat{M}\,\hat{L}\,(\bar{\theta}^T \bar{v} - \hat{L}^{-1}(u) + c_0^* (\hat{M}\hat{L})^{-1}(y_p) - \gamma \bar{v}^T \bar{v}\, e_1)$$

$$= y_p - \frac{1}{c_0^*} \hat{M}\,(u) + \frac{1}{c_0^*} \hat{M}\,\hat{L}\,(\bar{\theta}^T \hat{L}^{-1}(\bar{w}\,) - \gamma \bar{v}^T \bar{v}\, e_1) \qquad (3.3.23)$$

As before, the control signal is set equal to $u = \theta^T w = c_0^* r + \bar{\theta}^T \bar{w}$, and the equality is used to derive the error equation for the identifier

$$e_1 = y_p - \hat{M}\,(r) - \frac{1}{c_0^*} \hat{M}\,(\bar{\theta}^T \bar{w}\,) + \frac{1}{c_0^*} \hat{M}\hat{L}\,(\bar{\theta}^T \hat{L}^{-1}(\bar{w}\,) - \gamma \bar{v}^T \bar{v}\, e_1)$$

$$= y_p - y_m - \frac{1}{c_0^*} \hat{M}\hat{L}\,((\hat{L}^{-1}\bar{\theta}^T - \bar{\theta}^T \hat{L}^{-1})(\bar{w}\,) + \gamma \bar{v}^T \bar{v}\, e_1) \quad (3.3.24)$$

Again, the identifier error involves the output error $e_0 = y_p - y_m$. The additional term, which appeared starting with the work of Monopoli [1974], is denoted

$$y_a = \frac{1}{c_0^*} \hat{M}\hat{L}\,((\hat{L}^{-1}\bar{\theta}^T - \bar{\theta}^T \hat{L}^{-1})(\bar{w}\,) + \gamma \bar{v}^T \bar{v}\, e_1) \quad (3.3.25)$$

and the resulting error $e_1 = y_p - y_m - y_a$ is called the *augmented error*, in contrast to the original output error $e_0 = y_p - y_m$.

The error (3.3.22) is of the form of the modified SPR error equation of Chapter 2 provided that $\hat{M}\hat{L}$ is a strictly positive real transfer function. If this condition is satisfied, the properties of the identifier will follow and are the basis of the stability proof of Section 3.7.

Assumptions

The algorithm relies on assumptions (A1)–(A3) and the following assumption.

(A5) High-Frequency Gain and SPR Assumptions

Assume that k_p is known and that there exists $\hat{L}^{-1}$, a stable, minimum phase transfer function of relative degree $n - m - 1$, such that $\hat{M}\hat{L}$ is SPR.

The practical implementation of the algorithm is summarized hereafter.

Output Error Direct Adaptive Control Algorithm—Implementation

Assumptions

(A1)–(A3), (A5)

Data

n, m, k_p

Input

$r(t), y_p(t) \in \mathbb{R}$

Output

$u(t) \in \mathbb{R}$

Internal Signals

$\bar{w}(t) \in \mathbb{R}^{2n-1}$ $[w^{(1)}(t), w^{(2)}(t) \in \mathbb{R}^{n-1}]$

$\bar{\theta}(t) \in \mathbb{R}^{2n-1}$ $[(c(t), d(t) \in \mathbb{R}^{n-1}, d_0(t) \in \mathbb{R}]$

$\bar{v}(t) \in \mathbb{R}^{2n-1}$

$e_1(t), y_a(t), y_m(t) \in \mathbb{R}$

Initial conditions are arbitrary.

Design Parameters

Choose

- $\hat{M}$ (i.e. $k_m, \hat{n}_m, \hat{d}_m$) satisfying (A2) and (A5).
- $\Lambda \in \mathbb{R}^{n-1 \times n-1}$, $b_\lambda \in \mathbb{R}^{n-1}$, in controllable canonical form, such that $\det(sI - \Lambda)$ is Hurwitz, and contains the zeros of $\hat{n}_m(s)$.
- $\hat{L}^{-1}$ stable, minimum phase transfer function of relative degree $n - m - 1$, such that $\hat{M}\hat{L}$ is SPR.
- $g, \gamma > 0$.

Controller Structure

$$\dot{w}^{(1)} = \Lambda w^{(1)} + b_\lambda u$$

$$\dot{w}^{(2)} = \Lambda w^{(2)} + b_\lambda y_p$$

$$\bar{\theta}^T = (c^T, d_0, d^T)$$

$$\bar{w}^T = (w^{(1)^T}, y_p, w^{(2)^T})$$

$$c_0^* = k_m / k_p > 0$$

$$u = c_0^* r + \bar{\theta}^T \bar{w}$$

Identifier Structure

$$\bar{v}^T = \hat{L}^{-1}(\bar{w})$$

$$y_m = \hat{M}(r)$$

$$y_a = \frac{1}{c_0^*} \hat{M} \hat{L} (\hat{L}^{-1}(\bar{\theta}^T \bar{w}) - \bar{\theta}^T \hat{L}^{-1}(\bar{w}) - \gamma \bar{v}^T \bar{v} e_1)$$

$$e_1 = y_p - y_m - y_a$$

Gradient Algorithm

$$\dot{\bar{\theta}} = -g e_1 \bar{v}$$

□

Differences Between Input and Output Error

Traditionally, the starting point in the derivation of model reference adaptive control schemes has been the output error $e_0 = y_p - y_m$. Using the error between the plant and the reference model to update controller parameters is intuitive. However, stability proofs suggest that SPR conditions must be satisfied by the model and that an augmented error should be used when the relative degree of the plant is greater than 1. The derivation of the input error scheme shows that model reference adaptive control can in fact be achieved *without* formally involving the output error and without SPR conditions on the reference model.

Important differences should be noted between the input and output error schemes. The first is that the derivation of the equation error (3.3.24) from (3.3.22) relies on the input signal u being equal to the computed value $u = \theta^T w$, at all times. If the input saturates, updates of the identifier will be erroneous. When the input error scheme is used, this problem can be avoided, provided that the actual input entering the LTI plant is available and used in the identifier. This is because (3.3.19) is based on (3.3.17) and does not assume any particular value of u. If needed, the parameters used for identification and control can also be separated, and the identifier can be used "off-line."

A second difference appears between the input and output error schemes when the high-frequency gain k_p is unknown, and the relative degree of the plant is greater than 1. The error e_1 derived in (3.3.22) is not implementable if c_0^* is unknown. Although an SPR error equation can still be obtained in the unknown high-frequency gain case, the solution proposed by Morse [1980] (and also Narendra, Lin, & Valavani [1980]) requires an overparameterization of the identifier which excludes the possibility of asymptotic stability even when persistency of excitation (PE) conditions are satisfied (cf. Boyd & Sastry [1986], Anderson, Dasgupta, & Tsoi [1985]). In view of the recent examples due to Rohrs, and the connections between exponential convergence and robustness (see Chapter 5), this appears to be a major drawback of the algorithm.

Another advantage of the input error scheme is to lead to a linear error equation for which other identification algorithms, such as the least-squares algorithm, are available. These algorithms are an advantageous alternative to the gradient algorithm. Further, it was shown recently that there are advantages of the input error scheme in terms of robustness to unmodeled dynamics (cf. Bodson [1988]).

In some cases, the input error scheme requires more computations. This is because the observers for $w^{(1)}$, $w^{(2)}$ are on order n, instead of $n-1$ for the output error scheme. Also, the filter $\hat{L}^{-1}$ is one order higher. When the relative degree is 1, significant simplifications arise in the output error scheme, as discussed now.

Output Error Direct Adaptive Control—The Relative Degree 1 Case

The condition that $\hat{M}\hat{L}$ be SPR is considerably stronger than the condition that $\hat{M}\hat{L}$ simply be invertible (as required by the input error scheme and guaranteed by (A2)). The relative degree of $\hat{L}^{-1}$, however, is only required to be $n-m-1$, as compared to $n-m$ for proper invertibility. In the case when the relative degree $n-m$ of the model and of the plant is 1, $\hat{L}^{-1}$ is unnecessary along with the additional signal y_a. The output error direct adaptive control scheme then has a much simpler form, in which the error equation used for identification involves the output error $e_0 = y_p - y_m$ only. The simplicity of this scheme makes it attractive in that case. We assume therefore the following:

(A6) Relative Degree 1 and SPR Assumptions

$n-m = 1$, $\hat{M}$ is SPR.

Output Error Direct Adaptive Control Algorithm, Relative Degree 1—Implementation

Assumptions

(A1)–(A3), (A6)

Data

n, k_p

Input

$r(t), y_p(t) \in \mathbb{R}$

Output

$u(t) \in \mathbb{R}$

Internal Signals

$w(t) \in \mathbb{R}^{2n}$ $[w^{(1)}(t), w^{(2)}(t) \in \mathbb{R}^{n-1}]$

$\theta(t) \in \mathbb{R}^{2n}$ $[c_0(t), d_0(t) \in \mathbb{R}, c(t), d(t) \in \mathbb{R}^{n-1}]$

$y_m(t), e_0(t) \in \mathbb{R}$

Initial conditions are arbitrary.

Design Parameters

Choose

- $\hat{M}$ (i.e. k_m, $\hat{n}_m$, $\hat{d}_m$) satisfying (A2) and (A6).
- $\Lambda \in \mathbb{R}^{n-1 \times n-1}$, $b_\lambda \in \mathbb{R}^{n-1}$ in controllable canonical form and such that $\det(sI - \Lambda) = \hat{n}_m(s)$.
- $g > 0$.

Controller Structure

$$\dot{w}^{(1)} = \Lambda w^{(1)} + b_\lambda u$$
$$\dot{w}^{(2)} = \Lambda w^{(2)} + b_\lambda y_p$$
$$\theta^T = (c_0, c^T, d_0, d^T)$$
$$w^T = (r, w^{(1)^T}, y_p, w^{(2)^T})$$
$$u = \theta^T w$$

Identifier Structure

$$y_m = \hat{M}(r)$$
$$e_0 = y_p - y_m$$

Gradient Algorithm

$$\dot{\theta} = -g e_0 w$$

□

Comment

The identifier error equation is (3.3.24) and is the *basic* SPR error equation of Chapter 2. The high-frequency gain k_p (and consequently c_0) can be assumed to be unknown, but the sign of k_p must still be known to ensure that $c_0^* > 0$, so that $(1/c_0^*)\hat{M}$ is SPR.

3.3.3 Indirect Adaptive Control

In the indirect adaptive control scheme presented in this section, estimates of the plant parameters k_p, $\hat{n}_p$, and $\hat{d}_p$ are obtained using the standard equation error identifier of Chapter 2. The controller parameters c_0, $\hat{c}$ and $\hat{d}$ are then computed using the relationships resulting from the matching equality (3.2.6).

Note that the dimension of the signals $w^{(1)}$, $w^{(2)}$ used for identification in Chapter 2 is n, the order of the plant. For control, it is sufficient that this dimension be $n-1$. However, in order to share the observers for identification and control, we will let their dimension be n. Proposition 3.2.1 is still true then, but the degrees of the polynomials become, respectively, $\partial\hat{\lambda} = n$, $\partial\hat{\lambda}_0 = n - m$, $\partial\hat{q} = n - m$, $\partial\hat{d} = n - 1$, and $\partial\hat{c} = n - 1$. Since $\partial\hat{d} = n - 1$, it can be realized as $d^T(sI - \Lambda)^{-1}b$, without the direct gain d_0 from y_p. This a (minor) technical difference, and for simplicity, we will keep our previous notation. Thus, we define

$$\bar{\theta}^T := (c^T, d^T) \in \mathbb{R}^{2n} \qquad \bar{w}^T := (w^{(1)^T}, w^{(2)^T}) \in \mathbb{R}^{2n} \tag{3.3.26}$$

and

$$\theta^T := (c_0, \bar{\theta}^T) \in \mathbb{R}^{2n+1} \qquad w^T := (r, \bar{w}^T) \in \mathbb{R}^{2n+1} \tag{3.3.27}$$

The controller structure is otherwise completely identical to the controller structure described previously.

The identifier parameter is now different from the controller parameter θ. We will denote, in analogy with (2.2.17)

$$\begin{aligned}\pi^T &:= (a^T, b^T)\\ &:= (a_1, \ldots, a_{m+1}, 0, \ldots, b_1, \ldots, b_n) \in \mathbb{R}^{2n}\end{aligned} \tag{3.3.28}$$

Since the relative degree is assumed to be known, there is no need to update the parameters a_{m+2}—so that we let these parameters be zero in (3.3.28). The corresponding components of $\bar{w}$ are thus not used for identification. We let $\tilde{w}$ be equal to $\bar{w}$ except for those components which are not used and are thus set to zero, so that

$$\tilde{w}^T := (w_1^{(1)}, \ldots, w_{m+1}^{(1)}, 0, \ldots, w^{(2)^T}) \in \mathbb{R}^{2n} \tag{3.3.29}$$

A consequence (that will be used in the stability proof in Section 3.7) is that the relative degree of the transfer function from $u \to \tilde{w}$ is at least $n - m$.

The nominal value of the identifier parameter π^* can be found from the results of Chapter 2 through the polynomial equalities in (2.2.5), that is

$$\begin{aligned}\hat{a}^*(s) &= a_1^* + a_2^* s + \cdots + a_{m+1}^* s^m = k_p \hat{n}_p(s)\\ \hat{b}^*(s) &= b_1^* + b_2^* s + \cdots + b_n^* s^{n-1} = \hat{\lambda}(s) - \hat{d}_p(s)\end{aligned} \tag{3.3.30}$$

The *identifier parameter error* is now denoted

$$\psi := \pi - \pi^* \in \mathbb{R}^{2n} \tag{3.3.31}$$

The transformation $\pi \to \theta$ is chosen following a certainty equivalence principle to be the same as the transformation $\pi^* \to \theta^*$, as in (3.2.7)–(3.2.9). Note that our estimate of the high-frequency gain k_p is

a_{m+1}. Since $c_0^* = k_m / k_p$, we will let $c_0 = k_m / a_{m+1}$. The control input u will be unbounded if a_{m+1} goes to zero, and to avoid this problem, we make the following assumption.

(A7) Bound on the High-Frequency Gain

Assume $k_p \geq k_{\min} > 0$.

The practical implementation of the indirect adaptive control algorithm is summarized hereafter.

Indirect Adaptive Control Algorithm—Implementation

Assumptions

(A1)–(A3), (A7)

Data

$n, m, k_{\min}$

Input

$r(t), y_p(t) \in \mathbb{R}$

Output

$u(t) \in \mathbb{R}$

Internal Signals

$w(t) \in \mathbb{R}^{2n+1}$ $[w^{(1)}(t), w^{(2)}(t) \in \mathbb{R}^n]$

$\theta(t) \in \mathbb{R}^{2n+1}$ $[c_0(t) \in \mathbb{R}, c(t), d(t) \in \mathbb{R}^n]$

$\pi(t) \in \mathbb{R}^{2n}$ $[a(t), b(t) \in \mathbb{R}^n]$

$\tilde{w}(t) \in \mathbb{R}^{2n}$

$y_i(t), e_3(t) \in \mathbb{R}$

Initial conditions are arbitrary, except $a_{m+1}(0) > k_{\min}$.

Design Parameters

Choose

- $\hat{M}$ (i.e. $k_m, \hat{n}_m, \hat{d}_m$) satisfying (A2).
- $\Lambda \in \mathbb{R}^{n \times n}$, $b_\lambda \in \mathbb{R}^n$ in controllable canonical form, such that $\det(sI - \Lambda) = \hat{\lambda}(s)$ is Hurwitz and $\hat{\lambda}(s) = \hat{\lambda}_0(s)\hat{n}_m(s)$.
- $g, \gamma > 0$.

Controller Structure

$$\dot{w}^{(1)} = \Lambda w^{(1)} + b_\lambda u$$

$$\dot{w}^{(2)} = \Lambda w^{(2)} + b_\lambda y_p$$

$$\theta^T = (c_0, c^T, d^T) = (c_0, c_1, \ldots, c_n, d_1, \ldots, d_n)$$

$$w^T = (r, w^{(1)^T}, w^{(2)^T})$$
$$u = \theta^T w$$

Identifier Structure

$$\pi^T = (a^T, b^T) = (a_1, \ldots, a_{m+1}, 0, \ldots, b_1, \ldots, b_n)$$
$$\tilde{w} = (w_1^{(1)}, \ldots, w_{m+1}^{(1)}, 0, \ldots, w^{(2)^T})$$
$$y_i = \pi^T \tilde{w}$$
$$e_3 = \pi^T \tilde{w} - y_p$$

Normalized Gradient Algorithm with Projection

$$\dot{\pi} = -g \frac{e_3 \tilde{w}}{1 + \gamma \tilde{w}^T \tilde{w}}$$

If $a_{m+1} = k_{\min}$ and $\dot{a}_{m+1} < 0$, then let $\dot{a}_{m+1} = 0$.

Transformation Identifier Parameter → Controller Parameter

Let the polynomials with time-varying coefficients

$$\hat{a}(s) = a_1 + \cdots + a_{m+1} s^m \qquad \hat{c}(s) = c_1 + \cdots + c_n s^{n-1}$$
$$\hat{b}(s) = b_1 + \cdots + b_n s^{n-1} \qquad \hat{d}(s) = d_1 + \cdots + d_n s^{n-1}$$

Divide $\hat{\lambda}_0 \hat{d}_m$ by $(\hat{\lambda} - \hat{b})$, and let $\hat{q}$ be the quotient.

θ is given by the coefficients of the polynomials

$$\hat{c} = \hat{\lambda} - \frac{1}{a_{m+1}} \hat{q} \hat{a}$$
$$\hat{d} = \frac{1}{a_{m+1}} (\hat{q} \hat{\lambda} - \hat{q} \hat{b} - \hat{\lambda}_0 \hat{d}_m)$$

and by

$$c_0 = \frac{k_m}{a_{m+1}}$$

□

Transformation Identifier Parameter → Controller Parameter

We assumed that the transformation from the identifier parameter π to the controller parameter θ is performed *instantaneously*. Note that $\hat{\lambda} - \hat{b}$ is a monic polynomial, so that $\hat{q}$ is also a monic polynomial (of degree $n - m$). Its coefficients can be expressed as the sum of products of coefficients of $\hat{\lambda}_0 \hat{d}_m$ and $\hat{\lambda} - \hat{b}$. The same is true for $\hat{c}$, $\hat{d}$, and c_0 with an additional division by a_{m+1}. Therefore, given n and m, the transformation consists of a fixed number of multiplications, additions, and a division.

Note also that if the coefficients of $\hat{a}$ and $\hat{b}$ are bounded, and if a_{m+1} is bounded away from zero (as is guaranteed by the projection), then the coefficients of $\hat{q}$, $\hat{c}$, $\hat{d}$, and c_0 are bounded. Therefore, the transformation is also continuously differentiable and has bounded derivatives.

3.3.4 Alternate Model Reference Schemes

The input error scheme is closely related to the schemes presented in discrete time by Goodwin & Sin [1984], and in continuous time by Goodwin & Mayne [1987]. Their identifier structure is identical to the structure used here, but their controller structure is somewhat different. In our notation, Goodwin & Mayne choose

$$\hat{M}(s) = k_m \frac{\hat{n}(s)}{\hat{\lambda}(s)\hat{L}(s)} \tag{3.3.32}$$

where $\hat{n}$, $\hat{\lambda}$ and $\hat{L}$ are *polynomials* of degree $\leq n$, n, and $n-m$, respectively. The polynomials $\hat{\lambda}$, $\hat{L}$ are used for similar purposes as in the input error scheme. However, except for possible pole-zero cancellations, $\hat{\lambda}\hat{L}$ now also defines the model poles in (3.3.32). The filtered reference input

$$\bar{r} = k_m \frac{\hat{n}(s)}{\hat{\lambda}(s)}(r) \tag{3.3.33}$$

is used as input to the actual controller. Then, the transfer function $\bar{r} \to y_p$ is made to match $\hat{L}^{-1}$, so that the transfer function from $r \to y_p$ is $\hat{M}$. Thus, by prefiltering the input, the control problem of matching a transfer function $\hat{M}$ is altered to the problem of matching the arbitrary all-pole transfer function $\hat{L}^{-1}$.

The input error adaptive control scheme of Section 3.3.1 can be used to achieve this new objective and is represented in Figure 3.6. This scheme is the one obtained by Goodwin & Mayne (up to a small remaining difference described hereafter). Since the new model is $\hat{L}^{-1}$, the new transfer function $\hat{M}\hat{L}$ is equal to 1. Note that, in this instance, the input and output errors are identical and the input and output error schemes are very similar. The analysis is also considerably simplified.

Goodwin & Mayne's algorithms essentially control the plant by reducing the transfer function to an all-pole transfer function of relative degree $n-m$. The additional dynamics are provided by prefiltering the reference input. Thus, the input error scheme presented in Section 3.3.1 is a more general scheme, allowing for the placement of all the closed-

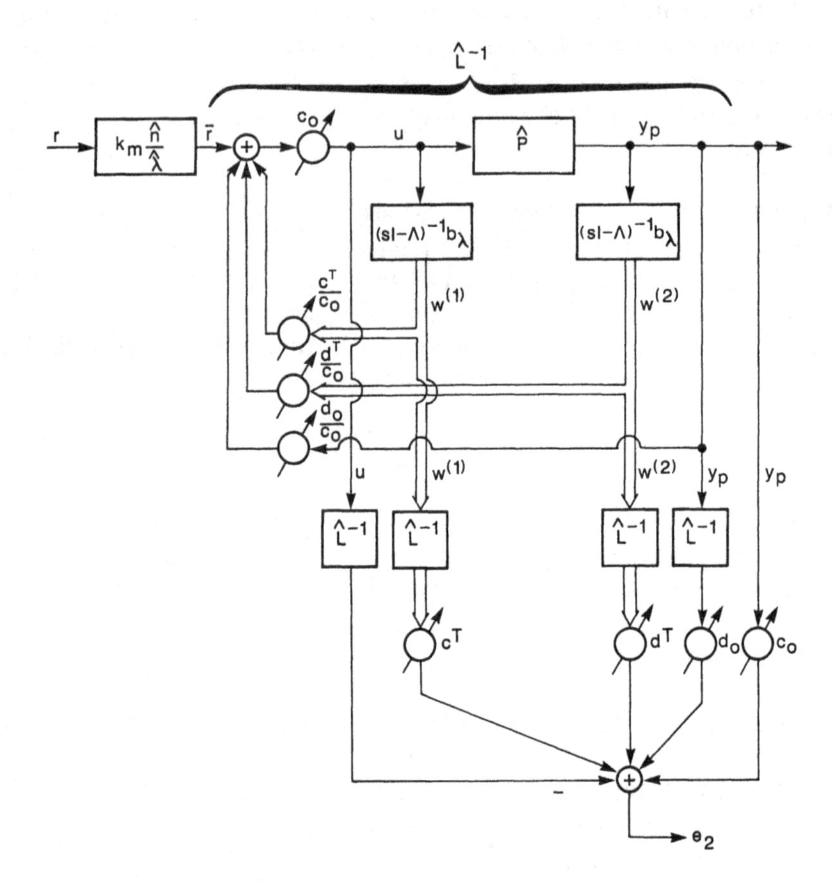

Figure 3.6: Alternate Input Error Scheme

loop poles directly at the desired locations without prefiltering.

Note that since identification and control can be separated in the input error scheme, we may identify $1/c_0$ and $\bar{\theta} / c_0$ rather than c_0 and $\bar{\theta}$. This is also shown in Figure 3.6. By dividing the identifier error e_2 by c_0, the appropriate linear error equation may be found and used for identification.

It is curious to note that the problems encountered are different depending whether we identify c_0 or $1/c_0$. If we identify $1/c_0$, as we did in the indirect scheme, the control input $u = c_0 r + \bar{\theta}^T \bar{w}$ will be unbounded if the estimate of $1/c_0$ goes to zero. To avoid the zero crossing, we require knowledge of the sign of $1/c_0$ (that is, of k_p), and of a lower bound on $1/c_0$, that is a *lower* bound on k_p to be used with the projection algorithm.

If we identify c_0 directly, as we did in the input error scheme, a different problem appears. If $c_0 = 0$ and $\bar{\theta} = 0$, then $u = 0$ and $e_2 = 0$ (cf. Figure 3.5). No adaptation will occur ($\dot{\phi} = 0$), although $y_p - y_m$ does not tend necessarily to zero and may even be unbounded. This is an identification problem, since we basically lose information in the regression vector. To avoid it, we require the knowledge of the sign of c_0 (i.e., of k_p) and a lower bound on c_0, therefore an *upper* bound on k_p, to be used by the projection algorithm.

3.3.5 Adaptive Pole Placement Control

The model reference adaptive control approach requires a minimum phase assumption on the plant $\hat{P}(s)$. This results from the necessity to cancel the plant zeros in order to replace them by the model zeros. One might consider the approach of letting the model be

$$\hat{M}(s) = k_m \frac{\hat{n}_p(s)}{\hat{d}_m(s)} \qquad i.e. \qquad \hat{n}_m(s) = \hat{n}_p(s) \tag{3.3.34}$$

that is, require that only the closed-loop poles be assigned. An adaptive control based on this idea is called an *adaptive pole placement control* algorithm. We now discuss some of the differences between this and the model reference adaptive control algorithms presented above.

First note that in (3.3.34), the reference model itself becomes adaptive, with $\hat{n}_p$ replaced by its estimated value. Since $\hat{n}_p$ is unknown *a priori* and it is not Hurwitz, it is impossible to choose $\hat{\lambda} = \hat{\lambda}_0 \hat{n}_m$ as before. Now, let $\hat{\lambda}$ be an arbitrary Hurwitz polynomial, and consider the same controller as previously, so that (3.2.4) is valid. The nominal values c_0^*, $\hat{c}^*$, $\hat{d}^*$ such that the closed-loop transfer function is equal to the $\hat{M}(s)$ defined in (3.3.34) must satisfy

$$(\hat{\lambda} - \hat{c}^*)\hat{d}_p - k_p \hat{n}_p \hat{d}^* = \left[c_0^* \frac{k_p}{k_m} \right] \hat{\lambda}\, \hat{d}_m \tag{3.3.35}$$

This equation is a *Diophantine equation*, that is a polynomial equation of the form

$$\hat{a}\hat{x} + \hat{b}\hat{y} = \hat{c} \tag{3.3.36}$$

A necessary condition for a solution $\hat{x}, \hat{y}$ to exist is that any common zero of $\hat{a}, \hat{b}$ is also a zero of $\hat{c}$. A sufficient condition is simply that $\hat{a}, \hat{b}$ be coprime, in this case, $\hat{n}_p$, $\hat{d}_p$ coprime (see lemma A6.2.3 in the

Appendix for the general solution in that case). Previously, (3.3.35) was replaced by (3.2.6), with $\hat{n}_p$ appearing on the right hand side, so that any common zero of $\hat{n}_p$, $\hat{d}_p$ was automatically a zero of the right hand side. Therefore, the solution always existed, although not unique when $\hat{n}_p$, $\hat{d}_p$ were not coprime.

An indirect adaptive pole placement control algorithm may be obtained in a similar way as the indirect model reference adaptive control algorithm of Section 3.3.3. Then $\hat{c}, \hat{d}$ are obtained by solving (3.3.35) with k_p, $\hat{n}_p$, $\hat{d}_p$ replaced by their estimates. A difficulty arises to guarantee that the estimates of $\hat{n}_p \hat{d}_p$ are coprime, since the solution of (3.3.35) will usually not exist otherwise. This was not necessary in the model reference case where the solution always existed.

Proving stability for an adaptive pole placement algorithm is somewhat complicated. It is often assumed that the input is sufficiently rich and that some procedure guarantees coprimeness of the estimates. Nevertheless, these algorithms have the significant advantage of not requiring minimum phase assumptions. A further discussion is presented in Section 6.2.2.

3.4 THE STABILITY PROBLEM IN ADAPTIVE CONTROL

Stability Definitions

Various definitions and concepts of stability have been proposed. A classical definition for systems of the form

$$\dot{x} = f(t,x) \tag{3.4.1}$$

is the stability in the sense of Lyapunov defined in Chapter 1.

The adaptive systems described so far are of the special form

$$\dot{x} = f(t,x,r(t)) \tag{3.4.2}$$

where r is the input to the system and x is the overall state of the system, including the plant, the controller, and the identifier. For practical reasons, stability in the sense of Lyapunov is not sufficient for adaptive systems. As we recall, this definition is a local property, guaranteeing that the trajectories will remain arbitrarily close to the equilibrium, *when started sufficiently close.* In adaptive systems, we do not have any control on how close initial conditions are to equilibrium values. A natural stability concept is then the *bounded-input bounded-state* stability (BIBS): for any $r(.)$ bounded, and $x_0 \in \mathbb{R}^n$, the solution $x(.)$ remains bounded. This is the concept of stability that will be used in this chapter.

The Problem of Proving Stability in Adaptive Control

The stability of the identifiers presented in Chapter 2 was assessed in theorem 2.4.5. There, the stability of the plant was assumed explicitly. In adaptive control, the stability of the controlled plant must be guaranteed by the identifier, which seriously complicates the problem. The stability of the *overall adaptive system*, which includes the plant, the controller, and the identifier, must then be considered.

To understand the nature of the problem, we will take a general approach in this section and consider the generic model reference adaptive control system shown in Figure 3.7.

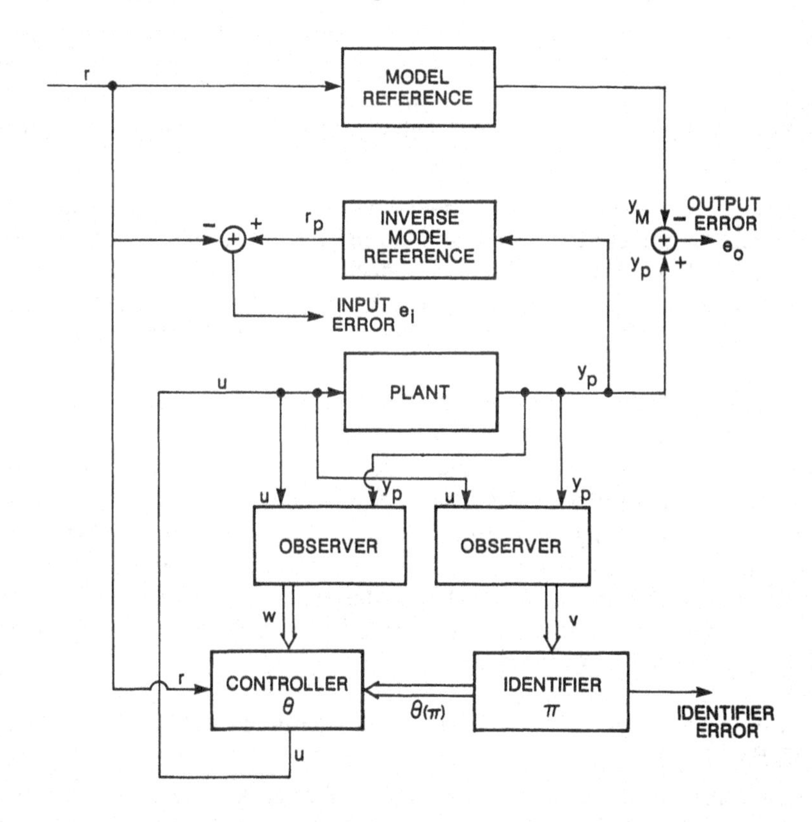

Figure 3.7: Generic Model Reference Adaptive Control System

The signals and systems defined previously can be recognized. θ is the controller parameter, and π is the identifier parameter. In the case of direct control, $\theta = \pi$, that is, the parameter being identified is directly the controller parameter. The identifier error may be the output error $e_0 = y_p - y_m$, the input error $e_i = r_p - r$, or any other error used for identification.

The problem of stability can be understood as follows. Intuitively, the plant with the control loop will be stable if θ is sufficiently close to the true value θ^*. However, as we saw in Chapter 2, the convergence of the identifier is dependent on the stability and persistent excitation of signals originating from the control loop.

To break this circular argument, we must first express properties of the identifier that are independent of the stability and persistency of excitation of these signals. Such properties were already derived in Chapter 2, and were expressed in terms of the identifier error. Recall that the identifier *parameter* error $\pi - \pi^*$ does not converge to zero but that only the identifier error converges to zero in some sense. Thus, we cannot argue that for t sufficiently large, the controller parameter θ will be arbitrarily close to the nominal value that stabilizes the plant-control loop.

Instead of relying on the convergence of θ to θ^* to prove stability, we can express the control signal as a nominal control signal—that makes the controlled plant match the reference model—plus a control error. The problem then is to transfer the properties of the identifier to the control loop, that is, the identifier error to the control error, and prove stability. Several difficulties are encountered here. First, the transformation $\theta(\pi)$ is usually nonlinear. In direct adaptive control, the transformation is the identity, and the proof is consequently simplified. Another difficulty arises however from the different signals v and w used for identification and control. A major step will be to transfer properties of the identifier involving v to properties of the controller involving w. Provided that the resulting control error is a "small" gain from plant signals, the proof of stability will basically be a *small gain theorem* type of proof, a generic proof to assess the stability of nonlinear time varying systems (cf. Desoer & Vidyasagar [1975]).

3.5 ANALYSIS OF THE MODEL REFERENCE ADAPTIVE CONTROL SYSTEM

We now return to the model reference adaptive control system presented in Sections 3.1–3.3. The results derived in this section are the basis for analyses presented in this and following chapters. Many identities involve signals which are not available in practice (since $\hat{P}$ is unknown) but are well defined for the analysis. Most results also rely on the control input being defined by

$$u = \theta^T w$$

$$\theta^T = (c_0, c^T, d_0, d^T)$$

$$w^T = (r, w^{(1)^T}, y_p, w^{(2)^T}) \tag{3.5.1}$$

Error Formulation

It will be useful to represent the adaptive system in terms of its deviation with respect to the ideal situation when $\theta = \theta^*$, that is, $\phi = 0$. This step is similar to transferring the equilibrium point of a differential equation as (3.4.1) to $x = 0$ by a change of coordinates.

Recall that we defined r_p in (3.3.1) as

$$r_p = \hat{M}^{-1}(y_p) \tag{3.5.2}$$

while

$$y_m = \hat{M}(r) \tag{3.5.3}$$

Applying $\hat{L}$ to (3.3.10), it follows, since θ^* is constant, that

$$u = c_0^* r_p + \bar{\theta}^{*^T} \bar{w} \tag{3.5.4}$$

and, since u is given by (3.5.1),

$$r_p = r + \frac{1}{c_0^*} \phi^T w \tag{3.5.5}$$

Further, applying $\hat{M}$ to both sides of (3.5.5)

$$y_p = y_m + \frac{1}{c_0^*} \hat{M}(\phi^T w) \tag{3.5.6}$$

The signal $\phi^T w$ will be called the *control error*. We note that the *input error* $e_i = r_p - r$ is directly proportional to the *control error* $\phi^T w$ (cf. (3.5.5)), while the *output error* $e_0 = y_p - y_m$ is related to the control error through the model transfer function $\hat{M}$ (cf. (3.5.6)).

Since $y_p = \hat{P}(u) = \hat{M}(r_p)$, the control input can also be expressed in terms of the control error as

$$u = \hat{P}^{-1}\hat{M}(r_p) = \hat{P}^{-1}\hat{M}(r + \frac{1}{c_0^*} \phi^T w) \tag{3.5.7}$$

and the vector $\bar{w}$ is similarly expressed as

$$\overline{w} = \begin{bmatrix} w^{(1)} \\ y_p \\ w^{(2)} \end{bmatrix} = \begin{bmatrix} (sI - \Lambda)^{-1} b_\lambda \hat{P}^{-1} \hat{M} \\ \hat{M} \\ (sI - \Lambda)^{-1} b_\lambda \hat{M} \end{bmatrix} (r + \frac{1}{c_0^*} \phi^T w) \tag{3.5.8}$$

while v (cf. (3.3.16)) is given by

$$v = \hat{L}^{-1} \begin{bmatrix} 1 \\ (sI - \Lambda)^{-1} b_\lambda \hat{P}^{-1} \hat{M} \\ \hat{M} \\ (sI - \Lambda)^{-1} b_\lambda \hat{M} \end{bmatrix} (r + \frac{1}{c_0^*} \phi^T w) \tag{3.5.9}$$

For the purpose of the analysis, we also define (cf. (3.3.10))

$$z := \hat{L}(v) = \begin{bmatrix} r_p \\ \cdot \\ \overline{w} \end{bmatrix} = \begin{bmatrix} 1 \\ (sI - \Lambda)^{-1} b_\lambda \hat{P}^{-1} \hat{M} \\ \hat{M} \\ (sI - \Lambda)^{-1} b_\lambda \hat{M} \end{bmatrix} (r + \frac{1}{c_0^*} \phi^T w) \tag{3.5.10}$$

Note that the transfer functions appearing in (3.5.6)–(3.5.10) are all stable (using assumptions (A1)–(A2) and the definitions of Λ and $\hat{L}^{-1}$).

Model Signals

The *model signals* are defined as the signals corresponding to the plant signals when $\theta = \theta^*$, that is, $\phi = 0$. As expected, the model signals corresponding to y_p and r_p are y_m and r, respectively (cf. (3.5.6) and (3.5.5)). Similarly, we define

$$\overline{w}_m := \begin{bmatrix} w_m^{(1)} \\ y_m \\ w_m^{(2)} \end{bmatrix} := \begin{bmatrix} (sI - \Lambda)^{-1} b_\lambda \hat{P}^{-1} \hat{M} \\ \hat{M} \\ (sI - \Lambda)^{-1} b_\lambda \hat{M} \end{bmatrix} (r)$$

$$:= \hat{H}_{\overline{w}_m r}(r) \tag{3.5.11}$$

and

$$v_m := \hat{L}^{-1}(z_m) := \hat{L}^{-1} \begin{bmatrix} 1 \\ (sI - \Lambda)^{-1} b_\lambda \hat{P}^{-1} \hat{M} \\ \hat{M} \\ (sI - \Lambda)^{-1} b_\lambda \hat{M} \end{bmatrix} (r) \tag{3.5.12}$$

By defining

$$w_m := \begin{bmatrix} r \\ \bar{w}_m \end{bmatrix} \tag{3.5.13}$$

we note the remarkable fact that

$$w_m = z_m \tag{3.5.14}$$

Since the transfer functions relating r to the model signals are all stable, and since r is bounded (assumption (A3)), it follows that all model signals are bounded functions of time. Consequently, if the differences between plant and model signals are bounded, the plant signals will be bounded.

State-Space Description

We now show how a state-space description of the overall adaptive system can be obtained. In particular, we will check that no cancellation of possibly unstable modes occurs when $\theta = \theta^*$.

The plant has a minimal state-space representation $[A_p, b_p, c_p^T]$ such that

$$\hat{P}(s) = k_p \frac{\hat{n}_p(s)}{\hat{d}_p(s)} = c_p^T (sI - A_p)^{-1} b_p \tag{3.5.15}$$

With the definitions of $w^{(1)}$, $w^{(2)}$ in (3.2.17)–(3.2.19), the plant with observer is described by

$$\begin{aligned} \dot{x}_p &= A_p x_p + b_p u \\ \dot{w}^{(1)} &= \Lambda w^{(1)} + b_\lambda u \\ \dot{w}^{(2)} &= \Lambda w^{(2)} + b_\lambda y_p = \Lambda w^{(2)} + b_\lambda c_p^T x_p \end{aligned} \tag{3.5.16}$$

The control input u can be expressed in terms of its desired value, plus the control error $\phi^T w$, as

$$u = \theta^T w = \theta^{*^T} w + \phi^T w \tag{3.5.17}$$

so that

$$\begin{bmatrix} \dot{x}_p \\ \dot{w}^{(1)} \\ \dot{w}^{(2)} \end{bmatrix} = \begin{bmatrix} A_p + b_p d_0^* c_p^T & b_p c^{*T} & b_p d^{*T} \\ b_\lambda d_0^* c_p^T & \Lambda + b_\lambda c^{*T} & b_\lambda d^{*T} \\ b_\lambda c_p^T & 0 & \Lambda \end{bmatrix} \cdot \begin{bmatrix} x_p \\ w^{(1)} \\ w^{(2)} \end{bmatrix}$$

$$+\begin{bmatrix} b_p \\ b_\lambda \\ 0 \end{bmatrix} \phi^T w + \begin{bmatrix} b_p \\ b_\lambda \\ 0 \end{bmatrix} c_0^* r$$

$$y_p = c_p^T x_p \tag{3.5.18}$$

Defining $x_{pw} \in \mathbb{R}^{3n-2}$ to be the total state of the plant and observer, this equation is rewritten as

$$\dot{x}_{pw} = A_m x_{pw} + b_m \phi^T w + b_m c_0^* r$$

$$y_p = c_m^T x_{pw} \tag{3.5.19}$$

where $A_m \in \mathbb{R}^{3n-2 \times 3n-2}$, $b_m \in \mathbb{R}^{3n-2}$ and $c_m \in \mathbb{R}^{3n-2}$ are defined through (3.5.18). Since the transfer function from $r \to y_p$ is $\hat{M}$ when $\phi = 0$, we must have that $c_m^T (sI - A_m)^{-1} b_m = (1/c_0^*)\, \hat{M}(s)$, that is, that $[A_m, b_m, c_m^T]$ is a representation of the model transfer function $\hat{M}$, divided by c_0^*. Therefore, we can also represent the model and its output by

$$\dot{x}_m = A_m x_m + b_m c_0^* r$$

$$y_m = c_m^T x_m \tag{3.5.20}$$

Note that although the transfer function $\hat{M}$ is stable, its representation is non-minimal, since the order of $\hat{M}$ is n, while the dimension of A_m is $3n - 2$. We can find where the additional modes are located by noting that the representation of the model is that of Figure 3.8. Using standard transfer function manipulations, but avoiding cancellations, we get

$$c_m^T (sI - A_m)^{-1} b_m = \frac{\dfrac{k_p \hat{n}_p \hat{\lambda}}{(\hat{\lambda} - \hat{c}^*)\hat{d}_p}}{1 - \dfrac{k_p \hat{n}_p \hat{\lambda}}{(\hat{\lambda} - \hat{c}^*)\hat{d}_p} \dfrac{\hat{d}^*}{\hat{\lambda}}}$$

$$= \frac{k_p \hat{\lambda} \hat{n}_p \hat{\lambda}_0 \hat{n}_m}{\hat{\lambda}((\hat{\lambda} - \hat{c}^*)\hat{d}_p - k_p \hat{n}_p \hat{d}^*)} \tag{3.5.21}$$

and, using the matching equality (3.2.6)

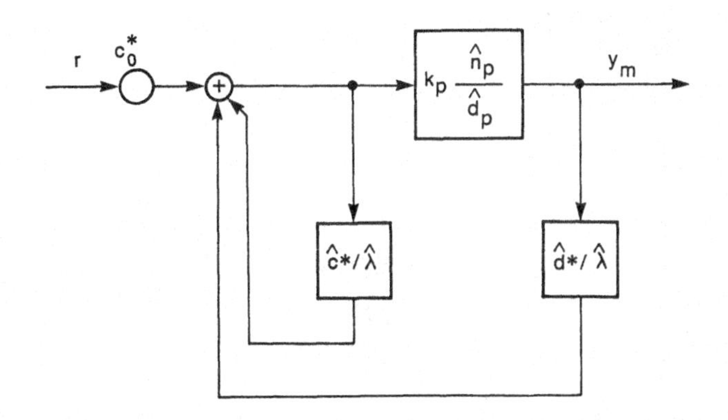

Figure 3.8: Representation of the Reference Model

$$c_m^T(sI - A_m)^{-1}b_m = \frac{1}{c_0^*}\, k_m \frac{\hat{n}_m}{\hat{d}_m} \frac{\hat{\lambda}\hat{\lambda}_0\hat{n}_p}{\hat{\lambda}\hat{\lambda}_0\hat{n}_p} = \frac{1}{c_0^*}\hat{M} \tag{3.5.22}$$

Thus, the additional modes are those of $\hat{\lambda}$, $\hat{\lambda}_0$, and $\hat{n}_p$, which are all stable by choice of $\hat{\lambda}$, $\hat{\lambda}_0$, and by assumption (A1). In other words, A_m is a stable matrix.

Since r is assumed to be bounded and A_m is stable, the state vector trajectory x_m is bounded. We can represent the plant states as their differences from the model states, letting the *state error* $e = x_{pw} - x_m \in \mathbb{R}^{3n-2}$, so that

$$\begin{aligned} \dot{e} &= A_m e + b_m \phi^T w \\ e_0 &= y_p - y_m = c_m^T e \end{aligned} \tag{3.5.23}$$

and

$$e_0 = y_p - y_m = \frac{1}{c_0^*}\hat{M}(\phi^T w) = \hat{M}\left(\frac{1}{c_0^*}\phi^T w\right) \tag{3.5.24}$$

which is equation (3.5.6), (derived above) through a somewhat shorter path.

Note that (3.5.23) is *not* a linear differential equation representing the plant with controller, because w depends on e. This can be resolved by expressing the dependence of w on e as

$$w = w_m + Qe \tag{3.5.25}$$

where

$$Q = \begin{bmatrix} 0 & 0 & 0 \\ 0 & I & 0 \\ c_p^T & 0 & 0 \\ 0 & 0 & I \end{bmatrix}$$

$$\in \begin{bmatrix} \mathbb{R}^{1 \times n} & \mathbb{R}^{1 \times n-1} & \mathbb{R}^{1 \times n-1} \\ \mathbb{R}^{n-1 \times n} & \mathbb{R}^{n-1 \times n-1} & \mathbb{R}^{n-1 \times n-1} \\ \mathbb{R}^{1 \times n} & \mathbb{R}^{1 \times n-1} & \mathbb{R}^{1 \times n-1} \\ \mathbb{R}^{n-1 \times n} & \mathbb{R}^{n-1 \times n-1} & \mathbb{R}^{n-1 \times n-1} \end{bmatrix} = \mathbb{R}^{2n \times 3n-2} \qquad (3.5.26)$$

A differential equation representing the plant with controller is then

$$\dot{e} = A_m e + b_m \phi^T w_m + b_m \phi^T Q e$$

$$e_1 = c_m^T e \qquad (3.5.27)$$

where w_m is an exogeneous, bounded input.

Complete Description—Output Error, Relative Degree 1 Case

To describe the adaptive system completely, one must simply add to this set of differential equations the set corresponding to the identifier. For example, in the case of the output error adaptive control scheme for relative degree 1 plants, the overall adaptive system (including the plant, controller and identifier) is described by

$$\dot{e} = A_m e + b_m \phi^T w_m + b_m \phi^T Q e$$

$$\dot{\phi} = -g c_m^T e w_m - g c_m^T e Q e \qquad (3.5.28)$$

As for most adaptive control schemes presented in this book, the adaptive control scheme is described by a nonlinear, time varying, ordinary differential equation. This specific case (3.5.28) will be used in subsequent chapters as a convenient example.

3.6 USEFUL LEMMAS

The following lemmas are useful to prove the stability of adaptive control schemes. Most lemmas are inspired from lemmas that are present in one form or another in existing stability proofs. In contrast with Sastry [1984] and Narendra, Annaswamy, & Singh [1985], we do not use any ordering of signals (order relations $o(.)$ and $O(.)$), but keep relationships between signals in terms of norm inequalities.

The systems considered in this section are of the general form

$$y = H(u) \qquad (3.6.1)$$

where $H : L_{pe} \to L_{pe}$ is a SISO *causal* operator, that is, such that

$$y_t = (H(u_t))_t \tag{3.6.2}$$

for all $u \in L_{pe}$ and for all $t \geq 0$. Lemmas 3.6.1–3.6.5 further restrict the attention to LTI systems with proper transfer functions $\hat{H}(s)$.

Lemma 3.6.1 is a standard result in linear system theory and relates the L_p norm of the output to the L_p norm of the input.

Lemma 3.6.1 Input/Output Lp Stability

Let $y = \hat{H}(u)$, where $\hat{H}$ is a proper, rational transfer function. Let h be the impulse response corresponding to $\hat{H}$.

If $\hat{H}$ is stable

Then for all $p \in [1, \infty]$ and for all $u \in L_p$

$$\| y\|_p \leq \| h\|_1 \| u\|_p + \| \epsilon\|_p \tag{3.6.3}$$

for all $u \in L_{\infty e}$

$$|y(t)| \leq \| h\|_1 \| u_t \|_\infty + |\epsilon(t)| \tag{3.6.4}$$

for all $t \geq 0$, where $\epsilon(t)$ is an exponentially decaying term due to the initial conditions.

Proof of Lemma 3.6.1 cf. Desoer & Vidyasagar [1975], p. 241.

It is useful, although not standard, to obtain a result that is the converse of lemma 3.6.1, that is, with u and y interchanged in (3.6.3)–(3.6.4). Such a lemma can be found in Narendra, Lin, & Valavani [1980], Narendra [1984], Sastry [1984], Narendra, Annaswamy, & Singh [1985], for $p = \infty$. Lemma 3.6.2 is a version that is valid for $p \in [1, \infty]$, with a completely different proof (see the Appendix).

Note that if $\hat{H}$ is minimum phase and has relative degree zero, then it has a proper and stable inverse, and the converse result is true by lemma 3.6.1. If $\hat{H}$ is minimum phase, but has relative degree greater than zero, then the converse result will be true provided that additional conditions are placed on the input signal u. This is the result of lemma 3.6.2.

Lemma 3.6.2 Output/Input Lp Stability

Let $y = \hat{H}(u)$, where $\hat{H}$ is a proper, rational transfer function. Let $p \in [1, \infty]$.

If $\hat{H}$ is minimum phase

For some $k_1, k_2 \geq 0$, for all $u, \dot{u} \in L_{pe}$ and for all $t \geq 0$

$$\| \dot{u}_t \|_p \leq k_1 \| u_t \|_p + k_2 \tag{3.6.5}$$

Then there exist $a_1, a_2 \geq 0$ such that

$$\| u_t \|_p \leq a_1 \| y_t \|_p + a_2 \tag{3.6.6}$$

for all $t \geq 0$.

Proof of Lemma 3.6.2 in Appendix.

It is also interesting to note the following equivalence, related to L_∞ norms. For all $a, b \in L_{\infty e}$

$$| a(t) | \leq k_1 \| b_t \|_\infty + k_2 \quad \text{iff} \quad \| a_t \|_\infty \leq k_1 \| b_t \|_\infty + k_2 \tag{3.6.7}$$

The same is true if the right-hand side of the inequalities is replaced by any positive, monotonically increasing function of time. Therefore, for $p = \infty$, the assumption (3.6.5) of lemma 3.6.2 is that u is regular (cf. definition in (2.4.14)). In particular, lemma 3.6.2 shows that if u is regular and y is bounded, then u is bounded. Lemma 3.6.2 therefore leads to the following corollary.

Corollary 3.6.3 Properties of Regular Signals

Let $y = \hat{H}(u)$, where $\hat{H}$ is a proper, rational transfer function. Let $\hat{H}$ be stable and minimum phase and $y \in L_{\infty e}$.

(a) *If* u is regular
Then $| u(t) | \leq a_1 \| y_t \|_\infty + a_2$ for all $t \geq 0$.

(b) *If* u is bounded and $\hat{H}$ is strictly proper
Then y is regular.

(c) *If* u is regular
Then y is regular.

The properties are also valid if u and y are vectors such that each component y_i of y is related to the corresponding u_i through $y_i = \hat{H}(u_i)$.

Proof of Corollary 3.6.3 in Appendix.

In Chapter 2, a key property of the identification algorithms was obtained in terms of a gain belonging to L_2. Lemma 3.6.4 is useful for such gains appearing in connection with systems with rational transfer function $\hat{H}$.

Lemma 3.6.4

Let $y = \hat{H}(u)$, where $\hat{H}$ is a proper, rational transfer function.

If $\hat{H}$ is stable, $u \in L_{\infty e}$, and for some $x \in L_{\infty e}$

$$|u(t)| \leq \beta_1(t) \| x_t \|_\infty + \beta_2(t) \tag{3.6.8}$$

for all $t \geq 0$ and for some $\beta_1, \beta_2 \in L_2$

Then there exist $\gamma_1, \gamma_2 \in L_2$ such that, for all $t \geq 0$

$$|y(t)| \leq \gamma_1(t) \| x_t \|_\infty + \gamma_2(t) \tag{3.6.9}$$

If in addition, either $\hat{H}$ is strictly proper, or $\beta_1, \beta_2 \in L_\infty$ and $\beta_1(t), \beta_2(t) \to 0$ as $t \to \infty$

Then $\gamma_1, \gamma_2 \in L_\infty$ and $\gamma_1(t), \gamma_2(t) \to 0$ as $t \to \infty$

Proof of Lemma 3.6.4 in Appendix.

The following lemma is the so-called *swapping lemma* (Morse [1980]), and is essential to the stability proofs presented in Section 3.7.

Lemma 3.6.5 Swapping Lemma

Let $\phi, w : \mathbb{R}_+ \to \mathbb{R}^n$ and ϕ be differentiable. Let $\hat{H}$ be a proper, rational transfer function.

If $\hat{H}$ is stable, with a minimal realization

$$\hat{H} = c^T (sI - A)^{-1} b + d \tag{3.6.10}$$

Then

$$\hat{H}(w^T \phi) - \hat{H}(w^T)\phi = \hat{H}_c(\hat{H}_b(w^T)\dot{\phi}) \tag{3.6.11}$$

where

$$\hat{H}_b = (sI - A)^{-1} b \qquad \hat{H}_c = -c^T (sI - A)^{-1} \tag{3.6.12}$$

Proof of Lemma 3.6.5 in Appendix.

Lemma 3.6.6 is the so-called *small gain theorem* (Desoer & Vidyasagar [1975]) and concerns general nonlinear time-varying systems connected as shown in Figure 3.9.

Roughly speaking, the small gain theorem states that the system of Figure 3.9, with inputs u_1, u_2 and outputs y_1, y_2, is BIBO stable, provided that H_1 and H_2 are BIBO stable and provided that the product of the gains of H_1 and H_2 is less than 1.

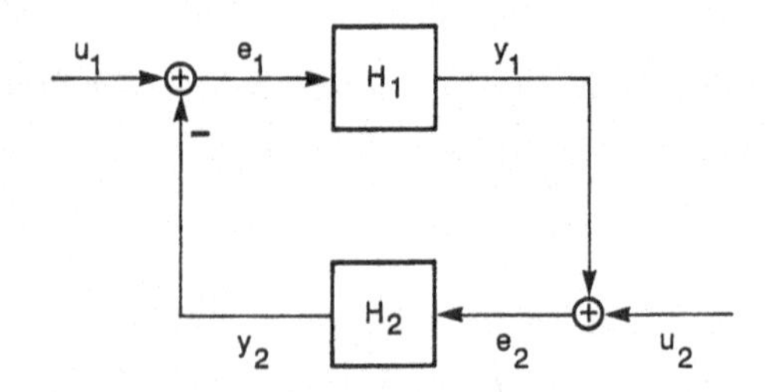

Figure 3.9: Feedback System for the Small-Gain Theorem

Lemma 3.6.6 Small Gain Theorem

Consider the system shown in Figure 3.9. Let $p \in [1, \infty]$. Let H_1, $H_2 : L_{pe} \to L_{pe}$ be causal operators. Let $e_1, e_2 \in L_{pe}$ and define u_1, u_2 by

$$\begin{aligned} u_1 &= e_1 + H_2(e_2) \\ u_2 &= e_2 - H_1(e_1) \end{aligned} \tag{3.6.13}$$

Suppose that there exist constants β_1, β_2 and $\gamma_1, \gamma_2 \geq 0$, such that

$$\begin{aligned} \| H_1(e_1)_t \| &\leq \gamma_1 \| e_{1_t} \| + \beta_1 \\ \| H_2(e_2)_t \| &\leq \gamma_2 \| e_{2_t} \| + \beta_2 \qquad \text{for all } t \geq 0 \end{aligned} \tag{3.6.14}$$

If $\gamma_1 \cdot \gamma_2 < 1$

Then

$$\begin{aligned} \| e_{1_t} \| &\leq (1 - \gamma_1\gamma_2)^{-1} (\| u_{1_t} \| + \gamma_2 \| u_{2_t} \| + \beta_2 + \gamma_2\beta_1) \\ \| e_{2_t} \| &\leq (1 - \gamma_1\gamma_2)^{-1} (\| u_{2_t} \| + \gamma_1 \| u_{1_t} \| + \beta_1 + \gamma_1\beta_2) \end{aligned} \tag{3.6.15}$$

for all $t \geq 0$.

If in addition, $u_1, u_2 \in L_p$

Then $e_1, e_2, y_1 = H_1(e_1), y_1 = H_2(e_2) \in L_p$ and (3.6.15) is valid with all subscripts t dropped.

Proof of Lemma 3.6.6 cf. Desoer & Vidyasagar [1975], p. 41.

3.7 STABILITY PROOFS

3.7.1 Stability–Input Error Direct Adaptive Control

The following theorem is the main stability theorem for the input error direct adaptive control scheme. It shows that, given any initial condition and any bounded input $r(t)$, the states of the adaptive system remain bounded (BIBS stability) and the output error tends to zero, as

$t \to \infty$. Further, the error is bounded by an L_2 function.

We also obtain that the difference between the regressor vector v and the corresponding model vector v_m tends to zero as $t \to \infty$ and is in L_2. This result will be useful to prove exponential convergence in Section 3.8.

We insist that initial conditions must be in some small B_h, because although the properties are valid for any initial conditions, the convergence of the error to zero and the L_2 bounds are not uniform globally. For example, there does not exist a fixed L_2 function that bounds the output error no matter how large the initial conditions are.

Theorem 3.7.1

Consider the input error direct adaptive control scheme described in Section 3.3.1, with initial conditions in an arbitrary B_h.

Then

(a) all states of the adaptive system are bounded functions of time.

(b) the output error $e_0 = y_p - y_m \in L_2$ and tends to zero as $t \to \infty$; the regressor error $v - v_m \in L_2$ and tends to zero as $t \to \infty$.

Comments

The proof of the theorem is organized to highlight the main steps that we described in Section 3.4.

Although the theorem concerns the adaptive scheme with the gradient algorithm, examination of the proof shows that it only requires the standard identifier properties resulting from theorems 2.4.1–2.4.4. Therefore, theorem 3.7.1 is also valid if the normalized gradient algorithm is replaced by the normalized least-squares (LS) algorithm with covariance resetting.

Proof of Theorem 3.7.1

(a) *Derive properties of the identifier that are independent of the boundedness of the regressor—Existence of the solutions.*

Properties obtained in theorems 2.4.1–2.4.4 led to

$$|\phi^T(t)\, v(t)| = \beta(t) \| v_t \|_\infty + \beta(t)$$

$$\beta \in L_2 \cap L_\infty$$

$$\phi \in L_\infty \qquad \dot{\phi} \in L_2 \cap L_\infty$$

$$c_0(t) \geq c_{min} > 0 \qquad \text{for all } t \geq 0 \tag{3.7.1}$$

The inequality for $c_0(t)$ follows from the use of the projection in the update law.

The question of the *existence* of the solutions of the differential equations describing the adaptive system may be answered as follows. The proof of (3.7.1) indicates that $\phi \in L_\infty$ as long as the solutions exist. In fact, $|\phi(t)| \leq |\phi(0)|$, so that $|\theta(t)| \leq \theta^* + |\phi(0)|$ for all $t \geq 0$. Therefore, the controlled system is a linear time invariant system with a linear time varying controller and *bounded* feedback gains. From proposition 1.4.1, it follows that all signals in the feedback loop, and therefore in the whole adaptive system, belong to $L_{\infty e}$.

(b) *Express the system states and inputs in term of the control error.*

This was done in Section 3.5 and led to the control error $\phi^T w$, with

$$r_p = r + \frac{1}{c_0^*}\phi^T w$$

$$u = \hat{P}^{-1}\hat{M}(r_p)$$

$$y_p = \hat{M}(r_p) = y_m + \frac{1}{c_0^*}\hat{M}(\phi^T w)$$

$$\overline{w} = \begin{bmatrix} (sI - \Lambda)^{-1} b_\lambda \hat{P}^{-1}\hat{M} \\ \hat{M} \\ (sI - \Lambda)^{-1} b_\lambda \hat{M} \end{bmatrix} (r_p) = \hat{H}_{\overline{w}_m r}(r_p)$$

$$= \overline{w}_m + \hat{H}_{\overline{w}_m r}\left(\frac{1}{c_0^*}\phi^T w\right) \tag{3.7.2}$$

where the transfer functions $\hat{M}$ and $\hat{H}_{\overline{w}_m r}$ are stable and strictly proper.

(c) *Relate the identifier error to the control error.*

The properties of the identifier are stated in terms of the error $\phi^T v = \phi^T L^{-1}(z)$, while the control error is $\phi^T w$. The relationship between the two can be examined in two steps.

(c1) Relate $\phi^T w$ to $\phi^T z$

Only the first component of w, namely r, is different from the first component of z, namely, r_p. The two can be related using (3.5.4), that is

$$u = c_0^* r_p + \overline{\theta}^{*T}\overline{w} \tag{3.7.3}$$

and using the fact that the control input

$$u = c_0 r + \bar{\theta}^T \bar{w} \tag{3.7.4}$$

to obtain

$$r_p = \frac{1}{c_0^*}(c_0 r + \bar{\phi}^T \bar{w}) = r + \frac{1}{c_0^*}\phi^T w \tag{3.7.5}$$

and

$$r = \frac{1}{c_0}(c_0^* r_p - \bar{\phi}^T \bar{w}) = r_p - \frac{1}{c_0}\phi^T z \tag{3.7.6}$$

It follows that

$$\frac{1}{c_0^*}\phi^T w = \frac{1}{c_0}\phi^T z \tag{3.7.7}$$

(c2) Relate $\phi^T z$ to $\phi^T v = \phi^T \hat{L}^{-1}(z)$

This relationship is obtained through the swapping lemma (lemma 3.6.5). We have, with notation borrowed from the lemma

$$\hat{L}^{-1}(\frac{1}{c_0}\phi^T z) = \frac{1}{c_0}\phi^T v + \hat{L}_c^{-1}(\hat{L}_b^{-1}(z^T)(\frac{\dot{\phi}}{c_0})) \tag{3.7.8}$$

and, using (3.7.7) with (3.7.8)

$$\frac{1}{c_0^*}\hat{M}(\phi^T w) = \hat{M}\hat{L}(\hat{L}^{-1}(\frac{1}{c_0^*}\phi^T w)) = \hat{M}\hat{L}(\hat{L}^{-1}(\frac{1}{c_0}\phi^T z))$$

$$= \hat{M}\hat{L}(\frac{1}{c_0}\phi^T v) + \hat{M}\hat{L}\hat{L}_c^{-1}(\hat{L}_b^{-1}(z^T)(\frac{\dot{\phi}}{c_0})) \tag{3.7.9}$$

With (3.7.2), this equation leads to Figure 3.10. It represents the plant as the model transfer function with the control error $\phi^T w$ in feedback. The control error has now been expressed as a function of the identifier error $\phi^T v$ using (3.7.9).

The gain ϕ^T operating on v is equal to the gain β operating on $\| v_t \|_\infty$, and this gain belongs to L_2. On the other hand, $\dot{\phi} \in L_2$, so that any of its component is in L_2. In particular $\dot{c}_0 \in L_2$. Also, $c_0(t) \geq c_{\min}$, so that $1/c_0 \in L_\infty$. Thus, $d/dt(\phi/c_0) \in L_2$. Therefore, in Figure 3.10, the controlled plant appears as a stable transfer function $\hat{M}$ with an L_2 feedback gain.

(d) *Establish the regularity of the signals.*

The need to establish the regularity of the signals can be understood from the following. We are not only concerned with the boundedness of

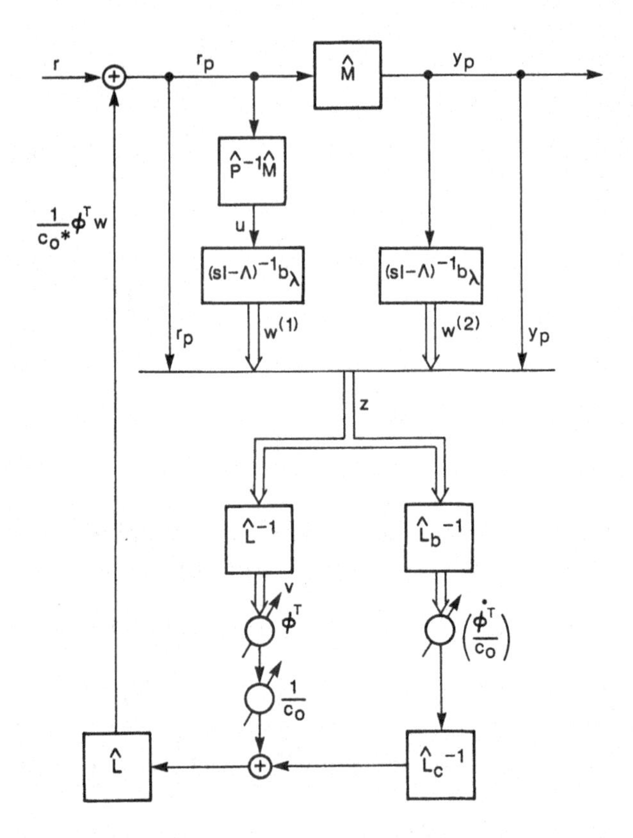

Figure 3.10: Representation of the Plant for the Stability Analysis

the output y_p but also of all the other signals present in the adaptive system. By ensuring the regularity of the signals in the loop, we guarantee, using lemma 3.6.2, that boundedness of one signal implies boundedness of all the others.

Now, note that since $\phi \in L_\infty$, the controller parameter θ is also bounded. It follows, from proposition 1.4.1, that all signals belong to $L_{\infty e}$.

Recall from (3.7.5) that

$$r_p = \frac{c_0}{c_0^*} r + \frac{1}{c_0^*} \bar{\phi}^T \bar{w} \tag{3.7.10}$$

Note that c_0 and r are bounded, by the results of (a) and by assumption (A3). $\bar{w}$ is related to r_p through a strictly proper, stable transfer function (cf. (3.7.2)). Therefore, with (3.7.10) and lemma 3.6.1

$$|\bar{w}| \leq k\| (\bar{\phi}^T \bar{w})_t \|_\infty + k$$

$$|\frac{d}{dt} \bar{w}| \leq k\| (\bar{\phi}^T \bar{w})_t \|_\infty + k \qquad (3.7.11)$$

for some constant $k \geq 0$. To prevent proliferation of constants, we will hereafter use the single symbol k, whenever such an inequality is valid for some positive constant.

Since ϕ is bounded, the last inequality implies that

$$|\frac{d}{dt} \bar{w}| \leq k\| \bar{w}_t \|_\infty + k \qquad (3.7.12)$$

that is, that $\bar{w}$ is regular.

Similarly, since ϕ and $\dot{\phi}$ are bounded and using (3.7.11)

$$|\frac{d}{dt} (\bar{\phi}^T \bar{w})| \leq |(\frac{d}{dt} \bar{\phi}^T)\bar{w}| + |\bar{\phi}^T (\frac{d}{dt} \bar{w})|$$

$$\leq k\| (\bar{\phi}^T \bar{w})_t \|_\infty + k \qquad (3.7.13)$$

so that $\bar{\phi}^T \bar{w}$ is also regular.

The output y_p is given by (using (3.7.10))

$$y_p = \hat{M}(r_p) = \frac{1}{c_0^*} \hat{M}(c_0 r) + \frac{1}{c_0^*} \hat{M}(\bar{\phi}^T \bar{w}) \qquad (3.7.14)$$

where $\hat{M}(c_0 r)$ is bounded. Using lemma 3.6.2, with the fact that $\bar{\phi}^T \bar{w}$ is regular and then (3.7.14)

$$|\bar{\phi}^T \bar{w}| \leq k\| (\hat{M}(\bar{\phi}^T \bar{w}))_t \|_\infty + k$$

$$\leq k \| y_{p_t} \|_\infty + k\| (\hat{M}(c_0 r))_t \|_\infty + k$$

$$\leq k\| y_{p_t} \|_\infty + k \qquad (3.7.15)$$

hence, with (3.7.10) and (3.7.11)

$$|r_p| \leq k\| (\bar{\phi}^T \bar{w})_t \|_\infty + k \leq k\| y_{p_t} \|_\infty + k$$

$$|\bar{w}| \leq k\| y_{p_t} \|_\infty + k \qquad (3.7.16)$$

Inequalities in (3.7.16) show that the boundedness of y_p implies the boundedness of $r_p, \bar{w}, u, \ldots$, and therefore of all the states of the adaptive system.

It also follows that v is regular, since it is the sum of two regular signals, specifically

$$v = \hat{L}^{-1}(z) = \begin{bmatrix} \hat{L}^{-1} r_p \\ \hat{L}^{-1} \overline{w} \end{bmatrix}$$

$$= \begin{bmatrix} \hat{L}^{-1} \dfrac{c_0}{c_0^*} r \\ 0 \end{bmatrix} + \begin{bmatrix} \hat{L}^{-1} \dfrac{1}{c_0^*} \overline{\phi}^T \overline{w} \\ \hat{L}^{-1} \overline{w} \end{bmatrix} \tag{3.7.17}$$

where the first term is the output of $\hat{L}^{-1}$ (a stable and strictly proper, minimum phase LTI system) with bounded input, while the second term is the output of $\hat{L}^{-1}$ with a regular input (cf. corollary 3.6.3).

(e) *Stability proof.*

Since v is regular, theorem 2.4.6 shows that $\beta \to 0$ as $t \to \infty$. From (3.7.2) and (3.7.9)

$$y_p = y_m + \frac{1}{c_0^*} \hat{M}(\phi^T w)$$

$$= y_m + \hat{M}\hat{L}\left(\frac{1}{c_0} \phi^T v\right) + \hat{M}\hat{L}\hat{L}_c^{-1}\left(\hat{L}_b^{-1}(z^T)\left(\frac{\dot{\phi}}{c_0}\right)\right) \tag{3.7.18}$$

We will now use the single symbol β in inequalities satisfied for some function satisfying the same conditions as β, that is $\beta \in L_2 \cap L_\infty$ and $\beta(t) \to 0$, as $t \to \infty$.

The transfer functions $\hat{M}\hat{L}$, $\hat{L}_b^{-1}$ and $\hat{L}_c^{-1}$, are all stable and the last two are strictly proper. The gain $\dfrac{1}{c_0}$ is bounded by (3.7.2), because of the projection in the update law. Therefore, using results obtained so far and lemmas 3.6.1 and 3.6.4

$$\begin{aligned} |y_p - y_m| &\le \beta \| v_t \|_\infty + \beta \| z_t \|_\infty + \beta \\ &\le \beta \| r_{p_t} \|_\infty + \beta \| \overline{w}_t \|_\infty + \beta \\ &\le \beta \| y_{p_t} \|_\infty + \beta \\ &\le \beta \| (y_p - y_m)_t \|_\infty + \beta \end{aligned} \tag{3.7.19}$$

Recall that since $\theta \in L_\infty$, all signals in the adaptive system belong to $L_{\infty e}$. On the other hand, for T sufficiently large, $\beta(t \ge T) < 1$.

Therefore, application of the small gain theorem (lemma 3.6.6) with (3.7.19) shows that $y_p - y_m$ is bounded for $t \geq T$. But since y_p, $y_m \in L_{\infty e}$, it follows that $y_p \in L_\infty$. Consequently, all signals belong to L_∞.

From (3.7.19), it also follows that $e_0 = y_p - y_m \in L_2$, and tends to zero as $t \to \infty$. Similarly, using (3.5.9), (3.5.12) and (3.7.9)

$$v = v_m + \begin{bmatrix} 1 \\ (sI - \Lambda)^{-1} b_\lambda \hat{P}^{-1} \hat{M} \\ \hat{M} \\ (sI - \Lambda)^{-1} b_\lambda \hat{M} \end{bmatrix}$$

$$\left(\frac{1}{c_0} \phi^T v + \hat{L}_c^{-1}\left(\hat{L}_b^{-1}(z^T)\left(\frac{\dot{\phi}}{c_0}\right)\right)\right) \tag{3.7.20}$$

so that $v - v_m$ also belongs to L_2 and tends to zero, as $t \to \infty$.

3.7.2 Stability—Output Error Direct Adaptive Control

Theorem 3.7.2

Consider the output error direct adaptive control scheme described in Section 3.3.2, with initial conditions in an arbitrary B_h.

Then

(a) all states of the adaptive system are bounded functions of time.

(b) the output error $e_0 = y_p - y_m \in L_2$ and tends to zero as $t \to \infty$ the regressor error $\hat{L}^{-1}(w) - \hat{L}^{-1}(w_m) \in L_2$ and tends to zero as $t \to \infty$.

Proof of Theorem 3.7.2

The proof is very similar to the proof for the input error scheme, and is just sketched here, following the steps of the proof of theorem 3.7.1.

(a) We now have, instead

$$e_1, \dot{\bar{\phi}} \in L_2$$

$$e_1, \bar{\phi} \in L_\infty \tag{3.7.21}$$

Note that these results are valid, although the realization of $\hat{M}$ is not minimal (but is stable).

(b) As in theorem 3.7.1.

(c) Since $c_0 = c_0^*$, (3.7.9) becomes

$$\frac{1}{c_0^*}\hat{M}(\bar{\phi}^T\bar{w})$$

$$= \frac{1}{c_0^*}\hat{M}\hat{L}(\bar{\phi}^T\bar{v}) + \frac{1}{c_0^*}\hat{M}\hat{L}(\hat{L}_c^{-1}(\hat{L}_b^{-1}(\bar{w}^T)\dot{\bar{\phi}})) \tag{3.7.22}$$

(d) As in theorem 3.7.1, it follows that $\bar{w}$ is regular—cf. (3.7.12). Unfortunately, $\bar{\phi}^T\bar{w}$ is not necessarily regular because $\dot{\bar{\phi}}$ is not bounded. However, we will show that (3.7.16) still holds, which is all we will need for the stability proof.

To prove (3.7.16), first note that lemma 3.6.2 may be modified as follows.
If there exists $z \in L_{pe}$, $k_3 \geq 0$ such that (3.6.5) is replaced by

$$\|\dot{u}_t\|_p \leq k_1 \|u_t\|_p + k_2 + k_3\|z_t\|_p \tag{3.7.23}$$

then lemma 3.6.2 is valid with (3.6.6) replaced by

$$\|u_t\|_p \leq a_1 \|y_t\|_p + a_2 + a_3\|z_t\|_p \tag{3.7.24}$$

for some $a_3 \geq 0$. We leave it to the reader to verify this new version of the lemma.

Now, recall that

$$\bar{w} = \begin{bmatrix} w^{(1)} \\ y_p \\ w^{(2)} \end{bmatrix} = \begin{bmatrix} (sI - \Lambda)^{-1} b_\lambda(u) \\ y_p \\ (sI - \Lambda)^{-1} b_\lambda(y_p) \end{bmatrix} \tag{3.7.25}$$

so that

$$\|w_t^{(2)}\| \leq k \|y_{p_t}\| + k \tag{3.7.26}$$

To apply the modified lemma 3.6.2, we note from (3.7.25) that

$$w^{(2)} = \hat{P}(w^{(1)}) \tag{3.7.27}$$

where $\hat{P}$ is minimum phase. Further

$$\dot{w}^{(1)} = \Lambda w^{(1)} + b_\lambda u = \Lambda w^{(1)} + b_\lambda c_0^* r + b_\lambda \bar{\theta}^T \bar{w} \tag{3.7.28}$$

It follows that

$$\|\dot{w}^{(1)}\|_\infty \leq k \|w_t^{(1)}\|_\infty + k + k(\|w_t^{(2)}\|_\infty + \|y_{p_t}\|_\infty) \tag{3.7.29}$$

and, applying the modified lemma 3.6.2

$$\|w_t^{(1)}\|_\infty \le k\,\|w_t^{(2)}\|_\infty + k\,\|y_{p_t}\|_\infty + k \tag{3.7.30}$$

Putting (3.7.30) with (3.7.25)–(3.7.26)

$$\begin{aligned}\|\overline{w}_t\|_\infty &\le \|w_t^{(1)}\|_\infty + \|y_{p_t}\|_\infty + \|w_t^{(2)}\|_\infty \\ &\le k\,\|y_{p_t}\|_\infty + k \end{aligned}\tag{3.7.31}$$

which is equivalent to (3.7.16) (cf. (3.6.7)).

(e) Recall, from (3.3.16) and the definition of the gradient update law, that

$$\begin{aligned}\frac{1}{c_0^*}\hat{M}\hat{L}(\overline{\phi}^T\overline{v}) &= e_1 + \frac{\gamma}{c_0^*}\hat{M}\hat{L}(\overline{v}^T\overline{v}\,e_1) \\ &= e_1 - \frac{\gamma}{gc_0^*}\hat{M}\hat{L}(\dot{\overline{\phi}}^T\overline{v})\end{aligned}\tag{3.7.32}$$

so that, with (3.7.22)

$$\begin{aligned}y_p - y_m &= \frac{1}{c_0^*}\hat{M}\hat{L}(\overline{\phi}^T\overline{v}) + \frac{1}{c_0^*}\hat{M}\hat{L}(\hat{L}_c^{-1}(\hat{L}_b^{-1}(\overline{w}^T))\dot{\overline{\phi}})) \\ &= e_1 - \frac{\gamma}{gc_0^*}\hat{M}\hat{L}(\dot{\overline{\phi}}\,\overline{v}) + \frac{1}{c_0^*}\hat{M}\hat{L}(\hat{L}_c^{-1}(\hat{L}_b^{-1}(\overline{w}^T)\dot{\overline{\phi}})\end{aligned}\tag{3.7.33}$$

Recall that e_1 is bounded (part (a)) and that $\hat{M}\hat{L}$ is strictly proper (in the output error scheme). The proof can then be completed as in theorem 3.7.1. □

3.7.3 Stability—Indirect Adaptive Control

Theorem 3.7.3

Consider the indirect adaptive control scheme described in Section 3.3.3, with initial conditions in an arbitrary B_h.

Then

(a) all states of the adaptive system are bounded functions of time.

(b) the output error $e_0 = y_p - y_m \in L_2$, and tends to zero as $t \to \infty$ the regressor error $\tilde{w} - \tilde{w}_m \in L_2$, and tends to zero as $t \to \infty$.

Proof of Theorem 3.7.3 in Appendix.

Comments

Compared with previous proofs, the proof of theorem 3.7.3 presents additional complexities due to the transformation $\pi \to \theta$. A major step is to relate the identification error $\psi^T \tilde{w}$ to the control error $\phi^T w$. We now discuss the basic ideas of the proof. The exact formulation is left to the Appendix.

To understand the approach of the proof, assume that the parameters π and θ are fixed in time and that k_p is known. For simplicity, let $k_p = a_{m+1} = k_m = 1$. The nominal values of the identifier parameters are then given by

$$\hat{a}^* = \hat{n}_p$$

$$\hat{b}^* = \hat{\lambda} - \hat{d}_p$$

The controller parameters are given as a function of the identifier parameters through

$$\hat{c} = \hat{\lambda} - \hat{q}\hat{a}$$

$$\hat{d} = \hat{q}\hat{\lambda} - \hat{q}\hat{b} - \hat{\lambda}_0 \hat{d}_m \tag{3.7.34}$$

while the nominal values are given by

$$\hat{c}^* = \hat{\lambda} - \hat{q}^*\hat{a}^* = \hat{\lambda} - \hat{q}^*\hat{n}_p$$

$$\hat{d}^* = \hat{q}^*\hat{\lambda} - \hat{q}^*\hat{b}^* - \hat{\lambda}_0 \hat{d}_m = \hat{q}^*\hat{d}_p - \hat{\lambda}_0 \hat{d}_m \tag{3.7.35}$$

It follows that

$$\begin{aligned} \hat{q}\hat{a} - \hat{q}\hat{a}^* &= (\hat{\lambda} - \hat{c}) - \hat{q}\hat{n}_p = -(\hat{c} - \hat{c}^*) + (\hat{\lambda} - \hat{c}^*) - \hat{q}\hat{n}_p \\ &= -(\hat{c} - \hat{c}^*) + (\hat{q}^* - \hat{q})\hat{n}_p \end{aligned} \tag{3.7.36}$$

and

$$\begin{aligned} \hat{q}\hat{b} - \hat{q}\hat{b}^* &= \hat{q}\hat{\lambda} - \hat{d} - \hat{\lambda}_0 \hat{d}_m - \hat{q}\hat{\lambda} + \hat{q}\hat{d}_p \\ &= -(\hat{d} - \hat{d}^*) + (-\hat{d}^* - \hat{\lambda}_0 \hat{d}_m + \hat{q}\hat{d}_p) \\ &= -(\hat{d} - \hat{d}^*) + (\hat{q} - \hat{q}^*)\hat{d}_p \end{aligned} \tag{3.7.37}$$

Therefore

$$\hat{q}\left[\frac{\hat{a} - \hat{a}^*}{\hat{\lambda}} + \frac{\hat{b} - \hat{b}^*}{\hat{\lambda}}\frac{\hat{n}_p}{\hat{d}_p}\right] = -\left[\frac{\hat{c} - \hat{c}^*}{\hat{\lambda}} + \frac{\hat{d} - \hat{d}^*}{\hat{\lambda}}\frac{\hat{n}_p}{\hat{d}_p}\right] \tag{3.7.38}$$

This equality of polynomial ratios can be interpreted as an operator equality in the Laplace transform domain, since we assumed that the parameters were fixed in time. If we apply the operator equality to the input u, it leads to (with the definitions of Section 3.3)

$$\hat{q}(\psi^T \tilde{w}) = -\bar{\phi}^T \bar{w} \tag{3.7.39}$$

and, consequently

$$y_p - y_m = -\frac{1}{c_0^*} \hat{M}\hat{q}(\psi^T \tilde{w}) \tag{3.7.40}$$

Since the degree of $\hat{q}$ is at most equal to the relative degree of the plant, the transfer function $\hat{M}\hat{q}$ is proper and stable. The techniques used in the proof of theorem 3.7.1 and the properties of the identifier would then lead to a stability proof.

Two difficulties arise when using this approach to prove the stability of the indirect adaptive system. The first is related to the unknown high-frequency gain, but only requires more complex manipulations. The real difficulty comes from the fact that the polynomials $\hat{q}$, $\hat{a}$, $\hat{b}$, $\hat{c}$, and $\hat{d}$ vary as functions of time. Equation (3.7.38) is still valid as a polynomial equality, but transforming it to an operator equality leading to (3.7.40) requires some care.

To make sense of time varying polynomials as operators in the Laplace transform domain, we define

$$\hat{s}_n = \begin{bmatrix} 1 \\ s \\ \cdot \\ \cdot \\ \cdot \\ s^{n-1} \end{bmatrix} \tag{3.7.41}$$

so that

$$\hat{a}(s) = a^T \hat{s}_n \qquad \frac{\hat{a}(s)}{\hat{\lambda}(s)} = a^T \left[\frac{\hat{s}_n}{\hat{\lambda}} \right] \tag{3.7.42}$$

Consider the following equality of polynomial ratios

$$\frac{\hat{a}(s)}{\hat{\lambda}(s)} = \frac{\hat{b}(s)}{\hat{\lambda}(s)} \tag{3.7.43}$$

where $\hat{a}$ and $\hat{b}$ vary with time but $\hat{\lambda}$ is a constant polynomial. Equality (3.7.43) implies the following operator equality

$$a^T \left[\frac{\hat{s}_n}{\hat{\lambda}}(.) \right] = b^T \left[\frac{\hat{s}_n}{\hat{\lambda}}(.) \right] \tag{3.7.44}$$

Similarly, consider the product

$$\frac{\hat{a}(s)}{\hat{\lambda}(s)} \cdot \frac{\hat{b}(s)}{\hat{\lambda}(s)} \tag{3.7.45}$$

This can be interpreted as an operator by multiplying the coefficients of the polynomials to lead to a ratio of higher order polynomials and then interpreting it as previously. We note that the product of polynomials can be expressed as

$$\hat{a}(s)\hat{b}(s) = a^T(\hat{s}_n \hat{s}_n^T)b \tag{3.7.46}$$

so that the operator corresponding to (3.7.45) is

$$a^T \left[\frac{\hat{s}_n}{\hat{\lambda}} \cdot \frac{\hat{s}_n^T}{\hat{\lambda}}(.) \right] b \tag{3.7.47}$$

i.e. by first operating the matrix transfer function on the argument and then multiplying by a and b in the time domain. Note that this operator is different from the operator

$$a^T \left[\frac{\hat{s}_n}{\hat{\lambda}} \left[\frac{\hat{s}_n^T}{\hat{\lambda}}(.)b \right] \right] \tag{3.7.48}$$

but the two operators can be related using the swapping lemma (lemma 3.6.5).

3.8 EXPONENTIAL PARAMETER CONVERGENCE

Exponential convergence of the identification algorithms under persistency of excitation conditions was established in Sections 2.5 and 2.6. Consider now the input error direct adaptive control scheme of Section 3.3.1. Using theorem 2.5.3, it would be straightforward to show that the parameters of the adaptive system converge exponentially to their nominal values, provided that the regressor v is persistently exciting. However, such result is useless, since the signal v is generated inside the adaptive system and is unknown *a priori*. Theorem 3.8.1 shows that it is sufficient for the *model* signal w_m to be persistently exciting to guarantee exponential convergence.

Note that in the case of adaptive control, we are not only interested in the convergence of the parameter error to zero, but also in the convergence of the errors between plant states and model states. In other words, we are concerned with the exponential stability of the overall adaptive system.

Theorem 3.8.1

Consider the input error direct adaptive control scheme of Section 3.3.1.

If w_m is PE

Then the adaptive system is exponentially stable in any closed ball.

Proof of Theorem 3.8.1

Since w_m, $\dot{w}_m$ are bounded, lemma 2.6.6 implies that $v_m = \hat{L}^{-1}(z_m) = \hat{L}^{-1}(w_m)$ is PE. In theorem 3.7.1, we found that $v - v_m \in L_2$. Therefore, using lemma 2.6.5, v_m PE implies that v is PE. Finally, since v is PE, by theorem 2.5.3, the parameter error ϕ converges exponentially to zero.

Recall that in Section 3.5, it was established that the errors between the plant and the model signals are the outputs of stable transfer functions with input $\phi^T w$. Since w is bounded (by theorem 3.7.1), $\phi^T w$ converges exponentially to zero. Therefore, all errors between plant and model signals converge to zero exponentially fast. □

Comments

Although theorem 3.8.1 establishes exponential stability in any closed ball, it does not prove global exponential stability. This is because $v - v_m$ is not bounded by a unique L_2 function for any initial condition. Results in Section 4.5 will actually show that the adaptive control system is not globally exponentially stable.

The various theorems and lemmas used to prove theorem 3.8.1 can be used to obtain estimates of the convergence rates of the parameter error. It is, however, doubtful that these estimates would be of any practical use, due to their complexity and to their conservatism. A more successful approach is that of Chapter 4, using averaging techniques.

The result of theorem 3.8.1 has direct parallels for the other adaptive control algorithms presented in Section 3.3.

Theorem 3.8.2

Consider the output error direct adaptive control scheme of Section 3.3.2 (or the indirect scheme of Section 3.3.3).

If w_m is PE ($\tilde{w}_m$ is PE)

Then the adaptive system is exponentially stable in any closed ball.

Proof of Theorem 3.8.2

The proof of theorem 3.8.2 is completely analogous to the proof of theorem 3.8.1 and is omitted here. □

3.9 CONCLUSIONS

In this chapter, we derived three model reference adaptive control schemes. All had a similar controller structure but different identification structures. The first two schemes were direct adaptive control schemes, where the parameters updated by the identifier were the same as those used by the controller. The third scheme was an indirect scheme, where the parameters updated by the identifier were the same as those of the basic identifier of Chapter 2. Then, the controller parameters were obtained from the identifier parameters through a nonlinear transformation resulting from the model reference control objective.

We investigated the connections between the adaptive control schemes and also with other known schemes. The difficulties related to the unknown high-frequency gain were also discussed. The stability of the model reference adaptive control schemes was proved, together with the result that the error between the plant and the reference model converged to zero as t approached infinity. We used a unified framework and an identical step-by-step procedure for all three schemes. We proved basic lemmas that are fundamental to the stability proofs and we emphasized a basic intuitive idea of the proof of stability, that was the existence of a *small loop gain* appearing in the adaptive system.

The exponential parameter convergence was established, with the additional assumption of the persistency of excitation of a model regressor vector. This condition was to be satisfied by an exogeneous model signal, influenced by the designer and was basically a condition on the reference input.

An interesting conclusion is that the stability and convergence properties are identical for all three adaptive control schemes. In particular, the indirect scheme had the same stability properties as the direct schemes. Further, the normalized gradient identification algorithm can be replaced by the least squares algorithm with projection without altering the results. Differences appear between the schemes however, in connection with the high-frequency gain and with other practical considerations.

The input error direct adaptive control scheme and the indirect scheme are attractive because they lead to linear error equations and do not involve SPR conditions. Another advantage is that they allow for a decoupling of identification and control useful in practice. The indirect scheme is quite more intuitive than the input error direct scheme, although more complex in implementation and especially as far as the analysis is concerned. The final result however shows that stability is not an argument to prefer one over the other.

The various model reference adaptive control schemes also showed that the model reference approach is not bound to the choice of a direct adaptive control scheme, to the use of the output error in the identification algorithm, or to SPR conditions on the reference model.

CHAPTER 4

PARAMETER CONVERGENCE USING AVERAGING TECHNIQUES

4.0 INTRODUCTION

Averaging is a method of analysis of differential equations of the form

$$\dot{x} = \epsilon f(t, x) \tag{4.0.1}$$

and relates properties of the solutions of system (4.0.1) to properties of the solutions of the so-called *averaged* system

$$\dot{x}_{av} = \epsilon f_{av}(x_{av}) \tag{4.0.2}$$

where

$$f_{av}(x) := \lim_{T \to \infty} \frac{1}{T} \int_{t_0}^{t_0 + T} f(\tau, x)\, d\tau \tag{4.0.3}$$

assuming that the limit exists and that the parameter ϵ is sufficiently small. The method was proposed originally by Bogoliuboff & Mitropolskii [1961], developed subsequently by Volosov [1962], Sethna [1973], Balachandra & Sethna [1975] and Hale [1980]; and stated in a geometric form in Arnold [1982] and Guckenheimer & Holmes [1983].

Averaging methods were introduced for the stability analysis of deterministic adaptive systems in the work of Astrom [1983], Astrom [1984], Riedle & Kokotovic [1985] and [1986], Mareels *et al* [1986], and Anderson *et al* [1986]. We also find early informal use of averaging in Astrom & Wittenmark [1973], and, in a stochastic context, in Ljung &

Soderstrom [1983] (the ODE approach).

Averaging is very valuable to assess the stability of adaptive systems in the presence of unmodeled dynamics and to understand mechanisms of instability. However, it is not only useful in stability problems, but in general as an *approximation* method, allowing one to replace a system of *nonautonomous* (time varying) differential equations by an *autonomous* (time invariant) system. This aspect was emphasized in Fu, Bodson, & Sastry [1986], Bodson *et al* [1986], and theorems were derived for one-time scale and two-time scale systems such as those arising in identification and control. These results are reviewed here, together with their application to the adaptive systems described in previous chapters. Our recommendation to the reader not familiar with these results is to derive the simpler versions of the theorems for linear periodic systems. In the following section, we present examples of averaging analysis which will help to understand the motivation of the methods discussed in this chapter.

4.1 EXAMPLES OF AVERAGING ANALYSIS

One-Time Scale Averaging

Consider the *linear* nonautonomous differential equation

$$\dot{x} = -\epsilon \sin^2(t)\, x \qquad x(0) = x_0 \tag{4.1.1}$$

where x is a *scalar*. This equation is a special case of the parameter error equation encountered in Chapter 2

$$\dot{\phi} = -g\, w(t)\, w^T(t)\, \phi \qquad \phi(0) = \phi_0 \tag{4.1.2}$$

and corresponds to the identification of a single constant θ^* from measurements of

$$y_p(t) = \theta^* \sin(t) \tag{4.1.3}$$

using a gradient update law. The general solution of a first order linear differential equation of the form

$$\dot{x} = a(t)\, x \qquad x(0) = x_0 \tag{4.1.4}$$

is known analytically, and is given by

$$x(t) = e^{\int_0^t a(\tau)\, d\tau}\, x_0 \tag{4.1.5}$$

In particular, the solution of (4.1.1) is simply

$$x(t) = e^{-\epsilon \int_0^t \sin^2(\tau)\, d\tau} x_0 = e^{-\epsilon \int_0^t (\frac{1}{2} - \frac{1}{2}\cos(2\tau))d\tau} x_0$$

$$= e^{-\epsilon \frac{t}{2} + \frac{\epsilon}{4}\sin(2t)} x_0 \tag{4.1.6}$$

Note that when we replaced $\sin^2(\tau)$ by $\frac{1}{2} - \frac{1}{2}\cos(2\tau)$ in (4.1.6), we separated the integrand into its average and periodic part. Indeed, for all t_0, x

$$\lim_{T\to\infty} \frac{1}{T} \int_{t_0}^{t_0+T} \sin^2(\tau) x \, d\tau = \frac{1}{2} x \tag{4.1.7}$$

Therefore, the averaged system defined by (4.0.2)–(4.0.3) is now given by

$$\dot{x}_{av} = -\frac{\epsilon}{2} x_{av} \qquad x_{av}(0) = x_0 \tag{4.1.8}$$

The solution of the averaged system is

$$x_{av}(t) = e^{-\epsilon \frac{t}{2}} x_0 \tag{4.1.9}$$

Let us now compare the solutions of the original system (4.1.6) and of the averaged system (4.1.9). The difference between the solutions, at a fixed t

$$|\, x(t) - x_{av}(t)| = e^{-\epsilon \frac{t}{2}} |\, e^{\frac{\epsilon}{4}\sin(2t)} - 1\,| \tag{4.1.10}$$

$$\to |\, \frac{\epsilon}{4} \sin(2t)| \qquad \text{as } \epsilon \to 0 \tag{4.1.11}$$

In other words, the solutions are arbitrarily close as $\epsilon \to 0$, so that we may *approximate* the original system by the averaged system. Also, both systems are *exponentially stable* (and if we were to change the sign in the differential equation, both would be unstable). As is now shown, the convergence rates are also identical.

Recall that the convergence rate of an exponentially stable system is the constant α such that the solutions satisfy

$$|\, x(t)| \le m\, e^{-\alpha(t - t_0)} |\, x(t_0)| \tag{4.1.12}$$

for all $x(t_0)$, $t_0 \ge 0$. A graphical representation may be obtained by plotting $\ln(|\, x(t)|^2)$, noting that

$$\ln(|x(t)|^2) \le \ln(m^2|x(t_0)|^2) - 2\alpha(t - t_0) \tag{4.1.13}$$

Therefore, the graph of $\ln(|x(t)|^2)$ is bounded by a straight line of slope -2α. In the above example, the original and the averaged system have identical convergence rate $\alpha = \frac{\epsilon}{2}$.

In this chapter, we will prove theorems stating similar results for more general systems. Then, the analytic solution of the original system is not available, and averaging becomes useful. The method of proof is completely different, but the results are essentially the same: closeness of the solutions, and closeness of the convergence rates as $\epsilon \to 0$. We devote the rest of this section to show how the averaged system may be calculated in more complex cases, using frequency-domain expressions.

One-Time Sale Averaging–Multiple Frequencies

Consider the system

$$\dot{x} = -\epsilon\, w^2(t)\, x \tag{4.1.14}$$

where $x \in \mathbb{R}$, and w contains multiple frequencies

$$w(t) = \sum_{k=1}^{n} a_k \sin(\omega_k t + \phi_k) \tag{4.1.15}$$

To define the averaged system, we need to calculate the average

$$\mathrm{AVG}\,(w^2(t)) := \lim_{T \to \infty} \frac{1}{T} \int_{t_0}^{t_0+T} w^2(\tau)\, d\tau \tag{4.1.16}$$

Expanding w^2 will give us a sum of product of sin's at the frequencies ω_k. However, a product of two sinusoids at different frequencies has zero average, so that

$$\mathrm{AVG}\,(w^2(t)) = \sum_{k=1}^{n} \frac{a_k^2}{2} \tag{4.1.17}$$

and the averaged system is

$$\dot{x}_{av} = -\epsilon \left[\sum_{k=1}^{n} \frac{a_k^2}{2} \right] x_{av} \tag{4.1.18}$$

The averaged system is exponentially stable as soon as w contains at least one sinusoid. Note also that the expression (4.1.18) is independent of the phases ϕ_k.

Two-Time Scale Averaging

Averaging may also be applied to systems of the form

$$\dot{x}(t) = -\epsilon\, w(t)\, y(t) \tag{4.1.19}$$

$$\dot{y}(t) = -a\, y(t) + b\, w(t)\, x(t) \tag{4.1.20}$$

where $x, y \in \mathbb{R}$, $a > 0$. In short, (4.1.20) is denoted

$$y = \frac{b}{s+a}(w\,x) = \hat{M}(w\,x) \tag{4.1.21}$$

where $\hat{M}$ is a stable transfer function. (4.1.19) becomes

$$\dot{x} = -\epsilon\, w\, \hat{M}(w\,x) \tag{4.1.22}$$

Equations (4.1.19)–(4.1.20) were encountered in model reference identification with x replaced by the parameter error ϕ, ϵ by the adaptation gain g, and y by the identifier error e_i.

When $\epsilon \to 0$, $x(t)$ varies slowly when compared to $y(t)$, and the time scales of their variations become separated. $x(t)$ is called the *slow state*, $y(t)$ the *fast state* and the system (4.1.19)–(4.1.20) a *two-time* scale system. In the limit as $\epsilon \to 0$, $x(t)$ may be considered frozen in (4.1.20) or (4.1.21), so that

$$\hat{M}(w\,x) = \hat{M}(w)\,x \tag{4.1.23}$$

The result of the averaging theory is that (4.1.22) may indeed be approximated by

$$\dot{x}_{av} = -\epsilon\, \mathrm{AVG}(w\, \hat{M}(w))\, x_{av} \tag{4.1.24}$$

Again, a frequency domain expression brings more interesting insight. Let w contain multiple sinusoids

$$w(t) = \sum_{k=1}^{n} a_k \sin(\omega_k t) \tag{4.1.25}$$

so that

$$\hat{M}(w) = \sum_{k=1}^{n} a_k\, |\hat{M}(j\omega_k)|\, \sin(\omega_k t + \phi_k) \tag{4.1.26}$$

where

$$\phi_k = \arg(\hat{M}(j\omega_k)) \tag{4.1.27}$$

The product $w\,\hat{M}(w)$ may be expanded as the sum of products of sinusoids. Further, $\sin(\omega_k t + \phi_k) = \sin(\omega_k t)\, \cos(\phi_k) + \cos(\omega_k t)$

$\sin(\phi_k)$. Now, products of sinusoids at different frequencies have zero average, as do products of sin's with cos's of any frequency. Therefore

$$\text{AVG}\left[w\, \hat{M}(w) \right] = \sum_{k=1}^{n} \frac{a_k^2}{2} \,|\, \hat{M}(j\omega_k) \,|\, \cos(\phi_k)$$

$$= \sum_{k=1}^{n} \frac{a_k^2}{2} \text{Re}\left[\hat{M}(j\omega_k) \right] \tag{4.1.28}$$

Using (4.1.28), a *sufficient* condition for the stability of the averaged system (4.1.24) is that

$$\text{Re}\, \hat{M}(j\omega) > 0 \qquad \text{for all } \omega > 0 \tag{4.1.29}$$

The condition is the familiar SPR condition obtained for the stability of the original system in the context of model reference identification. The averaging analysis brings this condition in evidence *directly in the frequency domain.* It is also evident that this condition is *necessary*, if one does not restrict the frequency content of the signal $w(t)$. Otherwise, it is sufficient that the ω_k's be concentrated in frequencies where $\text{Re}\, \hat{M}(j\omega) > 0$, so that the sum in (4.1.28) is positive.

Vector Case

In identification, we encountered (4.1.2), where ϕ was a vector. The solution (4.1.5) does not extend to the vector case, but the frequency domain analysis does, as will be shown in Section 4.3. We illustrate the procedure with the simple example of the identification of a first order system (cf. Section 2.0).

The regressor vector is given by

$$w = \begin{bmatrix} r \\ y_p \end{bmatrix} = \begin{bmatrix} r \\ \hat{P}(z) \end{bmatrix} \tag{4.1.30}$$

where $\hat{P} = k_p / (s + a_p)$. As before, we let the input r be periodic

$$r(t) = \sum_{k=1}^{n} r_k \quad \sin(\omega_k t) \tag{4.1.31}$$

so that the averaged system is given by (the gain g plays the role of ϵ)

$$\dot{\phi}_{av} = -g\, \text{AVG}(w\, w^T)\, \phi_{av}$$

$$= -g \begin{bmatrix} \text{AVG}(r\, r) & \text{AVG}(r\, \hat{P}(r)) \\ \text{AVG}(r\, \hat{P}(r)) & \text{AVG}(\hat{P}(r)\, \hat{P}(r)) \end{bmatrix} \phi_{av}$$

$$= -g \sum_{k=1}^{n} \frac{r_k^2}{2} \begin{bmatrix} 1 & \mathrm{Re}\,\hat{P}(j\omega_k) \\ \mathrm{Re}\,\hat{P}(j\omega_k) & |\hat{P}(j\omega_k)|^2 \end{bmatrix} \phi_{av} \tag{4.1.32}$$

$$= -g \sum_{k=1}^{n} \frac{r_k^2}{2} \begin{bmatrix} 1 & \dfrac{a_p k_p}{\omega_k^2 + a_p^2} \\ \dfrac{a_p k_p}{\omega_k^2 + a_p^2} & \dfrac{k_p^2}{\omega_k^2 + a_p^2} \end{bmatrix} \phi_{av} \tag{4.1.33}$$

The matrix above is symmetric and it may be checked to be positive semi-definite. Further, it is positive definite for all $\omega_k \neq 0$. Taking a Lyapunov function $v = \phi_{av}^T \phi_{av}$ shows that the averaged system is exponentially stable *as long as the input contains at least one sinusoid of frequency* $\omega \neq 0$. Thus, we directly recover a frequency-domain result obtained earlier for the original system through a much longer and laborious path.

Nonlinear Averaging

Analyzing *adaptive control* schemes using averaging is trickier because the schemes are usually nonlinear. This is the motivation for the derivation of nonlinear averaging theorems in this chapter. Note that it is possible to linearize the system around some nominal trajectory, or around the equilibrium. However, averaging allows us to approximate a nonautonomous system by an autonomous system, independently of the linearity or nonlinearity of the equations. Indeed, we will show that it is possible to keep the nonlinearity of the adaptive systems, and even obtain frequency domain results. The analysis is therefore not restricted to a neighborhood of some trajectory or equilibrium.

As an example, we consider the output error model reference adaptive control scheme for a first order system (cf. Section 3.0, with $k_p = k_m = c_0 = 1$)

$$\begin{aligned} \dot{y}_p &= -a_p y_p + u \\ \dot{y}_m &= -a_m y_m + r \end{aligned} \tag{4.1.34}$$

where $a_m > 0$, and a_p is unknown. The adaptive controller is defined by

$$\begin{aligned} u &= r + d_0 y_p \\ \dot{d}_0 &= -g e_0 y_p \end{aligned} \tag{4.1.35}$$

where $g > 0$ is the adaptation gain. The output error and the parameter error are given by

$$e_0 = y_p - y_m$$

$$\phi = d_0 - d_0^* = d_0 - (a_m - a_p) \tag{4.1.36}$$

The adaptive system is completely described by

$$\dot{e}_0 = -(a_m - \phi)\, e_0 + y_m \phi$$

$$\dot{\phi} = -g\, e_0\,(e_0 + y_m) \tag{4.1.37}$$

where

$$y_m = \frac{1}{s + a_m}\,(r) \tag{4.1.38}$$

When g is small (g takes the place of ϵ in the averaging analysis), ϕ varies slowly compared to r, y_m and e_0. The averaged system is defined by calculating $\mathrm{AVG}(e_0(e_0 + y_m))$, assuming that ϕ is fixed. In that case

$$e_0 = \frac{1}{s + a_m - \phi}\,(y_m)\;\phi$$

$$= \frac{1}{s + a_m - \phi}\left[\frac{1}{s + a_m}\,(r)\right]\phi \tag{4.1.39}$$

and

$$e_0 + y_m = \frac{1}{s + a_m - \phi}\,(y_m)\;\phi + y_m$$

$$= \frac{s + a_m}{s + a_m - \phi}\,(y_m) = \frac{1}{s + a_m - \phi}\,(r) \tag{4.1.40}$$

Note that $s + a_m - \phi$ is the *closed-loop polynomial*, that is the polynomial giving the closed-loop pole for ϕ fixed. Assume again that r is of the form

$$r = \sum_{k=1}^{n} r_k \sin(\omega_k t) \tag{4.1.41}$$

and it follows that

$$\mathrm{AVG}\Big[\, e_0(e_0 + y_m)\Big]_{\phi \text{ fixed}}$$

$$= \sum_{k=1}^{n} \frac{r_k^2}{2}\; \frac{1}{\omega_k^2 + (a_m - \phi)^2}\; \frac{a_m}{\omega_k^2 + a_m^2}\; \phi \tag{4.1.42}$$

so that the averaged system is given by

$$\dot{\phi}_{av} = -g \sum_{k=1}^{n} \frac{r_k^2}{2} \frac{1}{\omega_k^2 + (a_m - \phi_{av})^2} \frac{a_m}{\omega_k^2 + a_m^2} \phi_{av} \tag{4.1.43}$$

The averaged system is a scalar *nonlinear* system. Indeed, averaging did not alter the nonlinearity of the original system, only its time variation. Note that the averaged system is of the form

$$\dot{\phi}_{av} = -a(\phi_{av})\,\phi_{av} \tag{4.1.44}$$

where $a(\phi_{av})$ is a nonlinear function of ϕ_{av}. However, for all $h > 0$, there exists $\alpha > 0$ such that

$$a(\phi_{av}) \geq \alpha > 0 \qquad \text{for all } |\,\phi_{av}\,| \leq h \tag{4.1.45}$$

as long as r contains at least one sinusoid (including at $\omega = 0$). By taking a Lyapunov function $v = \phi_{av}^2$, it is easy to see that (4.1.43) is exponentially stable in B_h, with rate of convergence α. Since h is arbitrary, the system is not only locally exponentially stable, but also exponentially stable in any closed-ball. However, it is *not* globally exponentially stable, because α is not bounded below as $h \to \infty$.

Again, we recovered a result and a frequency domain analysis, obtained for the original system through a very different path. An advantage of the averaging analysis is to give us an expression (4.1.43) which may be used to predict parameter convergence quantitatively from frequency domain conditions.

The analysis of this section may be extended to the general identification and adaptive control schemes discussed in Chapter 2 and Chapter 3. We first present the averaging theory that supports the frequency-domain analysis.

4.2 AVERAGING THEORY–ONE-TIME SCALE

In this section, we consider differential equations of the form

$$\dot{x} = \epsilon f(t, x, \epsilon) \qquad x(0) = x_0 \tag{4.2.1}$$

where $x \in \mathbb{R}^n$, $t \geq 0$, $0 < \epsilon \leq \epsilon_0$, and f is piecewise continuous with respect to t. We will concentrate our attention on the behavior of the solutions in some closed ball B_h of radius h, centered at the origin.

For small ϵ, the variation of x with time is slow, as compared to the rate of time variation of f. The *method of averaging* relies on the assumption of the existence of the mean value of $f(t, x, 0)$ defined by the limit

$$f_{av}(x) = \lim_{T \to \infty} \frac{1}{T} \int_{t_0}^{t_0+T} f(\tau, x, 0)\, d\tau \qquad (4.2.2)$$

assuming that the limit exists uniformly in t_0 and x. This is formulated more precisely in the following definition.

Definition Mean Value of a Function, Convergence Function

The function $f(t, x, 0)$ is said to have mean value $f_{av}(x)$ if there exists a continuous function $\gamma(T)$: $\mathbb{R}_+ \to \mathbb{R}_+$, strictly decreasing, such that $\gamma(T) \to 0$ as $T \to \infty$ and

$$\left| \frac{1}{T} \int_{t_0}^{t_0+T} f(\tau, x, 0)\, d\tau - f_{av}(x) \right| \leq \gamma(T) \qquad (4.2.3)$$

for all $t_0 \geq 0$, $T \geq 0$, $x \in B_h$.

The function $\gamma(T)$ is called the *convergence function.*

Note that the function $f(t, x, 0)$ has mean value $f_{av}(x)$ if and only if the function

$$d(t, x) = f(t, x, 0) - f_{av}(x) \qquad (4.2.4)$$

has zero mean value.

It is common, in the literature on averaging, to assume that the function $f(t, x, \epsilon)$ is periodic in t, or almost periodic in t. Then, the existence of the mean value is guaranteed, without further assumption (Hale [1980], theorem 6, p. 344). Here, we do not make the assumption of (almost) periodicity, but consider instead the assumption of the existence of the mean value as the starting point of our analysis.

Note that if the function $d(t, x)$ is periodic in t and is bounded, then the integral of the function $d(t, x)$ is also a bounded function of time. This is equivalent to saying that there exists a convergence function $\gamma(T) = a/T$ (i.e., of the order of $1/T$) such that (4.2.3) is satisfied. On the other hand, if the function $d(t, x)$ is bounded, and is not required to be periodic but almost periodic, then the integral of the function $d(t, x)$ need not be a bounded function of time, even if its mean value is zero (Hale [1980], p. 346). The function $\gamma(T)$ is bounded (by the same bound as $d(t, x)$) and converges to zero as $T \to \infty$, but the convergence function need not be bounded by a/T as $T \to \infty$ (it may be of order $1/\sqrt{T}$ for example). In general, a zero mean function need not have a bounded integral, although the converse is true. In this book, we do not make the distinction between the periodic and the almost periodic case, but we do distinguish the bounded integral case from the general case and indicate the importance of the function $\gamma(T)$ in the subsequent developments.

System (4.2.1) will be called the *original system* and, assuming the existence of the mean value for the original system, the *averaged system* is defined to be

$$\dot{x}_{av} = \epsilon f_{av}(x_{av}) \qquad x_{av}(0) = x_0 \tag{4.2.5}$$

Note that the averaged system is autonomous and, for T fixed and ϵ varying, the solutions over intervals $[0, T/\epsilon]$ are identical, modulo a simple time scaling by ϵ.

We address the following two questions:

(a) the closeness of the response of the original and averaged systems on intervals $[0, T/\epsilon]$,

(b) the relationships between the stability properties of the two systems.

To compare the solutions of the original and of the averaged system, it is convenient to transform the original system in such a way that it becomes a *perturbed* version of the averaged system. An important lemma that leads to this result is attributed to Bogoliuboff & Mitropolskii [1961], p. 450 and Hale [1980], lemma 4, p. 346. We state a generalized version of this lemma.

Lemma 4.2.1 Approximate Integral of a Zero Mean Function

If $d(t, x) : \mathbb{R}_+ \times B_h \to \mathbb{R}^n$ is a bounded function, piecewise continuous with respect to t, and has zero mean value with convergence function $\gamma(T)$

Then there exists $\xi(\epsilon) \in K$ and a function $w_\epsilon(t, x) : \mathbb{R}_+ \times B_h \to \mathbb{R}^n$ such that

$$|\epsilon w_\epsilon(t, x)| \le \xi(\epsilon) \tag{4.2.6}$$

$$\left| \frac{\partial w_\epsilon(t, x)}{\partial t} - d(t, x) \right| \le \xi(\epsilon) \tag{4.2.7}$$

for all $t \ge 0$, $x \in B_h$. Moreover, $w_\epsilon(0, x) = 0$, for all $x \in B_h$.

If, moreover, $\gamma(T) = a/T^r$ for some $a \ge 0$, $r \in (0, 1]$

Then the function $\xi(\epsilon)$ can be chosen to be $2a\epsilon^r$.

Proof of Lemma 4.2.1 in Appendix.

Comments

The construction of the function $w_\epsilon(t, x)$ in the proof is identical to that in Bogoliuboff & Mitropolskii [1961], but the proof of (4.2.6), (4.2.7) is different and leads to the relationship between the convergence function $\gamma(T)$ and the function $\xi(\epsilon)$.

The main point of lemma 4.2.1 is that, although the exact integral of $d(t, x)$ may be an unbounded function of time, there exists a bounded function $w_\epsilon(t, x)$, whose first partial derivative with respect to t is arbitrarily close to $d(t, x)$. Although the bound on $w_\epsilon(t, x)$ may increase as $\epsilon \to 0$, it increases slower than $\xi(\epsilon)/\epsilon$, as indicated by (4.2.6).

It is necessary to obtain a function $w_\epsilon(t, x)$, as in lemma 4.2.1, that has some additional smoothness properties. A useful lemma is given by Hale ([1980], lemma 5, p. 349). At the price of additional assumptions on the function $d(t, x)$, the following lemma leads to stronger conclusions that will be useful in the sequel.

Lemma 4.2.2 Smooth Approximate Integral of a Zero Mean Function

If $d(t, x) : \mathbb{R}_+ \times B_h \to \mathbb{R}^n$ is piecewise continuous with respect to t, has bounded and continuous first partial derivatives with respect to x and $d(t, 0) = 0$, for all $t \geq 0$. Moreover, $d(t, x)$ has zero mean value, with convergence function $\gamma(T)|x|$ and $\dfrac{\partial d(t, x)}{\partial x}$ has zero mean value, with convergence function $\gamma(T)$

Then there exists $\xi(\epsilon) \in K$ and a function $w_\epsilon(t, x) : \mathbb{R}_+ \times B_h \to \mathbb{R}^n$, such that

$$|\epsilon w_\epsilon(t, x)| \leq \xi(\epsilon)|x| \tag{4.2.8}$$

$$\left|\frac{\partial w_\epsilon(t, x)}{\partial t} - d(t, x)\right| \leq \xi(\epsilon)|x| \tag{4.2.9}$$

$$\left|\epsilon \frac{\partial w_\epsilon(t, x)}{\partial x}\right| \leq \xi(\epsilon) \tag{4.2.10}$$

for all $t \geq 0$, $x \in B_h$. Moreover, $w_\epsilon(0, x) = 0$, for all $x \in B_h$.

If, moreover, $\gamma(T) = a/T^r$ for some $a \geq 0$, $r \in (0, 1]$,

Then the function $\xi(\epsilon)$ can be chosen to be $2a\epsilon^r$.

Proof of Lemma 4.2.2 in Appendix.

Comments

The difference between this lemma and lemma 4.2.1 is in the condition on the partial derivative of $w_\epsilon(t, x)$ with respect to x in (4.2.10) and the dependence on $|x|$ in (4.2.8), (4.2.9).

Note that if the original system is linear, i.e.

$$\dot{x} = \epsilon A(t)x \qquad x(0) = x_0 \tag{4.2.11}$$

for some $A(t) : \mathbb{R}_+ \to \mathbb{R}^{n \times n}$, then the main assumption of lemma 4.2.2 is that there exists A_{av} such that $A(t) - A_{av}$ has zero mean value.

The following assumptions will now be in effect.

Assumptions

For some $h > 0$, $\epsilon_0 > 0$

(A1) $x = 0$ is an equilibrium point of system (4.2.1), that is, $f(t, 0, 0) = 0$ for all $t \geq 0$. $f(t, x, \epsilon)$ is Lipschitz in x, that is, for some $l_1 \geq 0$

$$|f(t, x_1, \epsilon) - f(t, x_2, \epsilon)| \leq l_1 |x_1 - x_2| \tag{4.2.12}$$

for all $t \geq 0$, $x_1, x_2 \in B_h$, $\epsilon \leq \epsilon_0$.

(A2) $f(t, x, \epsilon)$ is Lipschitz in ϵ, linearly in x, that is, for some $l_2 \geq 0$

$$|f(t, x, \epsilon_1) - f(t, x, \epsilon_2)| \leq l_2 |x| \, |\epsilon_1 - \epsilon_2| \tag{4.2.13}$$

for all $t \geq 0$, $x \in B_h$, $\epsilon_1, \epsilon_2 \leq \epsilon_0$.

(A3) $f_{av}(0) = 0$ and $f_{av}(x)$ is Lipschitz in x, that is, for some $l_{av} \geq 0$

$$|f_{av}(x_1) - f_{av}(x_2)| \leq l_{av} |x_1 - x_2| \tag{4.2.14}$$

for all $x_1, x_2 \in B_h$.

(A4) the function $d(t, x) = f(t, x, 0) - f_{av}(x)$ satisfies the conditions of lemma 4.2.2.

Lemma 4.2.3 Perturbation Formulation of Averaging

If the original system (4.2.1) and the averaged system (4.2.5) satisfy assumptions (A1)–(A4)

Then there exist functions $w_\epsilon(t, x)$, $\xi(\epsilon)$ as in lemma 4.2.2 and $\epsilon_1 > 0$ such that the transformation

$$x = z + \epsilon w_\epsilon(t, z) \tag{4.2.15}$$

is a homeomorphism in B_h for all $\epsilon \leq \epsilon_1$ and

$$|x - z| \leq \xi(\epsilon) |z| \tag{4.2.16}$$

Under the transformation, system (4.2.1) becomes

$$\dot{z} = \epsilon f_{av}(z) + \epsilon p(t, z, \epsilon) \qquad z(0) = x_0 \tag{4.2.17}$$

where $p(t, z, \epsilon)$ satisfies

$$|p(t, z, \epsilon)| \le \psi(\epsilon)|z| \tag{4.2.18}$$

for some $\psi(\epsilon) \in K$. Further, $\psi(\epsilon)$ is of the order of $\epsilon + \xi(\epsilon)$.

Proof of Lemma 4.2.3 in Appendix.

Comments

a) A similar lemma can be found in Hale [1980] (lemma 3.2, p. 192). Inequality (4.2.18) is a Lipschitz type of condition on $p(t, z, \epsilon)$, which is not found in Hale [1980] and results from the stronger conditions and conclusions of lemma 4.2.2.

b) Lemma 4.2.3 is fundamental to the theory of averaging presented hereafter. It separates the error in the approximation of the original system by the averaged system $(x - x_{av})$ into two components: $x - z$ and $z - x_{av}$. The first component results from a pointwise (in time) transformation of variable. This component is guaranteed to be small by inequality (4.2.16). For ϵ sufficiently small ($\epsilon \le \epsilon_1$), the transformation $z \to x$ is invertible and as $\epsilon \to 0$, it tends to the identity transformation. The second component is due to the perturbation term $p(t, z, \epsilon)$. Inequality (4.2.18) guarantees that this perturbation is small as $\epsilon \to 0$.

c) At this point, we can relate the convergence of the function $\gamma(T)$ to the order of the two components of the error $x - x_{av}$ in the approximation of the original system by the averaged system. The relationship between the functions $\gamma(T)$ and $\xi(\epsilon)$ was indicated in lemma 4.2.1. Lemma 4.2.3 relates the function $\xi(\epsilon)$ to the error due to the averaging. If $d(t, x)$ has a bounded integral (i.e., $\gamma(T) \sim 1/T$), then both $x - z$ and $p(t, z, \epsilon)$ are of the order of ϵ with respect to the main term $f_{av}(z)$. It may indeed be useful to the reader to check the lemma in the linear periodic case. Then, the transformation (4.2.15) may be replaced by

$$x(t) = z(t) + \epsilon \left[\int_0^t (A(\tau) - A_{av})\, d\tau \right] z(t)$$

and $\psi(\epsilon)$, $\xi(\epsilon)$ are of the order of ϵ. If $d(t, x)$ has zero mean but unbounded integral, the perturbation terms go to zero as $\epsilon \to 0$, but possibly more slowly than linearly (as $\sqrt{\epsilon}$ for example). The proof of lemma 4.2.1 provides a direct relationship between the order of the convergence to the mean value and the order of the error terms.

We now focus attention on the approximation of the original system by the averaged system. Consider first the following assumption.

(A5) x_0 is sufficiently small so that, for fixed T and some $h' < h$, $x_{av}(t) \in B_{h'}$ for all $t \in [0, T/\epsilon]$ (this is possible, using the Lipschitz assumption (A3) and proposition 1.4.1).

Theorem 4.2.4 Basic Averaging Theorem

If the original system (4.2.1) and the averaged system (4.2.5) satisfy assumptions (A1)–(A5)

Then there exists $\psi(\epsilon)$ as in lemma 4.2.3 such that, given $T \geq 0$

$$|x(t) - x_{av}(t)| \leq \psi(\epsilon) b_T \tag{4.2.19}$$

for some $b_T \geq 0$, $\epsilon_T > 0$, and for all $t \in [0, T/\epsilon]$ and $\epsilon \leq \epsilon_T$.

Proof of Theorem 4.2.4

We apply the transformation of lemma 4.2.3, so that

$$|x - z| \leq \xi(\epsilon)|z| \leq \psi(\epsilon)|z| \tag{4.2.20}$$

for $\epsilon \leq \epsilon_1$. On the other hand, we have that

$$\frac{d}{dt}(z - x_{av}) = \epsilon \left[f_{av}(z) - f_{av}(x_{av}) \right] + \epsilon p(t, z, \epsilon) \tag{4.2.21}$$

$$z(0) - x_{av}(0) = 0$$

for all $t \in [0, T/\epsilon]$, $x_{av} \in B_{h'}$, $h' < h$.

We will now show that, on this time interval, and for as long as $x, z \in B_h$, the errors $(z - x_{av})$ and $(x - x_{av})$ can be made arbitrarily small by reducing ϵ. Integrating (4.2.21)

$$|z(t) - x_{av}(t)| \leq \epsilon l_{av} \int_0^t |z(\tau) - x_{av}(\tau)|\, d\tau + \epsilon \psi(\epsilon) \int_0^t |z(\tau)|\, d\tau \tag{4.2.22}$$

Using the Bellman-Gronwall lemma (lemma 1.4.2)

$$\begin{aligned} |z(t) - x_{av}(t)| &\leq \epsilon \psi(\epsilon) \int_0^t |z(\tau)|\, e^{\epsilon l_{av}(t - \tau)} d\tau \\ &\leq \psi(\epsilon) h \left[\frac{e^{\epsilon l_{av} T} - 1}{l_{av}} \right] \\ &:= \psi(\epsilon) a_T \end{aligned} \tag{4.2.23}$$

Combining (4.2.20), (4.2.23)

$$\begin{aligned} |x(t) - x_{av}(t)| &\leq |x(t) - z(t)| + |z(t) - x_{av}(t)| \\ &\leq \psi(\epsilon)|x_{av}(t)| + (1 + \psi(\epsilon))|z(t) - x_{av}(t)| \\ &\leq \psi(\epsilon)(h + (1 + \psi(\epsilon_1))a_T) \\ &:= \psi(\epsilon)b_T \end{aligned} \tag{4.2.24}$$

By assumption, $|x_{av}(t)| \leq h' < h$. Let ϵ_T (with $0 < \epsilon_T \leq \epsilon_1$) such that $\psi(\epsilon_T)b_T < h - h'$. It follows, from a simple contradiction argument, that $x(t) \in B_h$, and that the estimate in (4.2.24) is valid for all $t \in [0, T/\epsilon]$, whenever $\epsilon \leq \epsilon_T$. □

Comments

Theorem 4.2.4 establishes that the trajectories of the original system and of the averaged system are arbitrarily close on intervals $[0, T/\epsilon]$, when ϵ is sufficiently small. The error is of the order of $\psi(\epsilon)$, and the order is related to the order of convergence of $\gamma(T)$. If $d(t, x)$ has a bounded integral (i.e., $\gamma(T) \sim 1/T$), then the error is of the order of ϵ.

It is important to remember that, although the intervals $[0, T/\epsilon]$ are unbounded, theorem 4.2.4 does not state that

$$|x(t) - x_{av}(t)| \leq \psi(\epsilon)b \tag{4.2.25}$$

for all $t \geq 0$ and some b. Consequently, theorem 4.2.4 does not allow us to relate the stability of the original and of the averaged system. This relationship is investigated in theorem 4.2.5.

Theorem 4.2.5 Exponential Stability Theorem

If the original system (4.2.1) and the averaged system (4.2.5) satisfy assumptions (A1)–(A5), the function $f_{av}(x)$ has continuous and bounded first partial derivatives in x, and $x = 0$ is an exponentially stable equilibrium point of the averaged system

Then the equilibrium point $x = 0$ of the original system is exponentially stable for ϵ sufficiently small.

Proof of Theorem 4.2.5

The proof relies on the converse theorem of Lyapunov for exponentially stable systems (theorem 1.4.3). Under the hypotheses, there exists a function $v(x_{av}) : \mathbb{R}^n \to \mathbb{R}_+$ and strictly positive constants $\alpha_1, \alpha_2, \alpha_3, \alpha_4$ such that, for all $x_{av} \in B_h$,

$$\alpha_1 |x_{av}|^2 \leq v(x_{av}) \leq \alpha_2 |x_{av}|^2 \tag{4.2.26}$$

$$\dot{v}(x_{av})\Big|_{(4.2.5)} \le -\epsilon\alpha_3 |x_{av}|^2 \tag{4.2.27}$$

$$\left|\frac{\partial v}{\partial x_{av}}\right| \le \alpha_4 |x_{av}| \tag{4.2.28}$$

The derivative in (4.2.27) is to be taken along the trajectories of the averaged system (4.2.5).

The function v is now used to study the stability of the perturbed system (4.2.17), where $z(x)$ is defined by (4.2.15). Considering $v(z)$, inequalities (4.2.26) and (4.2.28) are still verified, with z replacing x_{av}. The derivative of $v(z)$ along the trajectories of (4.2.17) is given by

$$\dot{v}(z)\Big|_{(4.2.17)} = \dot{v}(z)\Big|_{(4.2.5)} + \left[\frac{\partial v}{\partial z}\right](\epsilon p(t, z, \epsilon)) \tag{4.2.29}$$

and, using previous inequalities (including those from lemma 4.2.3)

$$\dot{v}(z)\Big|_{(4.2.17)} \le -\epsilon\alpha_3 |z|^2 + \epsilon\alpha_4 \psi(\epsilon)|z|^2$$

$$\le -\epsilon\left[\frac{\alpha_3 - \psi(\epsilon)\alpha_4}{\alpha_2}\right] v(z) \tag{4.2.30}$$

for all $\epsilon \le \epsilon_1$. Let ϵ_2' be such that $\alpha_3 - \psi(\epsilon_2')\alpha_4 > 0$, and define $\epsilon_2 = \min(\epsilon_1, \epsilon_2')$. Denote

$$\alpha(\epsilon) := \frac{\alpha_3 - \psi(\epsilon)\alpha_4}{2\alpha_2} \tag{4.2.31}$$

Consequently, (4.2.30) implies that

$$v(z) \le v(z(t_0))\, e^{-2\epsilon\alpha(\epsilon)(t - t_0)} \tag{4.2.32}$$

and

$$|z(t)| \le \left[\frac{\alpha_2}{\alpha_1}\right]^{\frac{1}{2}} |z(t_0)|\, e^{-\epsilon\alpha(\epsilon)(t - t_0)} \tag{4.2.33}$$

Since $\alpha(\epsilon) > 0$ for all $\epsilon \le \epsilon_2$, system (4.2.17) is exponentially stable. Using (4.2.16), it follows that

$$|x(t)| \le \frac{1 + \xi(\epsilon)}{1 - \xi(\epsilon)} \left[\frac{\alpha_2}{\alpha_1}\right]^{\frac{1}{2}} |x(t_0)|\, e^{-\epsilon\alpha(\epsilon)(t - t_0)} \tag{4.2.34}$$

for all $t \ge t_0 \ge 0$, $\epsilon \le \epsilon_2$, and $x(t_0)$ sufficiently small that all signals remain in B_h. In other words, the original system is exponentially stable, with rate of convergence (at least) $\epsilon\alpha(\epsilon)$. □

Comments

a) Theorem 4.2.5 is a *local* exponential stability result. The original system will be *globally* exponentially stable if the averaged system is globally exponentially stable, and provided that *all* assumptions are valid globally.

b) The proof of theorem 4.2.5 gives a useful bound on the rate of convergence of the original system. As ϵ tends to zero, $\epsilon\alpha(\epsilon)$ tends to $\epsilon/2\ \alpha_3/\alpha_2$, which is the bound on the rate of convergence of the averaged system that one would obtain using (4.2.26)–(4.2.27). In other words, the proof provides a bound on the rate of convergence, and this bound gets arbitrarily close to the corresponding bound for the averaged system, provided that ϵ is sufficiently small. This is a useful conclusion because it is in general very difficult to obtain a guaranteed rate of convergence for the original, nonautonomous system. The proof assumes the existence of a Lyapunov function satisfying (4.2.26)–(4.2.28), but does not depend on the specific function chosen. Since the averaged system is autonomous, it is usually easier to find such a function for it than for the original system, and any such function will provide a bound on the rate of convergence of the original system for ϵ sufficiently small.

c) The conclusion of theorem 4.2.5 is quite different from the conclusion of theorem 4.2.4. Since both x and x_{av} go to zero exponentially with t, the error $x - x_{av}$ also goes to zero exponentially with t. Yet theorem 4.2.5 does not relate the bound on the error to ϵ. It is possible, however, to combine theorem 4.2.4 and theorem 4.2.5 to obtain a uniform approximation result, with an estimate similar to (4.2.25).

4.3 APPLICATION TO IDENTIFICATION

To apply the averaging theory to the identifier described in Chapter 2, we will study the case when $g = \epsilon > 0$ and the update law is given by (cf. (2.4.1))

$$\dot{\phi}(t) = -g\, e_1(t)\, w(t) \qquad \phi(0) = \phi_0 \tag{4.3.1}$$

The evolution of the parameter error is described by

$$\dot{\phi}(t) = -g\, w(t)\, w^T(t)\, \phi(t) \qquad \phi(0) = \phi_0 \tag{4.3.2}$$

In theorem 2.5.1, we found that system (4.3.2) is exponentially stable, provided that w is *persistently exciting*, i.e., there exist constants $\alpha_1, \alpha_2, \delta > 0$, such that

$$\alpha_2 I \ge \int_{t_0}^{t_0+\delta} w(\tau)\, w^T(\tau)\, d\tau \ge \alpha_1 I \qquad \text{for all } t_0 \ge 0 \tag{4.3.3}$$

On the other hand, the averaging theory presented above leads us to the limit

$$R_w(0) := \lim_{T \to \infty} \frac{1}{T} \int_{t_0}^{t_0+T} w(\tau) w^T(t+\tau)\, d\tau \quad \in \mathbb{R}^{n \times n} \tag{4.3.4}$$

where we used the notation of Section 1.6 for the *autocovariance* of w evaluated at 0. Recall that $R_w(t)$ may be expressed as the inverse Fourier transform of the positive *spectral measure* $S_w(d\omega)$

$$R_w(t) = \frac{1}{2\pi} \int_{-\infty}^{\infty} e^{j\omega t} S_w(d\omega) \tag{4.3.5}$$

Further, w is the output of a proper stable transfer function $\hat{H}_{wr}$ given by (cf. (2.2.16)–(2.2.17))

$$\hat{H}_{wr}(s) = \begin{bmatrix} (sI - \Lambda)^{-1} b_\lambda \\ (sI - \Lambda)^{-1} b_\lambda \hat{P}(s) \end{bmatrix} \quad \in \mathbb{R}^{2n}(s) \tag{4.3.6}$$

Therefore, if the input r is stationary, then w is also stationary. Its spectrum is related to the spectrum of r through

$$S_w(d\omega) = \hat{H}_{wr}^*(j\omega) \hat{H}_{wr}^T(j\omega) S_r(d\omega) \tag{4.3.7}$$

and, using (4.3.5) and (4.3.7), we have that

$$R_w(0) = \frac{1}{2\pi} \int_{-\infty}^{\infty} \hat{H}_{wr}^*(j\omega) \hat{H}_{wr}^T(j\omega) S_r(d\omega) \tag{4.3.8}$$

Since $S_r(d\omega)$ is an even function of ω, $R_w(0)$ is also given by

$$R_w(0) = \frac{1}{2\pi} \int_{-\infty}^{\infty} \mathrm{Re}\left[\hat{H}_{wr}^*(j\omega)\, \hat{H}_{wr}^T(j\omega) \right] S_r(d\omega)$$

It was shown in Section 2.7 (proposition 2.7.1) that when w is stationary, w is persistently exciting (PE) if and only if $R_w(0)$ is positive definite. It followed (proposition 2.7.2) that this is true if the support of $S_r(d\omega)$ is greater than or equal to $2n$ points (the dimension of w = the number of unknown parameters = $2n$). Note that a DC component in $r(t)$ contributes one point to the support of $S_r(d\omega)$, while a sinusoidal component contributes two points (at $+\omega$ and $-\omega$).

With these definitions, the averaged system corresponding to (4.3.2) is simply

$$\dot{\phi}_{av} = -g R_w(0) \phi_{av} \qquad \phi_{av}(0) = \phi_0 \tag{4.3.9}$$

This system is particularly easy to study, since it is linear.

Convergence Analysis

When w is persistently exciting, $R_w(0)$ is a positive definite matrix. A natural Lyapunov function for (4.3.9) is

$$v(\phi_{av}) = \frac{1}{2} | \phi_{av} |^2 = \frac{1}{2} \phi_{av}^T \phi_{av} \tag{4.3.10}$$

and

$$-g \lambda_{min}(R_w(0)) | \phi_{av} |^2 \leq -\dot{v}(\phi_{av}) \leq -g \lambda_{max}(R_w(0)) | \phi_{av} |^2 \tag{4.3.11}$$

where λ_{min} and λ_{max} are, respectively, the minimum and maximum eigenvalues of $R_w(0)$. Thus, the rate of exponential convergence of the averaged system is at least $g \lambda_{min}(R_w(0))$ and at most $g \lambda_{max}(R_w(0))$. We can conclude that the rate of convergence of the original system for g small enough is close to the interval $[g \lambda_{min}(R_w(0)), g \lambda_{max}(R_w(0))]$.

Equation (4.3.8) gives an interpretation of $R_w(0)$ in the frequency domain, and also a mean of computing an estimate of the rate of convergence of the adaptive algorithm, given the spectral content of the reference input. If the input r is periodic or almost periodic

$$r(t) = \sum_k r_k \sin(\omega_k t) \tag{4.3.12}$$

then the integral in (4.3.8) may be replaced by a summation

$$R_w(0) = \sum_k \frac{r_k^2}{2} \mathrm{Re}\left[\hat{H}_{wr}^*(j\omega_k) \hat{H}_{wr}^T(j\omega_k)\right] \tag{4.3.13}$$

Since the transfer function $\hat{H}_{wr}$ depends on the unknown plant being identified, the use of (4.3.11) to determine the rate of convergence is limited. With knowledge of the plant, it could be used to determine the spectral content of the reference input that will optimize the rate of convergence of the identifier, given the physical constraints on r. Such a procedure is very reminiscent of the procedure indicated in Goodwin & Payne [1977] (Chapter 6) for the design of input signals in identification. The autocovariance matrix defined here is similar to the *average information matrix* defined in Goodwin & Payne [1977] (p. 134). Our interpretation is, however, in terms of rates of parameter convergence of the averaged system rather than in terms of parameter error covariance in a stochastic framework.

Note that the proof of exponential stability of theorem 2.5.1 was based on the Lyapunov function of theorem 1.4.1 that was an average of the norm along the trajectories of the system. In this chapter, we averaged the *differential equation* itself and found that the norm becomes a Lyapunov function to prove exponential stability.

It is also interesting to compare the convergence rate obtained through averaging with the convergence rate obtained in Chapter 2. We found, in the proof of exponential convergence of theorem 2.5.1, that the estimate of the convergence rate tends to $g\alpha_1/\delta$ when the adaptation gain g tends to zero. The constants α_1, δ resulted from the PE condition (2.5.3), i.e., (4.3.3). By comparing (4.3.3) and (4.3.4), we find that the estimates provided by direct proof and by averaging are essentially identical for $g = \epsilon$ small.

Example

To illustrate the conclusions of this section, we consider the following example

$$\hat{P}(s) = \frac{k_p}{s + a_p} \tag{4.3.14}$$

The filter is chosen to be $\hat{\lambda}(s) = (s + l_2)/l_1$ (where $l_1 = 10.05$, $l_2 = 10$ are arbitrarily chosen such that $|\hat{\lambda}(j1)| = 1$). Although $\hat{\lambda}$ is not monic, the gain l_1 can easily be taken into account.

Since the number of unknown parameters is 2, parameter convergence will occur when the support of $S_r(d\omega)$ is greater than or equal to 2 points. We consider an input of the form $r = r_0 \sin(\omega_0 t)$, so that the support consists of exactly 2 points.

The averaged system can be found by using (4.3.9), (4.3.13)

$$\dot{\phi}_{av} = -g\frac{r_0^2}{2}\frac{l_1^2}{l_2^2 + \omega_0^2}\begin{bmatrix} 1 & \dfrac{a_p k_p}{\omega_0^2 + a_p^2} \\ \dfrac{a_p k_p}{\omega_0^2 + a_p^2} & \dfrac{k_p^2}{\omega_0^2 + a_p^2}\end{bmatrix} \cdot \phi_{av} \tag{4.3.15}$$

with $\phi_{av}(0) = \phi_0$. When $r_0 = 1$, $\omega_0 = 1$, $a_p = 1$, $k_p = 2$, the eigenvalues of the averaged system (4.3.15) are computed to be $-\frac{3+\sqrt{5}}{4}g = -1.309\,g$, and $-\frac{3-\sqrt{5}}{4}g = -0.191\,g$. The nominal parameter $\theta^{*T} = (k_p/l_1,\ (l_2 - a_p)/l_1)$. We let $\theta(0) = 0$, so that

$\phi^T(0) = (-0.199, -0.9)$.

Figures 4.1 to 4.4 show the plots of the parameter errors ϕ_1 and ϕ_2, for both the original and averaged systems, and with two different adaptation gains $g = 1$, and $g = 0.1$.

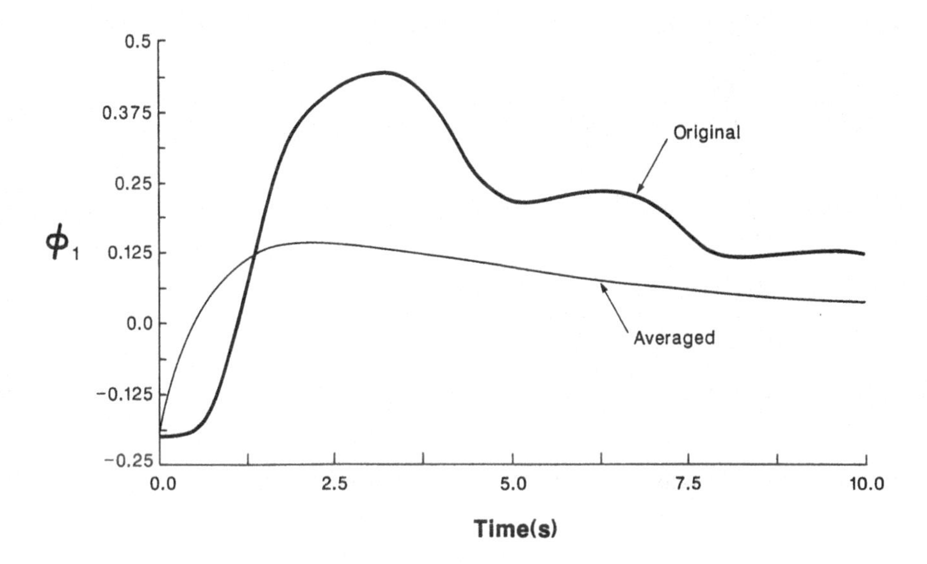

Figure 4.1: Parameter Error ϕ_1 $(g = 1)$

We notice the closeness of the approximation for $g = 0.1$.

Figures 4.5 and 4.6 are plots of the Lyapunov function (4.3.10) for $g = 1$ and $g = 0.1$, using a logarithmic scale. We observe the two slopes, corresponding to the two eigenvalues. The closeness of the estimate of the convergence rate by the averaged system can also be appreciated from these figures.

Figure 4.7 represents the two components of ϕ, one as a function of the other when $g = 0.1$. It shows the two subspaces corresponding to the small and large eigenvalues: the parameter error first moves fast along the direction of the eigenvector corresponding to the large eigenvalue. Then, it slowly moves along the direction corresponding to the small eigenvalue.

4.4 AVERAGING THEORY—TWO-TIME SCALES

We now consider a more general class of differential equations arising in the adaptive control schemes presented in Chapter 3.

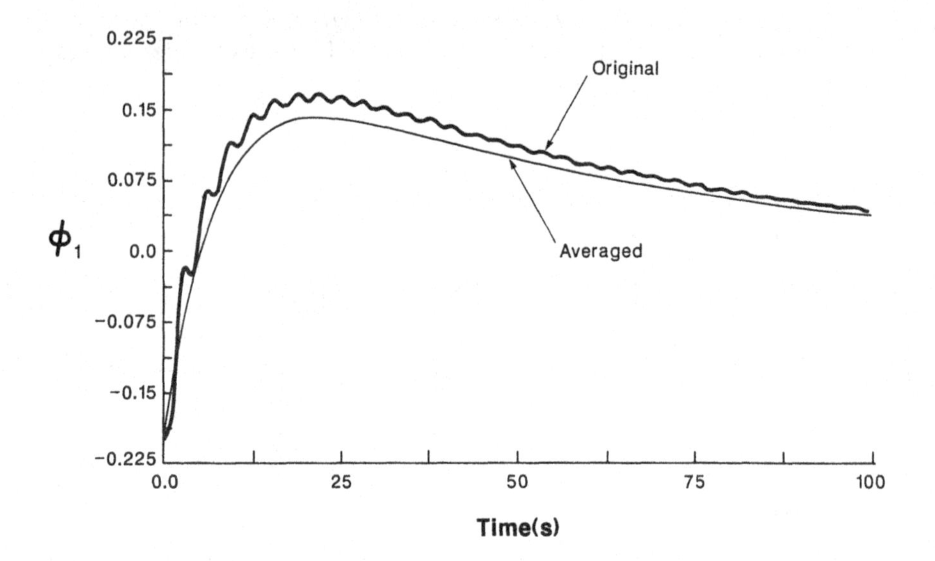

Figure 4.2: Parameter Error ϕ_1 ($g = 0.1$)

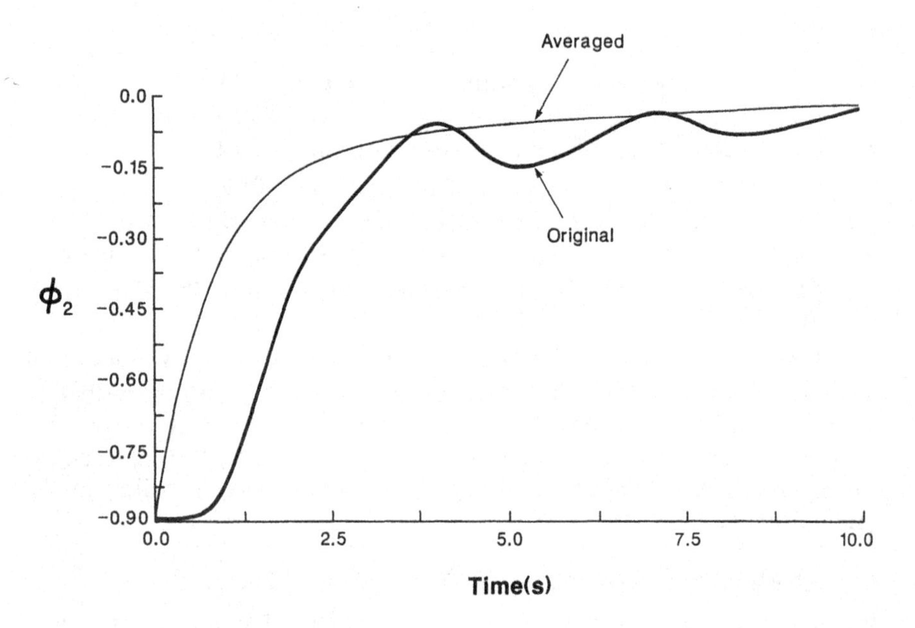

Figure 4.3: Parameter Error ϕ_2 ($g = 1$)

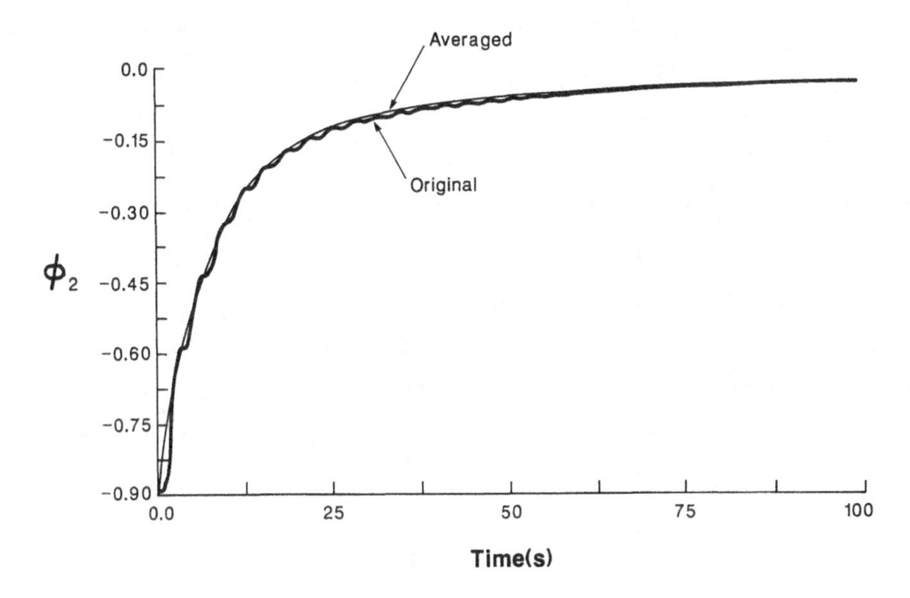

Figure 4.4: Parameter Error ϕ_2 (g = 0.1)

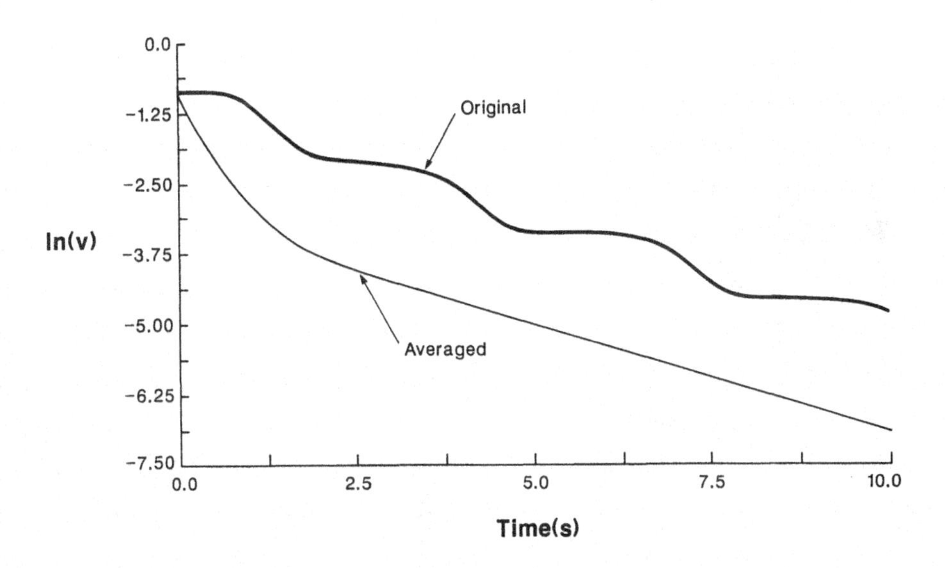

Figure 4.5: Logarithm of the Lyapunov Function (g = 1)

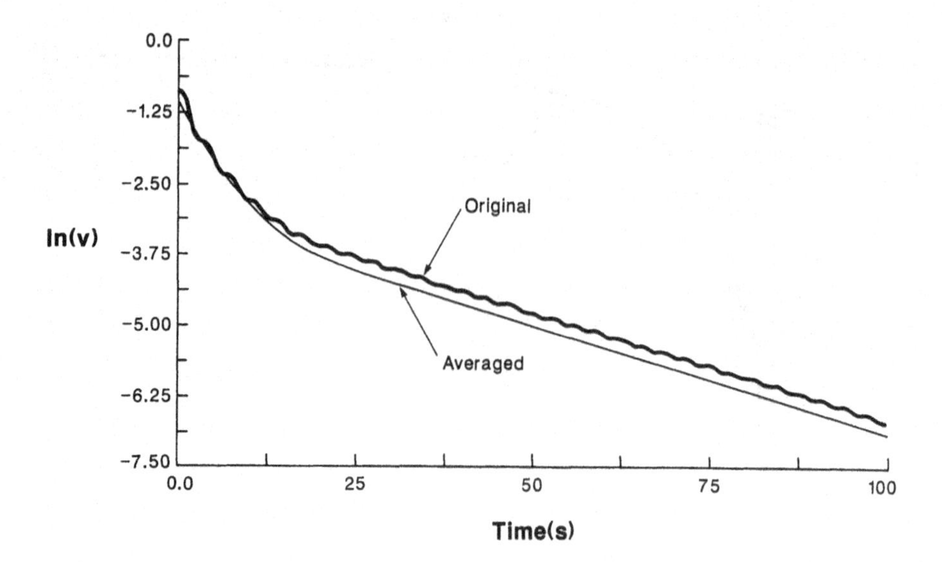

Figure 4.6: Logarithm of the Lyapunov Function ($g = 0.1$)

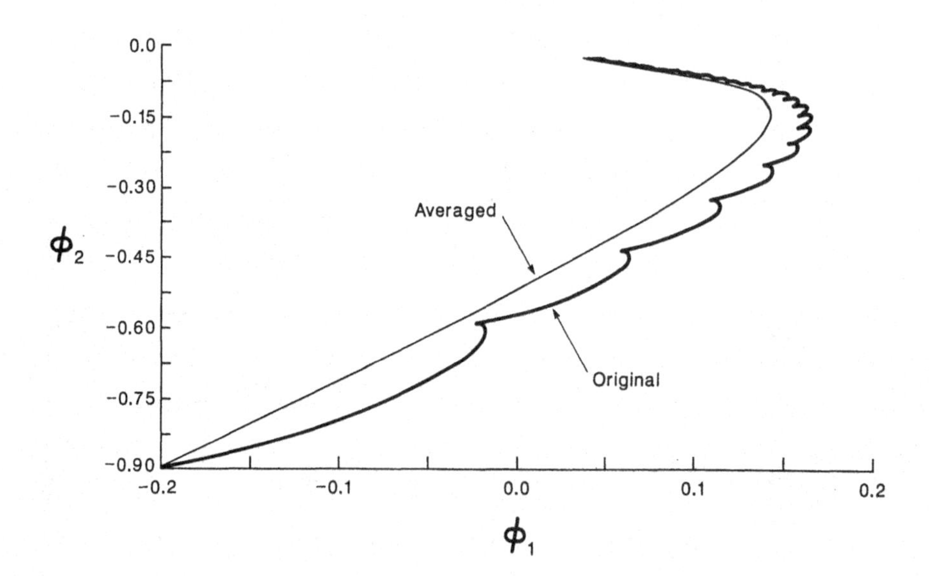

Figure 4.7: Parameter Error $\phi_2(\phi_1)$ ($g = 0.1$)

4.4.1 Separated Time Scales

We first consider the system of differential equations

$$\dot{x} = \epsilon f(t, x, y) \tag{4.4.1}$$

$$\dot{y} = A(x)y + \epsilon g(t, x, y) \tag{4.4.2}$$

where $x(0) = x_0$, $y(0) = y_0$, $x \in \mathbb{R}^n$, and $y \in \mathbb{R}^m$.

The state vector is divided into a fast state vector y and a slow state vector x, whose dynamics are of the order of ϵ with respect to the fast dynamics. The dominant term in (4.4.2) is linear in y, but is itself allowed to vary as a function of the slow state vector.

As previously, we define

$$f_{av}(x) = \lim_{T \to \infty} \frac{1}{T} \int_{t_0}^{t_0+T} f(\tau, x, 0)\, d\tau \tag{4.4.3}$$

and the system

$$\dot{x}_{av} = f_{av}(x_{av}) \qquad x_{av}(0) = x_0 \tag{4.4.4}$$

is the *averaged system* corresponding to (4.4.1)–(4.4.2). We make the following additional assumption.

Definition Uniform Exponential Stability of a Family of Square Matrices

The family of matrices $A(x) \in \mathbb{R}^{m \times m}$ is *uniformly exponentially stable for all* $x \in B_h$, if there exist m, λ, m', $\lambda' > 0$, such that for all $x \in B_h$ and $t \geq 0$

$$m' e^{-\lambda' t} \leq \| e^{A(x)t} \| \leq m\, e^{-\lambda t} \tag{4.4.5}$$

Comments

This definition is equivalent to require that the solutions of the system $\dot{y} = A(x)y$ are bounded above and below by decaying exponentials, independently of the parameter x.

It is also possible to show that the definition is equivalent to requiring that there exist p_1, p_2, q_1, $q_2 > 0$, such that for all $x \in B_h$, there exists $P(x)$ satisfying $p_1 I \leq P(x) \leq p_2 I$, and $-q_2 I \leq A^T(x)P(x) + P(x)A(x) \leq -q_1 I$.

We will make the following assumptions.

Assumptions

For some $h > 0$

(B1) The functions f and g are piecewise continuous functions of time, and continuous functions of x and y. Moreover, $f(t, 0, 0) = 0$, $g(t, 0, 0) = 0$ for all $t \geq 0$, and for some l_1, l_2, $l_3, l_4 \geq 0$

$$|f(t, x_1, y_1) - f(t, x_2, y_2)| \leq l_1 |x_1 - x_2| + l_2 |y_1 - y_2|$$

$$|g(t, x_1, y_1) - g(t, x_2, y_2)| \leq l_3 |x_1 - x_2| + l_4 |y_1 - y_2| \quad (4.4.6)$$

for all $t \geq 0$, x_1, $x_2 \in B_h$, y_1, $y_2 \in B_h$. Also assume that $f(t, x, 0)$ has continuous and bounded first partial derivatives with respect to x, for all $t \geq 0$, and $x \in B_h$.

(B2) The function $f(t, x, 0)$ has average value $f_{av}(x)$. Moreover, $f_{av}(0) = 0$, and $f_{av}(x)$ has continuous and bounded first partial derivatives with respect to x, for all $x \in B_h$, so that for some $l_{av} \geq 0$

$$|f_{av}(x_1) - f_{av}(x_2)| \leq l_{av} |x_1 - x_2| \quad (4.4.7)$$

for all x_1, $x_2 \in B_h$.

(B3) Let $d(t, x) = f(t, x, 0) - f_{av}(x)$, so that $d(t, x)$ has zero average value. Assume that the convergence function can be written as $\gamma(T)|x|$, and that $\dfrac{\partial d(t, x)}{\partial x}$ has zero average value, with convergence function $\gamma(T)$.

(B4) A(x) is uniformly exponentially stable for all $x \in B_h$ and, for some $k_a \geq 0$

$$\left\| \frac{\partial A(x)}{\partial x} \right\| \leq k_a \qquad \text{for all } x \in B_h \quad (4.4.8)$$

(B5) For some $h' < h$, $|x_{av}(t)| \in B_{h'}$ on the time intervals considered, and for some h_0, $y_0 \in B_{h_0}$ (where h', h_0 are constants to be defined later). This assumption is technical, and will allow us to guarantee that all signals remain in B_h.

As for one-time scale systems, we first obtain the following preliminary lemma, similar to lemma 4.2.3.

Lemma 4.4.1 Perturbation Formulation of Averaging—Two-Time Scales

If the original system (4.4.1)–(4.4.2) and the averaged system (4.4.4) satisfy assumptions (B1)–(B3)

Then there exist functions $w_\epsilon(t, x)$, $\xi(\epsilon)$ as in lemma 4.2.2 and $\epsilon_1 > 0$, such that the transformation

$$x = z + \epsilon w_\epsilon(t, z) \tag{4.4.9}$$

is a homeomorphism in B_h for all $\epsilon \le \epsilon_1$, and

$$|x - z| \le \xi(\epsilon)|z| \tag{4.4.10}$$

Under the transformation, system (4.4.1) becomes

$$\dot{z} = \epsilon f_{av}(z) + \epsilon p_1(t, z, \epsilon) + \epsilon p_2(t, z, y, \epsilon) \tag{4.4.11}$$

$$z(0) = x_0$$

where

$$|p_1(t, z, \epsilon)| \le \xi(\epsilon) k_1 |z|$$

$$|p_2(t, z, y, \epsilon)| \le k_2 |y| \tag{4.4.12}$$

for some k_1, k_2 depending on l_1, l_2, l_{av}.

Proof of Lemma 4.4.1 in Appendix.

We are now ready to state the averaging theorems concerning the differential system (4.4.1)–(4.4.2). Theorem 4.4.2 is an approximation theorem similar to theorem 4.2.4 and guarantees that the trajectories of the original and averaged system are arbitrarily close on compact intervals, when ϵ tends to zero. Theorem 4.4.3 is an exponential stability theorem, similar to theorem 4.2.5.

Theorem 4.4.2 Basic Averaging Theorem

If the original system (4.4.1)–(4.4.2) and the averaged system (4.4.4) satisfy assumptions (B1)–(B5)

Then there exists $\psi(\epsilon)$ as in lemma 4.2.3 such that, given $T \ge 0$

$$|x(t) - x_{av}(t)| \le \psi(\epsilon) b_T \tag{4.4.13}$$

for some $b_T \ge 0$, $\epsilon_T > 0$ and for all $t \in [0, T/\epsilon]$, and $\epsilon \le \epsilon_T$.

Theorem 4.4.3 Exponential Stability Theorem

If the original system (4.4.1)–(4.4.2) and the averaged system (4.4.4) satisfy assumptions (B1)–(B5), the function $f_{av}(x)$ has continuous and bounded first partial derivatives in x, and $x = 0$ is an exponentially stable equilibrium point of the averaged system

Then the equilibrium point $x = 0$, $y = 0$ of the original system is exponentially stable for ϵ sufficiently small.

Comments

As for theorem 4.2.5, the proof of theorem 4.4.3 gives a useful bound on the rate of convergence of the nonautonomous system. As $\epsilon \to 0$, the rate tends to the bound on the rate of convergence of the averaged system that one would obtain using the Lyapunov function for the averaged system. Since the averaged system is autonomous, it is usually easier to obtain such a Lyapunov function for the averaged system than for the original nonautonomous system, and conclusions about its exponential convergence can be applied to the nonautonomous system for ϵ sufficiently small.

4.4.2 Mixed Time Scales

We now discuss a more general class of two-time scale systems, arising in adaptive control

$$\dot{x} = \epsilon f'(t, x, y') \tag{4.4.14}$$

$$\dot{y}' = A(x)y' + h(t, x) + \epsilon g'(t, x, y') \tag{4.4.15}$$

We will show that system (4.4.14)–(4.4.15) can be transformed into the system (4.4.1)–(4.4.2). In this case, x is a slow variable, but y' has both a fast and a slow component.

The averaged system corresponding to (4.4.14), (4.4.15) is obtained as follows. Define the function

$$v(t, x) := \int_0^t e^{A(x)(t-\tau)} h(\tau, x)\, d\tau \tag{4.4.16}$$

and assume that the following limit exists uniformly in t and x

$$f_{av}(x) = \lim_{T \to \infty} \frac{1}{T} \int_{t_0}^{t_0+T} f'(\tau, x, v(\tau, x))\, d\tau \tag{4.4.17}$$

Intuitively, $v(t, x)$ represents the steady-state value of the variable y' with x frozen and $\epsilon = 0$ in (4.4.15). Then, f is averaged with $v(t, x)$ replacing y' in (4.4.14).

Consider now the transformation

$$y = y' - v(t, x) \tag{4.4.18}$$

Since $v(t, x)$ satisfies

$$\frac{\partial}{\partial t} v(t, x) = A(x) v(t, x) + h(t, x) \qquad v(t, 0) = 0 \tag{4.4.19}$$

we have that

$$\dot{y} = A(x)y + \epsilon\left[-\frac{\partial v(t,x)}{\partial x}f'(t,x,y+v(t,x)) + g'(t,x,y+v(t,x))\right] \tag{4.4.20}$$

so that (4.4.14), (4.4.20) is of the form of (4.4.1), (4.4.2) when

$$f(t,x,y) = f'(t,x,y+v(t,x)) \tag{4.4.21}$$

$$g(t,x,y) = -\frac{\partial v(t,x)}{\partial x}f'(t,x,y+v(t,x)) + g'(t,x,y+v(t,x)) \tag{4.4.22}$$

The averaged system is obtained by averaging the right-hand side of (4.4.21) with $y = 0$, so that the definitions (4.4.17), and (4.4.3) (with f given by (4.4.21)) agree.

To apply theorems 4.4.2 and 4.4.3, we require Assumptions (B1)–(B5) to be satisfied. In particular, we assume similar Lipschitz conditions on f', g', and the following assumption on $h(t,x)$

(B6) $h(t,0) = 0$ for all $t \geq 0$, and $\left\|\frac{\partial h(t,x)}{\partial x}\right\|$ is bounded for all $t \geq 0$, $x \in B_h$.

This new assumption implies that $v(t,0) = 0$. It also implies that $\left\|\frac{\partial v(t,x)}{\partial x}\right\|$ is bounded for all $t \geq 0$, $x \in B_h$, since

$$\frac{\partial v(t,x)}{\partial x_i} = \int_0^t \left[e^{A(x)(t-\tau)}\frac{\partial h(\tau,x)}{\partial x_i} + \frac{\partial}{\partial x_i}\left[e^{A(x)(t-\tau)}\right]h(\tau,x)\right]d\tau \tag{4.4.23}$$

and using the fact that $e^{A(x)(t-\tau)}$ and $\frac{\partial}{\partial x}e^{A(x)(t-\tau)}$ are bounded by exponentials ((4.4.5), and (A4.4.30) in the proof of theorem 4.4.3).

4.5 APPLICATIONS TO ADAPTIVE CONTROL

For illustration, we apply the previous results to the output error direct adaptive control algorithm for the relative degree 1 case.

We established the complete description of the adaptive system in Section 3.5 with (3.5.28), i.e.,

$$\dot{e}(t) = A_m e(t) + b_m \phi^T(t) w_m(t) + b_m \phi^T(t) Q\, e(t)$$

$$\dot{\phi}(t) = -g\, c_m^T e(t)\, w_m(t) - g\, c_m^T e(t)\, Q\, e(t) \tag{4.5.1}$$

where g is the adaptation gain. With the exception of the last terms (quadratic in e and ϕ), (4.5.1) is a set of linear time varying differential equations. They describe the adaptive control system, linearized around the equilibrium $e = 0$, $\phi = 0$. We first study these equations, then turn to the nonlinear equations.

4.5.1 Output Error Scheme—Linearized Equations

The linearized equations, describing the adaptive system for small values of e and ϕ, are

$$\begin{aligned} \dot{e}(t) &= A_m e(t) + b_m\, w_m^T(t)\, \phi(t) \\ \dot{\phi}(t) &= -g\, w_m(t)\, c_m{}^T e(t) \end{aligned} \tag{4.5.2}$$

Since w_m is bounded, it is easy to see that (4.5.2) is of the form of (4.4.14), (4.4.15) with the functions f' and h satisfying the conditions of Section 4.4. Recall that A_m is a stable matrix.

The function $v(t,\phi)$ defined in (4.4.16) is now

$$v(t,\phi) = \left[\int_0^t e^{A_m(t-\tau)}\, b_m w_m^T(\tau)\, d\tau\right] \phi \tag{4.5.3}$$

and f_{av} is given by

$$f_{av}(\phi) = -\lim_{T\to\infty} \frac{1}{T} \int_{t_0}^{t_0+T} w_m(t)\, c_m^T \cdot \left[\int_0^t e^{A_m(t-\tau)}\, b_m w_m^T(\tau)\, d\tau\right] dt\ \phi \tag{4.5.4}$$

Frequency Domain Analysis

To derive frequency domain expressions, we assume that r is stationary. Since the transfer function from $r \to w_m$ is stable, this implies that w_m is stationary. The spectral measure of w_m is related to that of r by

$$S_{w_m}(d\omega) = \hat{H}^*_{w_m r}(j\omega)\, \hat{H}^T_{w_m r}(j\omega)\, S_r(d\omega) \tag{4.5.5}$$

where the transfer function from $r \to w_m$ is given by (using (3.5.11))

$$\hat{H}_{w_m r} = \begin{bmatrix} 1 \\ (sI - \Lambda)^{-1} b_\lambda \hat{P}^{-1} \hat{M} \\ \hat{M} \\ (sI - \Lambda)^{-1} b_\lambda \hat{M} \end{bmatrix} \tag{4.5.6}$$

which is a stable transfer function.

Define a filtered version of w_m to be

$$w_{mf}(t) = \int_0^t c_m^T e^{A_m(t-\tau)} b_m w_m(\tau)\, d\tau$$

$$= \frac{1}{c_0^*} \hat{M}(w_m) \tag{4.5.7}$$

where the last equality follows from (3.5.22). Note that the signal w_f was also used in the direct proof of exponential convergence in Chapter 2 (cf. (2.6.34)).

Since $c_m{}^T(sI - A_m)^{-1} b_m = \dfrac{1}{c_0^*} \hat{M}(s)$ is stable, $w_{mf}(t)$ is stationary. We let

$$R_{w_m w_{mf}}(0) = \lim_{T \to \infty} \frac{1}{T} \int_{t_0}^{t_0 + T} w_m(t)\, w_{mf}^T(t)\, dt \tag{4.5.8}$$

which was called the *cross correlation* between w_m and w_{mf} (evaluated at 0) in Section 1.6. Consequently, we may use (4.5.7) and (4.5.8) to obtain a frequency domain expression for $R_{w_m w_{mf}}(0)$ as

$$R_{w_m w_{mf}}(0) = \frac{1}{2\pi c_0^*} \int_{-\infty}^{\infty} \hat{H}^*_{w_m r}(j\omega) \hat{H}^T_{w_m r}(j\omega) \hat{M}(j\omega) S_r(d\omega) \tag{4.5.9}$$

Since r is a scalar, $S_r(d\omega)$ is *real* and consequently an even function of ω (cf. Section 1.6). This may be used to show that

$$R_{w_m w_{mf}}(0) = \frac{1}{2\pi c_0^*} \int_{-\infty}^{\infty} \mathrm{Re}\left[\hat{H}^*_{w_m r}(j\omega) \hat{H}^T_{w_m r}(j\omega) \right] \mathrm{Re}\, \hat{M}(j\omega)\, S_r(d\omega)$$

$$+ \frac{1}{2\pi c_0^*} \int_{-\infty}^{\infty} \mathrm{Im}\left[\hat{H}^*_{w_m r}(j\omega) \hat{H}^T_{w_m r}(j\omega) \right] \mathrm{Im}\, \hat{M}(j\omega)\, S_r(d\omega)$$

and that the first matrix in the right-hand side is symmetric, while the second in antisymmetric.

With (4.5.7) and (4.5.8), (4.5.4) shows that the averaged system is a LTI system

$$\dot{\phi}_{av} = -g\, R_{w_m w_{mf}}(0)\, \phi_{av} \qquad \phi_{av}(0) = \phi_0 \tag{4.5.10}$$

Convergence Analysis

Since $\hat{M}(s)$ is strictly positive real, the matrix $R_{w_m w_{mf}}(0)$ is a positive semidefinite matrix. Unlike the matrix $R_w(0)$ of Section 4.3, $R_{w_m w_{mf}}(0)$ need not be symmetric, so that its eigenvalues need not be real. However, the real parts are guaranteed to be positive, and a natural Lyapunov function is again

$$v(\phi_{av}) = |\phi_{av}|^2 = \phi_{av}^T \phi_{av} \tag{4.5.11}$$

and

$$-\dot{v}(\phi_{av}) = g\, \phi_{av}^T \left[R_{w_m w_{mf}}(0) + R_{w_m w_{mf}}^T(0) \right] \phi_{av} \tag{4.5.12}$$

The matrix in parentheses is symmetric positive semidefinite. As previously, it is positive definite if w_m is PE.

When the reference input r is periodic or almost periodic, i.e.,

$$r(t) = \sum_k r_k \sin(\omega_k t) \tag{4.5.13}$$

an expression for $R_{w_m w_{mf}}(0)$ is

$$\begin{aligned} R_{w_m w_{mf}}(0) = \frac{1}{c_0^*} \sum_k & \left[\frac{r_k^2}{2} \operatorname{Re}\left[\hat{H}_{w_m r}^*(j\omega_k)\, \hat{H}_{w_m r}^T(j\omega_k) \right] \cdot \operatorname{Re} \hat{M}(j\omega_k) \right] \\ + \frac{1}{c_0^*} \sum_k & \left[\frac{r_k^2}{2} \operatorname{Im}\left[\hat{H}_{w_m r}^*(j\omega_k)\, \hat{H}_{w_m r}^T(j\omega_k) \right] \cdot \operatorname{Im} \hat{M}(j\omega_k) \right] \end{aligned} \tag{4.5.14}$$

Example

As an illustration of the preceding results, we consider the following example of a first order plant with an unknown pole and an unknown gain

$$\hat{P}(s) = \frac{k_p}{s + a_p} \tag{4.5.15}$$

We will choose values of the parameters corresponding to the Rohrs examples (Rohrs *et al* [1982], see also Section 5.2) , when no unmodeled dynamics are present.

The adaptive process is to adjust the feedforward gain c_0 and the feedback gain d_0 so as to make the closed-loop transfer function match the model transfer function

$$\hat{M}(s) = \frac{k_m}{s + a_m} \tag{4.5.16}$$

To guarantee persistency of excitation, we use a sinusoidal input signal of the form

$$r(t) = r_0 \sin(\omega_0 t) \tag{4.5.17}$$

Thus, (4.5.2) becomes

$$\begin{aligned} \dot{e}_0(t) &= -a_m e_0(t) + k_p(\phi_r(t) r(t) + \phi_y(t) y_m(t)) \\ \dot{\phi}_r(t) &= -g\, e_0(t) r(t) \\ \dot{\phi}_y(t) &= -g\, e_0(t) y_m(t) \end{aligned} \tag{4.5.18}$$

where

$$\begin{aligned} \phi_r(t) &= c_0(t) - c_0^* \\ \phi_y(t) &= d_0(t) - d_0^* \end{aligned} \tag{4.5.19}$$

It can be checked, using (4.5.14), that the averaged system defined in (4.5.10) is now

$$\dot{\phi}_{av} = -g \frac{r_0^2}{2} \frac{k_p}{k_m} \begin{bmatrix} \dfrac{a_m k_m}{(a_m^2 + \omega_0^2)} & \dfrac{k_m^2 (a_m^2 - \omega_0^2)}{(a_m^2 + \omega_0^2)^2} \\ \dfrac{k_m^2}{(a_m^2 + \omega_0^2)} & \dfrac{a_m k_m^3}{(a_m^2 + \omega_0^2)^2} \end{bmatrix} \phi_{av} \tag{4.5.20}$$

With $a_m = 3$, $k_m = 3$, $a_p = 1$, $k_p = 2$, $r_0 = 1$, $\omega_0 = 1$, $g = 1$, the two eigenvalues of the averaged system are computed to be $-0.0163g$ and $-0.5537g$, and are both real negative. The nominal parameter $\theta^{*^T} = (k_m/k_p, (a_p - a_m)/k_p)$. We let $\theta(0) = 0$, so that $\phi^T(0) = (-1.5, 1)$.

Figures 4.8, 4.9 and 4.10 show the plots of the parameter errors $\phi_y(\phi_r)$ for the original and averaged system, with three different frequencies ($\omega_0 = 1, 3, 5$).

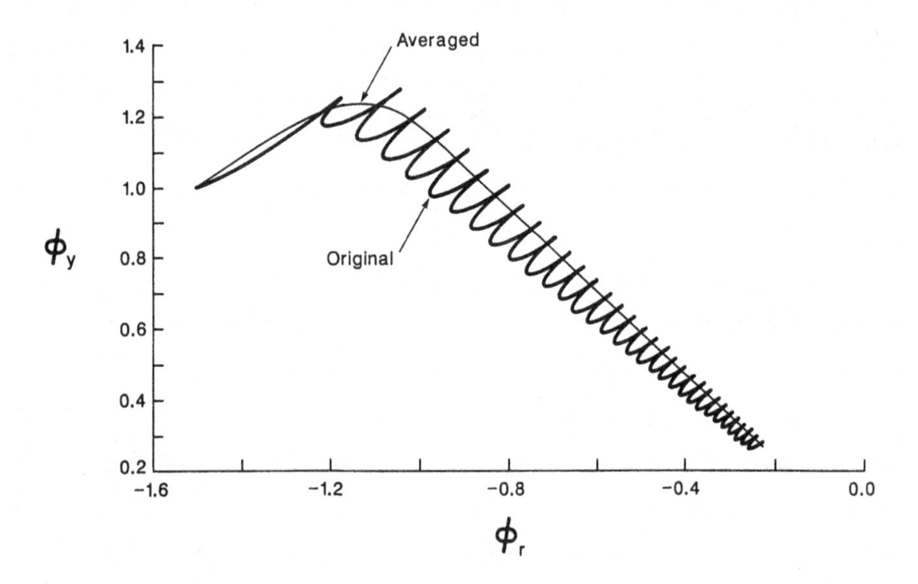

Figure 4.8: Parameter Error $\phi_y(\phi_r)$ ($r = \sin t$)

Figure 4.10 corresponds to a frequency of the input signal $\omega_0 = 5$, such that the eigenvalues of the matrix $R_{w_m w_{mf}}(0)$ are complex: $(-0.0553 \pm j\,0.05076)g$. This explains the oscillatory behavior of the original and averaged systems observed in the figure, which did not exist in the previous examples of Section 4.3.

4.5.2 Output Error Scheme—Nonlinear Equations

We now return to the complete, nonlinear differential equations

$$\dot{e}(t) = A_m e(t) + b_m \phi^T(t) w_m(t) + b_m \phi^T(t) Q\, e(t)$$

$$\dot{\phi}(t) = -g\, w_m(t) c_m^T e(t) - g\, Q\, e(t) c_m^T e(t) \tag{4.5.21}$$

From (4.4.45)

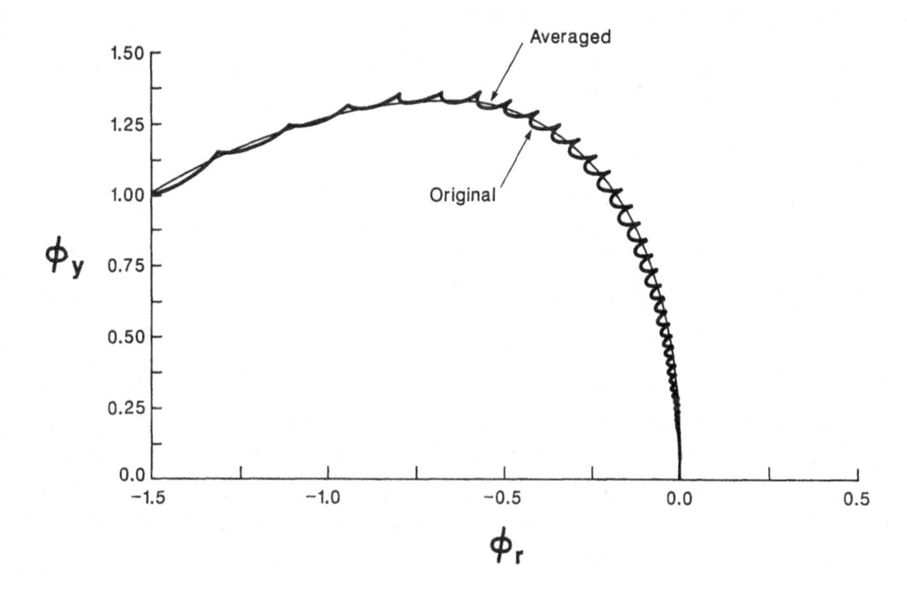

Figure 4.9: Parameter Error $\phi_y(\phi_r)$ $(r = \sin 3t)$

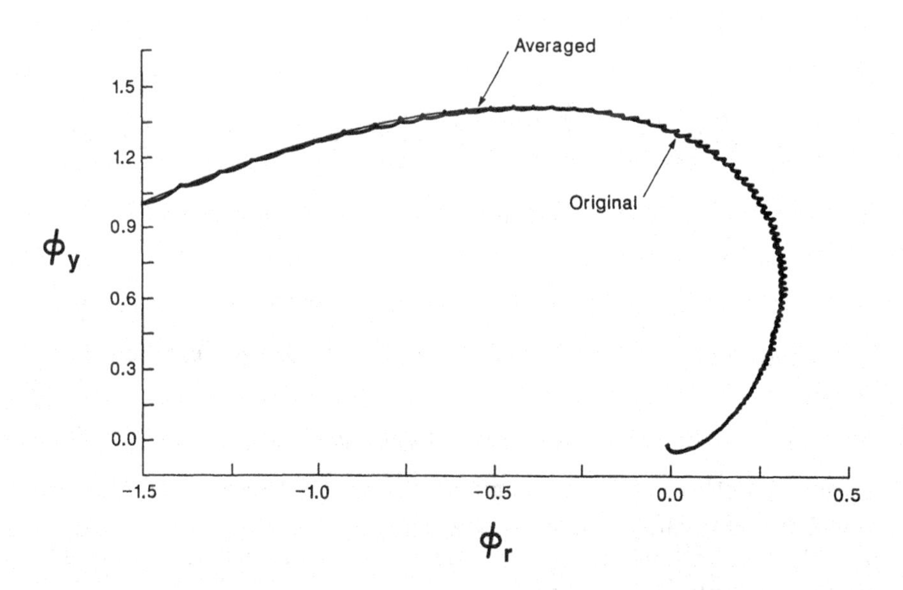

Figure 4.10: Parameter Error $\phi_y(\phi_r)$ $(r = \sin 5t)$

$$v(t,\phi) = \int_0^t e^{(A_m + b_m \phi^T Q)(t-\tau)} b_m \phi^T w_m(\tau)\, d\tau \tag{4.5.22}$$

so that the averaged system is

$$\dot{\phi}_{av} = g f_{av}(\phi_{av}) \qquad \phi_{av}(0) = \phi(0) \tag{4.5.23}$$

where f_{av} is defined by the limit

$$f_{av}(\phi) = -\lim_{T \to \infty} \frac{1}{T} \int_{t_0}^{t_0+T} \Big[w_m(t) c_m^T v(t,\phi) + Q\, v(t,\phi) c_m^T v(t,\phi) \Big]\, dt \tag{4.5.24}$$

The assumptions of the theorems will be satisfied if the limit in (4.5.24) is uniform in the sense of (B3) and provided that the matrix $A_m + b_m \phi^T Q$ is uniformly exponentially stable for $\phi \in B_h$. This means that if the controller parameters are frozen at any point of the trajectory the resulting time invariant system must be closed-loop stable.

Frequency Domain Analysis

The expression of f_{av} in (4.5.24) can be translated into the frequency domain, noting that w_m is related to r through the vector transfer function $\hat{H}_{w_m r}$

$$f_{av}(\phi) = -\frac{1}{2\pi} \int_{-\infty}^{\infty} \Big[\hat{H}_{w_m r}(-j\omega) + Q(-j\omega I - A_m - b_m \phi^T Q)^{-1} b_m \phi^T \hat{H}_{w_m r}(-j\omega) \Big] \Big[c_m^T (j\omega I - A_m - b_m \phi^T Q)^{-1} b_m \phi^T \hat{H}_{w_m r}(j\omega) \Big]\, S_r(d\omega) \tag{4.5.25}$$

where $S_r(d\omega)$ is the spectral measure of r. Note that f_{av} can be factored as

$$f_{av}(\phi) = -A_{av}(\phi) \cdot \phi \tag{4.5.26}$$

where $A_{av} : \mathbb{R}^{2n} \to \mathbb{R}^{2n \times 2n}$ is similar to $R_{w_m w_{mf}}(0)$ in Section 4.5.1, but now depends nonlinearly on ϕ. The expression in (4.5.25) is more complex than in the linear case, but some manipulations will allow us to obtain a more interesting result.

Recall that (4.5.21) was obtained from the differential equation

$$\begin{aligned} \dot{e}(t) &= A_m e(t) + b_m \phi^T(t) w(t) \\ \dot{\phi}(t) &= -g\, w(t) c_m^T e(t) \end{aligned} \qquad (4.5.27)$$

by noting that $w(t) = w_m(t) + Q\, e(t)$. In general, (4.5.27) is of limited use, precisely because w depends on e. The signal w is not an external signal, but depends on internal variables. On the other hand, w_m is an exogeneous signal, related to r through a stable transfer function.

In the context of averaging, the differential equation describing the fast variable (i.e., e) is averaged, assuming that the slow variable (i.e., ϕ) is constant. However, when ϕ is constant, w *is* related to r through a linear time invariant system, with a transfer function depending on ϕ. If $\det(sI - A_m - b_m \phi^T Q)$ is Hurwitz (as we assume to apply averaging), this transfer function is stable. Therefore, *assuming that ϕ is fixed*, we can write

$$\hat{w} = \hat{H}_{wr}(s, \phi)\ \hat{r} \qquad (4.5.28)$$

so that using (4.5.27), (4.5.25) can be replaced by (4.5.26), with an expression similar to the expression of $R_{w_m w_{mf}}(0)$ in (4.5.9), i.e.

$$A_{av}(\phi) = \frac{1}{2\pi c_0^*} \int_{-\infty}^{\infty} \hat{H}_{wr}^*(j\omega, \phi) \hat{H}_{wr}^T(j\omega, \phi) \hat{M}(j\omega) S_r(d\omega) \qquad (4.5.29)$$

Explicit Expression for $\hat{\mathbf{H}}_{\mathbf{wr}}(\mathbf{s}, \phi)$

Recall that $\bar{w}_m$ is related to r through the transfer function $\hat{H}_{\bar{w}_m r}$, whose poles are the zeros of $\det(sI - A_m)$. Let

$$\hat{\chi}_m(s) = \det(sI - A_m) \qquad (4.5.30)$$

and write the transfer function $\hat{H}_{\bar{w}_m r}$ as the ratio of a vector polynomial $\hat{n}(s)$, and a characteristic polynomial $\hat{\chi}_m(s)$, i.e.,

$$\hat{H}_{\bar{w}_m r}(s) = \frac{\hat{n}(s)}{\hat{\chi}_m(s)} \qquad (4.5.31)$$

We found in Section 3.5 (cf. (3.5.8), (3.5.11)) that

$$\bar{w} = \frac{\hat{n}(s)}{\hat{\chi}_m(s)} \left[r + \frac{1}{c_0^*} \phi^T w \right] \qquad (4.5.32)$$

Denote $\phi_r = c_0 - c_0^*$, so that $\phi^T w = \phi_r r + \bar{\phi}^T \bar{w}$. Assuming that ϕ is constant, (4.5.32) becomes

$$\overline{w} = \left[\hat{\chi}_m(s) \cdot \mathrm{I} - \frac{1}{c_0^*} \hat{n}(s) \overline{\phi}^T \right]^{-1} \hat{n}(s) \left[\left[1 + \frac{\phi_r}{c_0^*} \right] r \right]$$

$$= \frac{\hat{n}(s)}{\hat{\chi}_m(s) - \frac{1}{c_0^*} \overline{\phi}^T \hat{n}(s)} \left[\left[1 + \frac{\phi_r}{c_0^*} \right] r \right] \tag{4.5.33}$$

Denote

$$\hat{\chi}_\phi(s) := \hat{\chi}_m(s) - \frac{1}{c_0^*} \overline{\phi}^T \hat{n}(s) \tag{4.5.34}$$

$\hat{\chi}_\phi(s)$ is closed-loop characteristic polynomial, giving the poles of the adaptive system with feedback θ, that is, the poles of the model transfer function with feedback ϕ. Therefore, $\hat{\chi}_\phi(s)$ is also given by

$$\hat{\chi}_\phi(s) = \det(sI - A_m - b_m \phi^T Q) \tag{4.5.35}$$

With this notation, (4.5.33) can be written

$$\overline{w} = \frac{\hat{\chi}_m}{\hat{\chi}_\phi} \cdot \hat{H}_{\overline{w}_m r} \left[r + \frac{\phi_r}{c_0^*} r \right]$$

$$= \frac{\hat{\chi}_m}{\hat{\chi}_\phi} (\overline{w}_m) + \frac{\hat{\chi}_m}{\hat{\chi}_\phi} \left[\frac{\phi_r}{c_0^*} \overline{w}_m \right] \tag{4.5.36}$$

On the other hand

$$r = \frac{\hat{\chi}_m}{\hat{\chi}_\phi} \left[1 - \frac{\overline{\phi}^T \hat{n}}{c_0^* \hat{\chi}_m} \right] \cdot (r)$$

$$= \frac{\hat{\chi}_m}{\hat{\chi}_\phi} (r) - \frac{\hat{\chi}_m}{\hat{\chi}_\phi} \left[\frac{\overline{\phi}^T}{c_0^*} \cdot \overline{w}_m \right] \tag{4.5.37}$$

Define

$$B(\phi) := \begin{bmatrix} \mathrm{O} & -\dfrac{1}{c_0^*}\,\bar{\phi}^T \\ \mathrm{O} & \dfrac{\phi_r}{c_0^*}\cdot \mathrm{I} \end{bmatrix} \in \begin{bmatrix} \mathbb{R}^{1\times 1} & \mathbb{R}^{1\times 2n-1} \\ \mathbb{R}^{2n-1\times 1} & \mathbb{R}^{2n-1\times 2n-1} \end{bmatrix} \tag{4.5.38}$$

i.e.,

$$B(\phi) \in \mathbb{R}^{2n\times 2n}$$

so that (4.5.36)–(4.5.37) can be written

$$w = \begin{bmatrix} r \\ \bar{w} \end{bmatrix} = \frac{\hat{\chi}_m}{\hat{\chi}_\phi}\begin{bmatrix} r \\ \bar{w}_m \end{bmatrix} + \frac{\hat{\chi}_m}{\hat{\chi}_\phi} B(\phi)\cdot \begin{bmatrix} r \\ \bar{w}_m \end{bmatrix} \tag{4.5.39}$$

The vector transfer function $\hat{H}_{wr}$ can therefore be expressed in terms of the vector transfer function $\hat{H}_{w_m r}$ by

$$\hat{H}_{wr}(s,\phi) = \frac{\hat{\chi}_m(s)}{\hat{\chi}_\phi(s)}(I + B(\phi))\hat{H}_{w_m r}(s) \tag{4.5.40}$$

and, as expected

$$\hat{H}_{wr}(s,0) = \hat{H}_{w_m r}(s) \tag{4.5.41}$$

Convergence Analysis

With (4.5.40), A_{av} can be written

$$A_{av}(\phi) = \frac{1}{2\pi c_0^*}\int_{-\infty}^{\infty}\left|\frac{\hat{\chi}_m(j\omega)}{\hat{\chi}_\phi(j\omega)}\right|^2 (I + B(\phi))\hat{H}^*_{w_m r}(j\omega)$$

$$\cdot\ \hat{H}^T_{w_m r}(j\omega)(I + B^T(\phi))\hat{M}(j\omega)\,S_r(d\omega) \tag{4.5.42}$$

Consider now the trajectories of the averaged system and let $v(\phi_{av}) = |\phi_{av}|^2 = \phi_{av}^T\phi_{av}$. Note that by (4.5.38), it follows that

$$\phi^T\cdot B(\phi) = 0 \qquad \text{for all } \phi \tag{4.5.43}$$

Denote

$$R(\phi_{av}) := \frac{1}{2\pi c_0^*}\int_{-\infty}^{\infty}\left|\frac{\hat{\chi}_m(j\omega)}{\hat{\chi}_{\phi_{av}}(j\omega)}\right|^2 \hat{H}^*_{w_m r}(j\omega)$$

$$\cdot\ \hat{H}^T_{w_m r}(j\omega)\,\hat{M}(j\omega)\,S_r(d\omega) \tag{4.5.44}$$

It follows that the derivative of v is given by

$$-\dot{v}(\phi_{av}) = g\,\phi^T_{av}(R(\phi_{av}) + R^T(\phi_{av}))\,\phi_{av} \tag{4.5.45}$$

which is identical to the expression for the linear case (4.5.12), provided that $R(\phi_{av})$ given in (4.5.44) replaces $R_{w_m w_{mf}}(0)$ given in (4.5.9). It is remarkable that this result differs from the expression obtained by linearization followed by averaging in Section 4.5.1 only by the *scalar* weighting factor $|\hat{\chi}_m / \hat{\chi}_\phi|^2$. Recall that $\hat{\chi}_m(s)$ defines the nominal closed-loop poles (i.e. when $\phi = 0$, while $\hat{\chi}_\phi$ defines the closed-loop poles with feedback gains $\theta = \phi + \theta^*$. The term $|\hat{\chi}_m / \hat{\chi}_\phi|^2$ is strictly positive, given any ϕ bounded, and it approaches unity continuously as ϕ approaches zero.

Since $\hat{M}(s)$ is strictly positive real, $R(\phi_{av})$ is at least positive semidefinite. As in the linearized case, it is positive definite if w_m is persistently exciting. Using the Lyapunov function $v(\phi_{av})$, this argument itself constitutes a proof of exponential stability of the averaged system, using (4.5.45). By theorem 4.4.3, the exponential stability of the original system is also guaranteed for g sufficiently small.

Rates of convergence can also be determined, using the Lyapunov function $v(\phi_{av})$, so that

$$\begin{aligned} -\dot{v} &= g\,\phi^T_{av}(R(\phi_{av}) + R^T(\phi_{av}))\,\phi_{av} \\ &\geq g \inf_{\phi_{av} \in B_h} (\lambda_{\min}(R(\phi_{av}) + R^T(\phi_{av})))\,v := 2g\alpha v \end{aligned} \tag{4.5.46}$$

and the guaranteed rate of parameter convergence of the averaged adaptive system is $g\alpha$. The rate of convergence of the original system can be estimated by the same value, for g sufficiently small.

It is interesting to note that, as $|\phi_{av}|$ increases, $\lambda_{\min}(R(\phi_{av}) + R^T(\phi_{av}))$ tends to zero in some directions. This indicates that the adaptive control system may *not* be globally exponentially stable.

Example

We consider the previous two parameter example. The adaptive system is described by

$$\dot{e}_0(t) = -a_m e_0(t) + k_p(\phi_r(t)\,r(t) + \phi_y(t)\,e_0(t) + \phi_y(t)\,y_m(t))$$

$$\dot{\phi}_r(t) = -g\, e_0(t)\, r(t)$$

$$\dot{\phi}_y(t) = -g\, e_0^2(t) - g\, e_0(t)\, y_m(t) \tag{4.5.47}$$

Consider the case when $r = r_0 \sin(\omega_0 t)$. The averaged system can be computed using (4.5.42). We can also verify the expression using (4.5.47) and the definition of the averaged system (4.5.22). After some manipulations, we obtain, for the averaged system (dropping the "av" subscripts for simplicity)

$$\dot{\phi}_r = -g\, k_p \frac{r_0^2}{2} \frac{1}{\omega_0^2 + (a_m - k_p \phi_y)^2} \Bigg[(a_m - k_p \phi_y)\, \phi_r + \left[\frac{a_m^2 - \omega_0^2}{\omega_0^2 + a_m^2} k_m \right] \phi_y - \frac{k_p\, a_m\, k_m}{\omega_0^2 + a_m^2} \phi_y^2 \Bigg] \tag{4.5.48}$$

$$\dot{\phi}_y = -g\, k_p \frac{r_0^2}{2} \frac{1}{\omega_0^2 + (a_m - k_p \phi_y)^2} \Bigg[k_m \phi_r + \frac{a_m k_m^2}{\omega_0^2 + a_m^2} \phi_y + k_p \phi_r^2 + \frac{k_p\, a_m\, k_m}{\omega_0^2 + a_m^2} \phi_r \phi_y \Bigg] \tag{4.5.49}$$

Using this result, or using (4.5.42)–(4.5.43), we find that for $v = \phi^T \phi$

$$-\dot{v} = 2g \left[\frac{\omega_0^2 + a_m^2}{\omega_0^2 + (a_m - k_p \phi_y)^2} \right] \frac{r_0^2}{2} \frac{k_p}{k_m} \cdot \phi^T \begin{bmatrix} \dfrac{a_m k_m}{\omega_0^2 + a_m^2} & \dfrac{k_m^2 (a_m^2 - \omega_0^2)}{(\omega_0^2 + a_m^2)^2} \\ \dfrac{k_m^2}{\omega_0^2 + a_m^2} & \dfrac{a_m k_m^3}{(\omega_0^2 + a_m^2)^2} \end{bmatrix} \phi \tag{4.5.50}$$

It can easily be checked that when the first term in brackets is equal to 1 (i.e. with ϕ_y replaced by zero), the result is the same as the result obtained by first linearizing the system, then averaging it (cf. (4.5.20)). In fact, it can be seen, from the expressions of the averaged systems

((4.5.10) with (4.5.9), and (4.5.23) with (4.5.26), (4.5.38) and (4.5.42)), that the system obtained by linearization followed by averaging is *identical* to the system obtained by averaging followed by linearization. Also, given any prescribed B_h (but such that det $(sI - A_m - b_m \phi^T Q)$ is Hurwitz), (4.5.50) can be used to obtain estimates of the rates of convergence of the *nonlinear* system.

We reproduce here simulations for the following values of the parameters: $a_m = 3$, $k_m = 3$, $a_p = 1$, $k_p = 2$, $r_0 = 1$, $\omega_0 = 1$, $g = 1$. The first set of figures is a simulation for initial conditions $\phi_r(0) = -0.5$ and $\phi_y(0) = 0.5$. Figure 4.11 represents the time variation of the function $\ln(v = \phi^T \phi)$ for the original, averaged, and linearized-averaged systems (the minimum slope of the curve gives the rate of convergence).

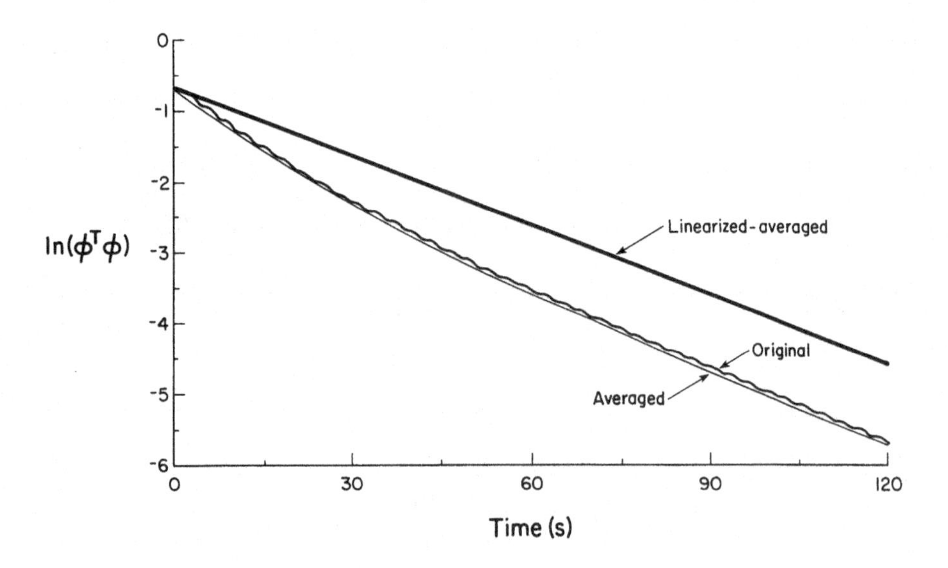

Figure 4.11: Logarithm of the Lyapunov Function

It shows the close approximation of the original system by the averaged system. The slope for the linearized-averaged system is asymptotically identical to that of the averaged system, since parameters eventually get arbitrarily close to their nominal values. Figures 4.12 and 4.13 show the approximation of the trajectories of ϕ_r and ϕ_y.

Figure 4.14 represents the logarithm of the Lyapunov function for a simulation with identical parameters, but initial conditions $\phi_r(0) = 0.5$, $\phi_y(0) = -0.5$. Due to the change of sign in $\phi_y(0)$, the rate of convergence of the nonlinear system is less now than the rate of the linearized

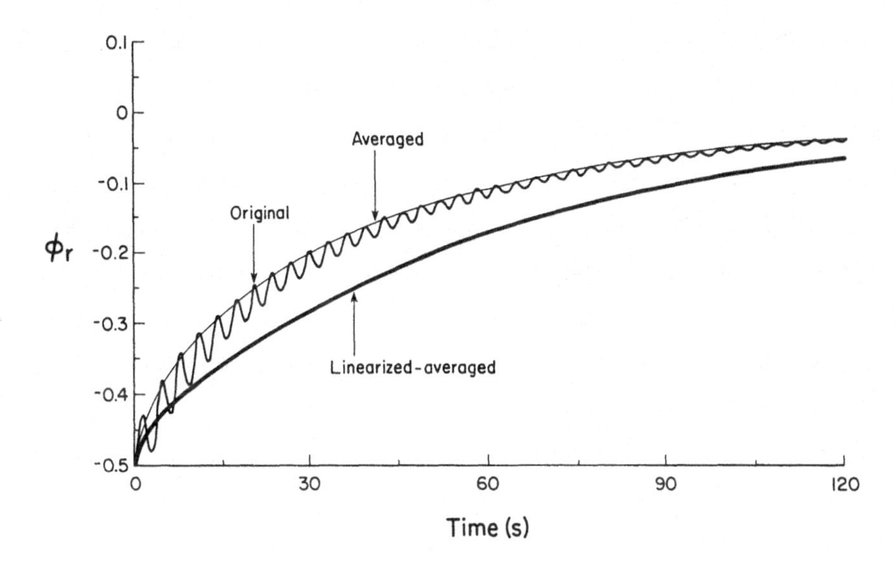

Figure 4.12: Parameter Error ϕ_r

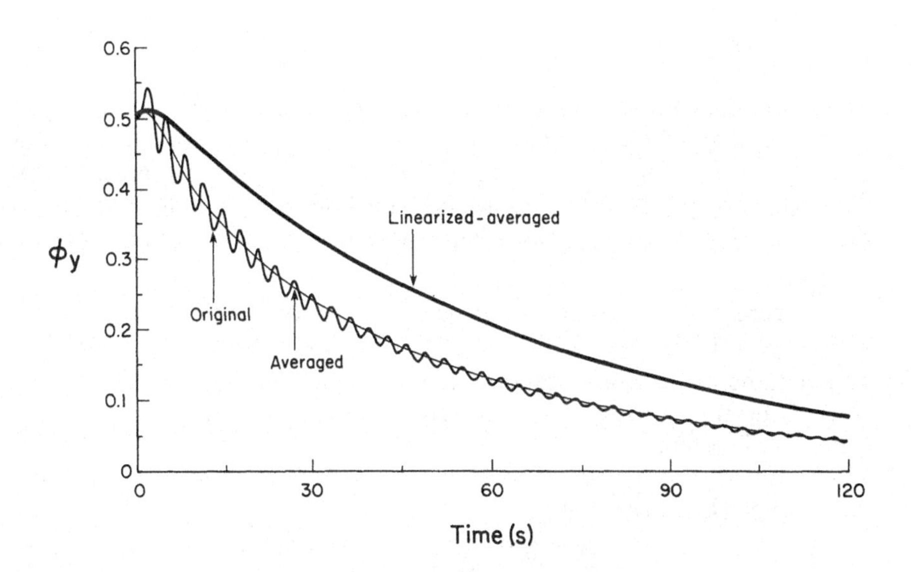

Figure 4.13: Parameter Error ϕ_y

system, while it was larger in the previous case.

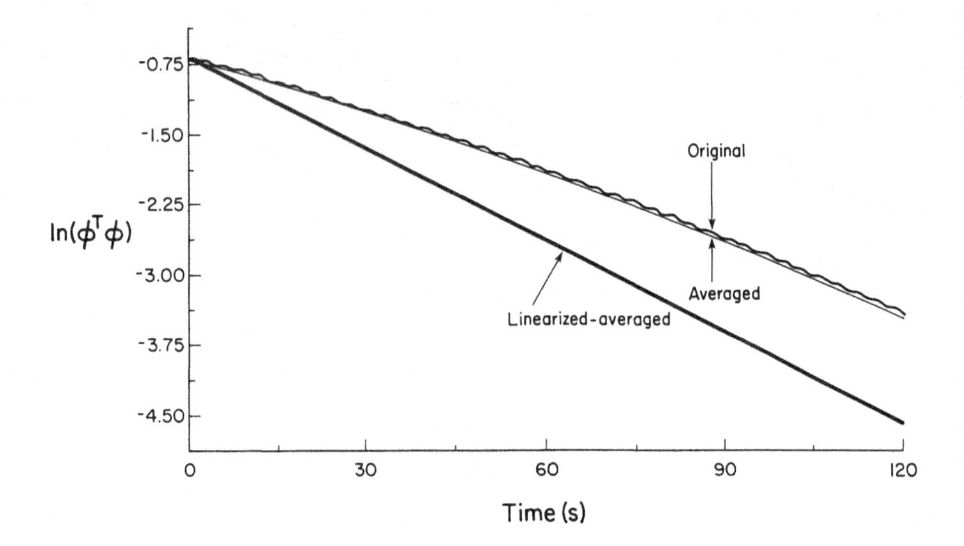

Figure 4.14: Logarithm of the Lyapunov Function

These simulations demonstrate the close approximation by the averaged system, and it should be noted that this is achieved despite an adaptation gain g equal to 1. This shows that the averaging method is useful for values of g which are not necessarily infinitesimal (i.e. not necessarily for very slow adaptation), but for values which are often practical ones.

Figure 4.15 shows the state-space trajectory $\phi_y(\phi_r)$, corresponding to Figure 4.10, that is with initial conditions $\phi_r(0) = -1.5, \phi_y(0) = 1$, and parameters as above except $\omega_0 = 5$. Figure 4.15 shows the distortion of the trajectories in the state-space, due to the nonlinearity of the differential system.

4.5.3 Input Error Scheme

An expression for the averaged system corresponding to the input error scheme may also be obtained. We consider the scheme for arbitrary relative degree. For simplicity, however, we neglect the normalization factor in the gradient algorithm and the projection for the feedforward gain c_0.

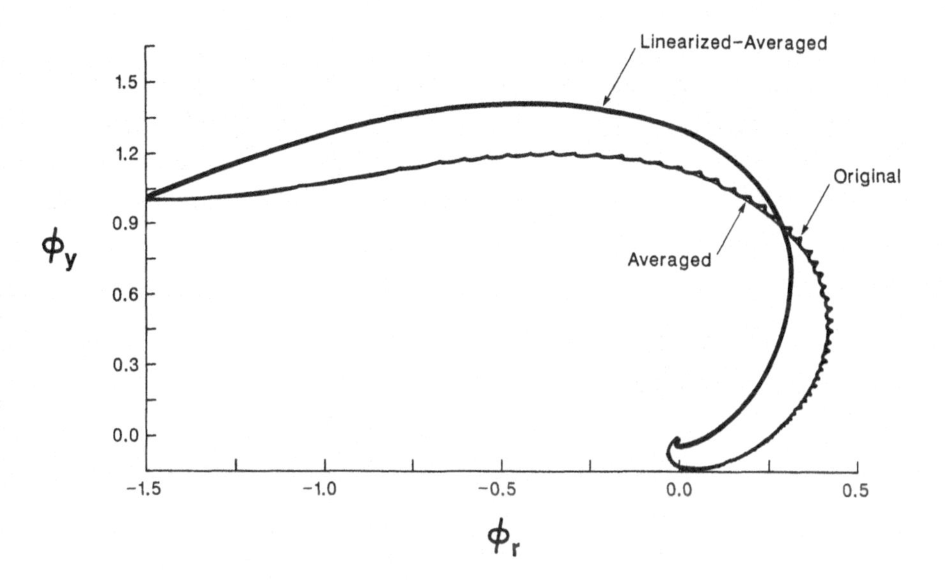

Figure 4.15: Parameter Error $\phi_y(\phi_r)(r = \sin 5t)$

The equation describing the parameter update is then simply

$$\dot{\phi}(t) = -g\, v(t) v^T(t) \phi(t) \tag{4.5.51}$$

so that the averaged system is again of the form

$$\dot{\phi}_{av} = -g A_{av}(\phi_{av})\phi_{av} \tag{4.5.52}$$

where $A_{av}(\phi_{av})$ is the autocovariance of the vector v at $t = 0$. It depends on ϕ_{av} because v is obtained from a closed-loop system with feedback depending on the parameter error ϕ_{av}. Within the framework of averaging, the system is a linear time invariant system, so that we may write (as in (4.5.28))

$$\hat{v} = \hat{H}_{vr}(s, \phi) \cdot \hat{r} \tag{4.5.53}$$

Explicit Expression of $\hat{\mathbf{H}}_{vr}(s, \phi)$
Recall that (cf. (3.5.10))

$$v = \hat{L}^{-1}(z) = \hat{L}^{-1}\begin{bmatrix} r_p \\ \overline{w} \end{bmatrix} = \hat{L}^{-1}\begin{bmatrix} r + \dfrac{1}{c_0^*}\phi^T w \\ \overline{w} \end{bmatrix} \tag{4.5.54}$$

Since the controller is the same as for the output error scheme, we may use (4.5.36)–(4.5.37). First, rewrite (4.5.36) using $\phi_r = c_0 - c_0^*$ so that

$$\overline{w} = \frac{c_0}{c_0^*}\frac{\hat{\chi}_m}{\hat{\chi}_\phi}(\overline{w}_m) \tag{4.5.55}$$

and

$$\begin{aligned} r + \frac{1}{c_0^*}\phi^T w &= \frac{c_0}{c_0^*} r + \frac{1}{c_0^*}\overline{\phi}^T \overline{w} \\ &= \frac{c_0}{c_0^*} r + \frac{c_0}{c_0^*}\frac{\hat{\chi}_m}{\hat{\chi}_\phi}\left(\frac{\overline{\phi}^T}{c_0^*}\overline{w}_m\right) \end{aligned} \tag{4.5.56}$$

With (4.5.37), (4.5.56) simply becomes

$$r + \frac{1}{c_0^*}\phi^T w = \frac{c_0}{c_0^*}\frac{\hat{\chi}_m}{\hat{\chi}_\phi}(r) \tag{4.5.57}$$

Therefore

$$v = \frac{c_0}{c_0^*}\hat{L}^{-1}\frac{\hat{\chi}_m}{\hat{\chi}_\phi}\begin{bmatrix} r \\ \overline{w}_m \end{bmatrix} = \frac{c_0}{c_0^*}\hat{L}^{-1}\frac{\hat{\chi}_m}{\hat{\chi}_\phi}(w_m) \tag{4.5.58}$$

and

$$\hat{H}_{vr}(s,\phi) = \frac{c_0}{c_0^*}\hat{L}^{-1}(s)\frac{\hat{\chi}_m(s)}{\hat{\chi}_\phi(s)}\hat{H}_{w_m r}(s) \tag{4.5.59}$$

which is the equivalent of (4.5.40) for the input error scheme.

Convergence Analysis

Using the foregoing result, we may express $A_{av}(\phi)$ in the frequency domain as

$$A_{av}(\phi) = \frac{1}{2\pi}\left(\frac{c_0}{c_0^*}\right)^2 \int_{-\infty}^{\infty} |\hat{L}^{-1}(j\omega)|^2 \left|\frac{\hat{\chi}_m(j\omega)}{\hat{\chi}_\phi(j\omega)}\right|^2 \hat{H}^*_{w_m r}(j\omega)$$

$$\cdot\ \hat{H}^T_{w_m r}\,(j\omega)\ S_r\,(d\omega) \tag{4.5.60}$$

Note that the matrix $A_{av}(\phi)$ is now symmetric and is a positive semidefinite matrix for all ϕ. It is positive definite if the input r is sufficiently rich. Again, parameter convergence rates may be estimated from the preceding expression. Although the convergence properties are quite similar, the symmetry of $A_{av}(\phi)$ guarantees that around the equilibrium, the linearized system is described by a linear time invariant system with only real eigenvalues. Therefore, the oscillatory behavior of the output error scheme is not observed for the input error scheme.

Example

We consider once again the example of Section 4.5.2, but for the input error scheme. The model transfer function is $\hat{M} = k_m / (s + a_m)$, and we choose $\hat{L} = (s + l_2) / l_1$. Note that $(\hat{M}\hat{L})^{-1}$ may be expressed as

$$(\hat{M}\hat{L})^{-1} = \frac{(s + a_m) l_1}{k_m (s + l_2)} = \frac{l_1}{k_m} + \frac{a_m - l_2}{k_m}\,\frac{l_1}{s + l_2} \tag{4.5.61}$$

The equations describing the overall adaptive system with the input error scheme are

$$\dot{y}_p = -a_p y_p + k_p u$$

$$\dot{y}_m = -a_m y_m + k_m r$$

$$u = c_0 r + d_0 y_p$$

$$\dot{x}_1 = -l_2 x_1 + l_1 u \qquad \text{i.e.} \quad x_1 = \hat{L}^{-1}(u)$$

$$\dot{x}_2 = -l_2 x_2 + l_1 y_p \qquad \text{i.e.} \quad x_2 = \hat{L}^{-1}(y_p)$$

$$x_3 = \frac{l_1}{k_m}(y_p) + \frac{a_m - l_2}{k_m}(x_2) \qquad \text{i.e}\ x_3 = (\hat{M}\hat{L})^{-1}(y_p)$$

$$e_2 = c_0 x_3 + d_0 x_2 - x_1$$

$$\dot{c}_0 = -g\, e_2 x_3 \qquad \phi_r = c_0 - c_0^* = c_0 - k_m / k_p$$

$$\dot{d}_0 = -g\, e_2 x_2 \qquad \phi_y = d_0 - d_0^* = d_0 - \frac{(a_p - a_m)}{k_p}$$

Again, we neglected the normalization factor and the projection in the update law for simplicity.

When $r = r_0 \sin(\omega_0 t)$, the averaged system is

$$\begin{bmatrix} \dot{\phi}_r \\ \dot{\phi}_y \end{bmatrix} = -g \frac{r_0^2}{2} \frac{l_1^2}{\omega_0^2 + l_2^2} \left[\frac{\phi_r}{c_0^*} + 1 \right]^2 \frac{\omega_0^2 + a_m^2}{\omega_0^2 + (a_m - k_p \phi_y)^2}$$

$$\begin{bmatrix} 1 & \dfrac{k_m a_r}{a_m^2 + \omega_0^2} \\ \dfrac{k_m a_m}{a_m^2 + \omega_0^2} & \dfrac{k_m^2}{a_m^2 + \omega_0^2} \end{bmatrix} \begin{bmatrix} \phi_r \\ \phi_y \end{bmatrix} \tag{4.5.62}$$

When $\omega_0 = 5$, Figure 4.16 shows that trajectories of the output scheme exhibit an oscillatory type of response. Figure 4.17 shows the response for the input error scheme under comparable conditions.

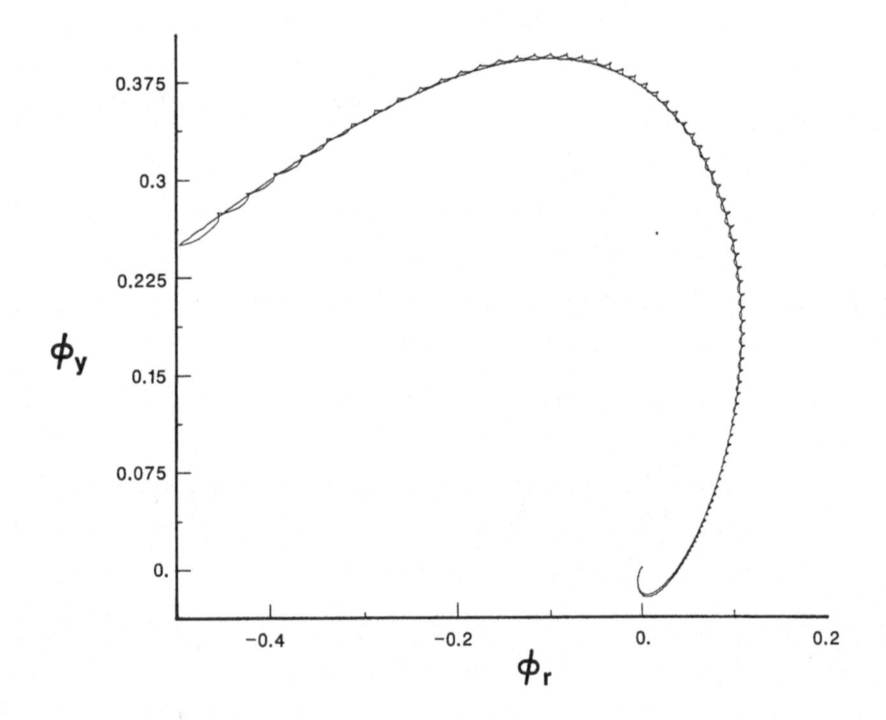

Figure 4.16: Parameter Error $\phi_y(\phi_r)$–Output Error Scheme

The parameters are $k_m = 3$, $a_m = 3$, $a_p = 1$, $k_p = 2$, $r_0 = 1$, $\omega_0 = 5$, $g = 1$, $l_1 = 10.05$, $l_2 = 10$, $\phi_r(0) = -0.5$, $\phi_y(0) = 0.25$. As may be observed, the

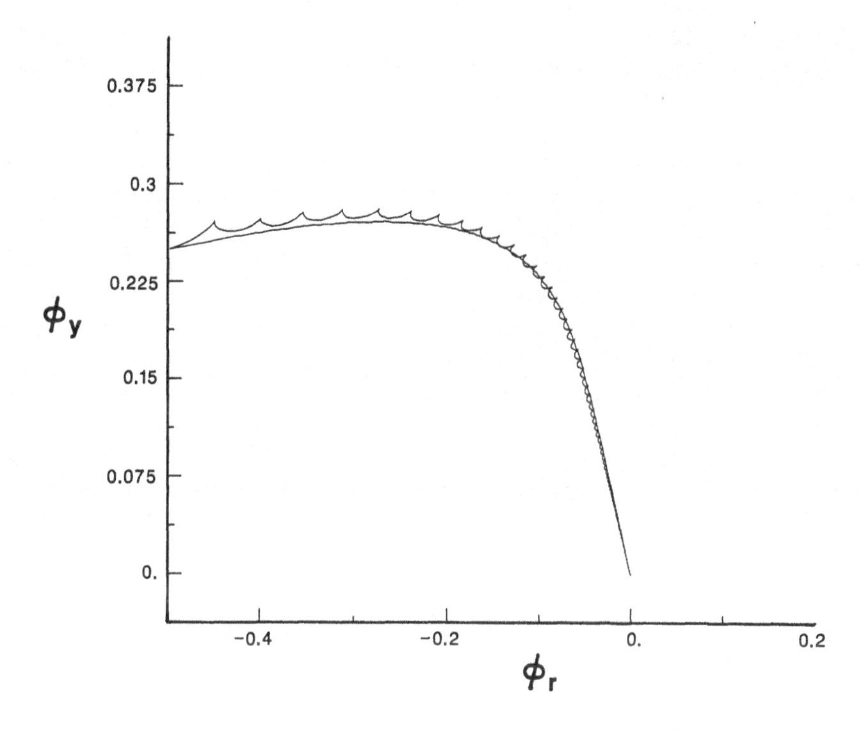

Figure 4.17: Parameter Error $\phi_y\,(\phi_r)$–Input Error Scheme

trajectories do not exhibit oscillatory behavior, reflecting the fact that the matrix above is symmetric and, therefore, has only real eigenvalues.

4.6 CONCLUSIONS

Averaging is a powerful tool to approximate nonautonomous differential equations by autonomous differential equations. In this chapter, we introduced averaging as a method of analysis of adaptive systems. The approximation of parameter convergence rates using averaging was justified by general results concerning a class of systems including the adaptive systems described in Chapters 2 and 3. The analysis had the interesting feature of considering nonlinear differential equations as well as linear ones. Therefore, the application was not restricted to linear or linearized systems, but extended to all adaptive systems considered in this work, including adaptive control systems.

The application to adaptive systems included useful parameter convergence rates estimates for identification and adaptive control systems. The rates depended strongly on the reference input and a frequency domain analysis related the frequency content of the reference input to the convergence rates, even in the nonlinear adaptive control case. These results are useful for the optimum design of reference input. They have the limitation of depending on unknown plant parameters, but an approximation of the complete parameter trajectory is obtained and the understanding of the dynamical behavior of the parameter error is considerably increased using averaging. For example, it was found that the trajectory of the parameter error corresponding to the linear error equation could be approximated by an LTI system with real negative eigenvalues, while for the strictly positive real (SPR) error equation it had possibly complex eigenvalues.

Besides requiring stationarity of input signals, averaging also required slow parameter adaptation. We showed however, through simulations, that the approximation by the averaged system was good for values of the adaptation gain that were close to 1 (that is, not necessarily infinitesimal) and for acceptable time constants in the parameter variations. In fact, it appeared that a basic condition is simply that parameters vary more slowly than do other states and signals of the adaptive system.

CHAPTER 5
ROBUSTNESS

5.1 STRUCTURED AND UNSTRUCTURED UNCERTAINTY

In a large number of control system design problems, the designer does not have a detailed state-space model of the plant to be controlled, either because it is too complex, or because its dynamics are not completely understood. Even if a detailed high-order model of the plant is available, it is usually desirable to obtain a reduced order controller, so that part of the plant dynamics must be neglected. We begin discussing the representation of such uncertainties in plant models, in a framework similar to Doyle & Stein [1981].

Consider the kind of prior information available to control a *stable* plant, and obtained for example by performing input-output experiments, such as sinusoidal inputs. Typically, Bode diagrams of the form shown in Figures 5.1 and 5.2 are obtained. An inspection of the diagrams shows that the data obtained beyond a certain frequency ω_H is unreliable because the measurements are poor, corrupted by noise, and so on. They may also correspond to the high-order dynamics that one wishes to neglect. What is available, then, is essentially no phase information, and only an "envelope" of the magnitude response beyond ω_H. The dashed lines in the magnitude and phase response correspond to the approximation of the plant by a finite order model, assuming that there are no dynamics at frequencies beyond ω_H. For frequencies below ω_H, it is easy to guess the presence of a zero near ω_1, poles in the neighborhood of ω_2, ω_3, and complex pole pairs in the neighborhood of ω_4, ω_5.

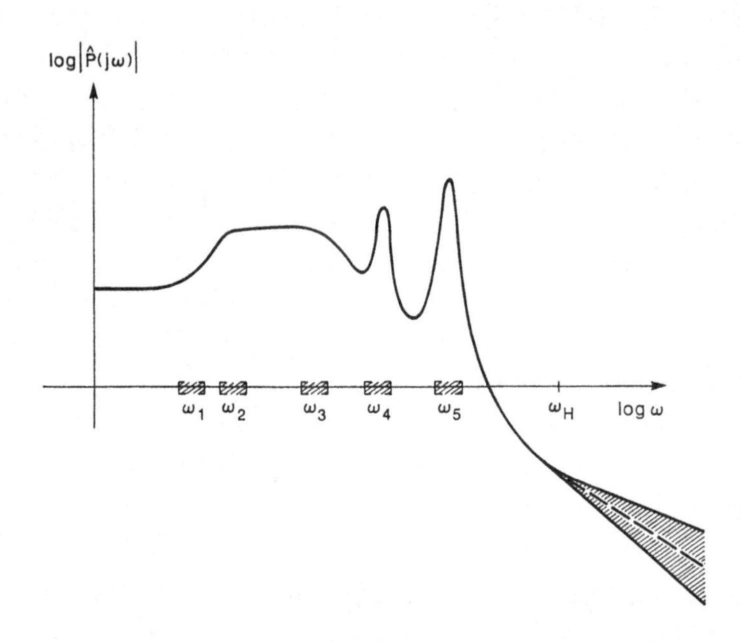

Figure 5.1: Bode Plot of the Plant (Gain)

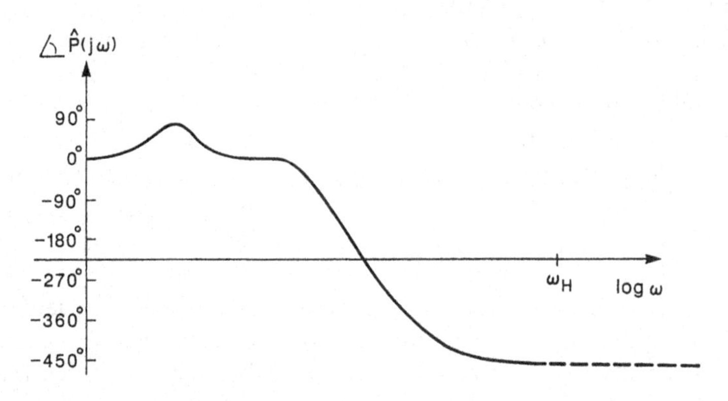

Figure 5.2: Bode Plot of the Plant (Phase)

To keep the design goal specific and consistent with our previous analysis, we will assume that the designer's goal is *model following*: the designer is furnished with a desired closed-loop response and selects an appropriate reference model with transfer function $\hat{M}(s)$. The problem is to design a control system to get the plant output $y_p(t)$ to track the model output $y_m(t)$ in response to reference signals $r(t)$ driving the model. This is shown in Figure 5.3.

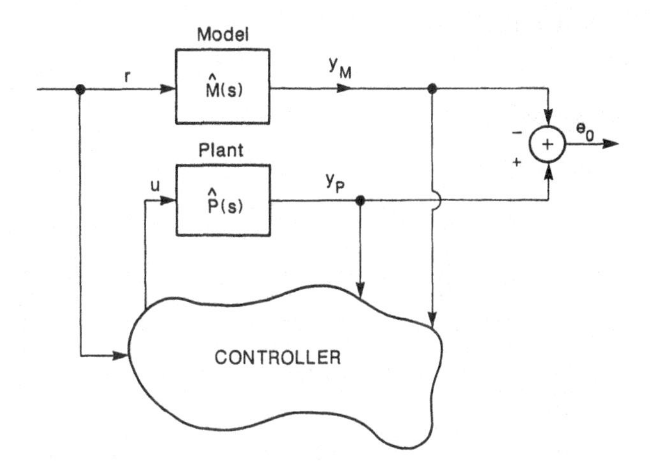

Figure 5.3: Model Following Control System

The controller generates the input $u(t)$ of the plant, using $y_m(t)$, $y_p(t)$ and $r(t)$ so that the error between the plant and model output $e_0(t) := y_p(t) - y_m(t)$ tends to zero asymptotically.

Two options are available to the designer at this point.

Non-Adaptive Robust Control. The designer uses as model for the plant the nominal transfer function $\hat{P}^*(s)$

$$\hat{P}^*(s) = \frac{k_p(s+\omega_1)}{(s+\omega_2)\,(s+\omega_3)((s+\nu_4)^2+(\omega_4)^2)((s+\nu_5)^2+(\omega_5)^2)} \tag{5.1.1}$$

The gain k_p in (5.1.1) is obtained from the nominal high-frequency asymptote of Figure 5.1 (i.e. the dashed line). The modeling errors due to inaccuracies in the pole-zero locations, and to poor data at high frequencies may be taken into account by assuming that the actual plant transfer function is of the form

$$\hat{P}(s) = \hat{P}^*(s) + \hat{H}_a(s) \tag{5.1.2}$$

or

$$\hat{P}(s) = \hat{P}^*(s)\,(1+\hat{H}_m(s)) \tag{5.1.3}$$

where $\hat{H}_a(s)$ is referred to as the *additive uncertainty* and $\hat{H}_m(s)$ as the *multiplicative uncertainty*. Of course, $|\hat{H}_a(j\omega)|$ and $|\hat{H}_m(j\omega)|$ are unknown, but magnitude bounds may be determined from input-output measurements and other available information. A typical bound for $|\hat{H}_m(j\omega)|$ is shown in Figure 5.4.

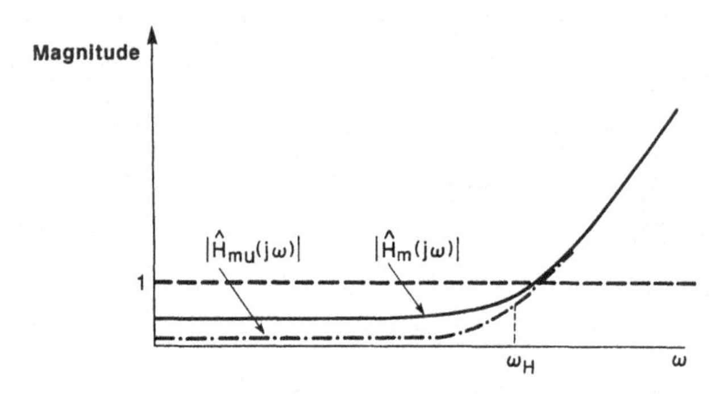

Figure 5.4: Typical Plot of Uncertainty $|\hat{H}_m(j\omega)|$ and $|\hat{H}_{mu}(j\omega)|$

Given the desired transfer function $\hat{M}(s)$, one attempts to build a *linear, time-invariant* controller of the form shown in Figure 5.5, with feedforward compensator $\hat{C}(s)$ and feedback compensator $\hat{F}(s)$, so that the nominal closed-loop transfer function approximately matches the reference model, that is,

$$\hat{P}^*(s)\hat{C}(s)\left[I + \hat{F}(s)\hat{P}^*(s)\hat{C}(s)\right]^{-1} \sim \hat{M}(s) \qquad (5.1.4)$$

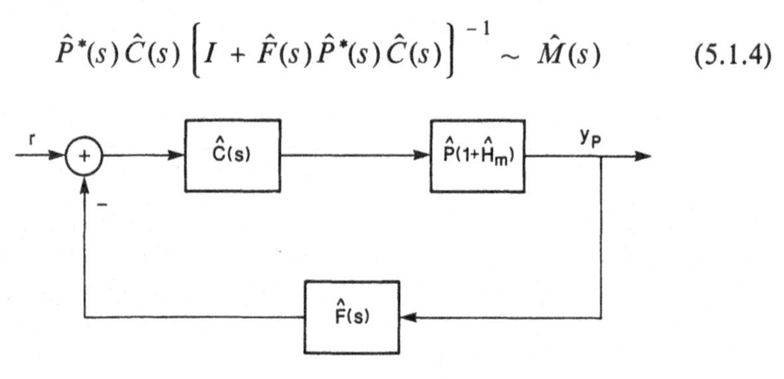

Figure 5.5: Non-adaptive Controller Structure

over the frequency range of interest (the frequency range of r). Further, $\hat{C}(s)$ and $\hat{F}(s)$ are chosen so as to at least *preserve stability* and also *reduce sensitivity* of the actual closed-loop transfer function to the modeling errors represented by $\hat{H}_a$, or $\hat{H}_m$ within some given bounds.

Adaptive Control. The designer makes a distinction between the two kinds of uncertainty present in the description of Figures 5.1–5.2: the *parametric* or *structured* uncertainty in the pole and zero locations and the *inherent* or *unstructured* uncertainty due to additional dynamics beyond ω_H. Rather than postulate a transfer function for the plant, the designer decides to identify the pole-zero locations *on-line*, i.e. during the operation of the plant. This on-line "tune-up" is for the purpose of reduction of the structured uncertainty during the course of plant operation. The aim is to obtain a better match between $\hat{M}(s)$ and the controlled plant for frequencies below ω_H. A key feature of the on-line tuning approach is that the controller is generally nonlinear and time-varying. The added complexity of adaptive control is made worthwhile when the performance achieved by non-adaptive control is inadequate.

The plant model for adaptive control is given by

$$\hat{P}(s) = \hat{P}_{\theta^*}(s) + \hat{H}_{au}(s) \tag{5.1.5}$$

or

$$\hat{P}(s) = \hat{P}_{\theta^*}(s)(1 + \hat{H}_{mu}(s)) \tag{5.1.6}$$

where $\hat{P}_{\theta^*}(s)$ stands for the plant indexed by the parameters θ^* and $\hat{H}_{au}(s)$ and $\hat{H}_{mu}(s)$ are the additive and multiplicative uncertainties respectively. The difference between (5.1.2)–(5.1.3) and (5.1.5)–(5.1.6) lies in the on-line tuning of the parameter θ^* to reduce the uncertainty, so that it only consists of the *unstructured uncertainty* due to high-frequency unmodeled dynamics.

When the plant is *unstable*, a frequency response curve as shown in Figures 5.1–5.2 is not available, and a certain amount of off-line identification and detailed modeling needs to be performed. As before, however, the plant model will have both *structured* and *unstructured* uncertainty, and the design options will be the same as above. The difference only arises in the representation of uncertainty. Consider, for example, the multiplicative uncertainty in the nonadaptive and adaptive cases. Previously, $\hat{H}_m(s)$ was stable. However, when the plant is unstable, since the nominal locations of the unstable poles may not be chosen exactly, $\hat{H}_m(s)$ may be an unstable transfer function. For adaptive control, we require merely that *all unstable poles of the system be parameterized* (of course, their exact location is not essential!), so that the description for the uncertainty is still given by (5.1.6), with $\hat{H}_{mu}(s)$ stable, even though $\hat{P}_{\theta^*}(s)$ may not be.

A simple example illustrates this: consider a plant with transfer function

$$\hat{P}(s) = \frac{m}{(s - 1 + \epsilon)(s + m)} \tag{5.1.7}$$

with $\epsilon > 0$ small and $m > 0$ large.

For non-adaptive control, the nominal plant is chosen to be $1/s - 1$, so that

$$\hat{H}_{mu}(s) = \frac{-s^2 + s - \epsilon(s + m)}{(s - 1 + \epsilon)(s + m)} \quad \text{(unstable)} \tag{5.1.8}$$

For adaptive control on the other hand, $\hat{P}_{\theta^*}(s) = 1/(s + \theta^*)$ is chosen with

$$\hat{H}_{mu}(s) = -\frac{s}{s + m} \quad \text{(stable)} \tag{5.1.9}$$

and $\theta^* = -1 + \epsilon$.

In the preceding chapters, we only considered the adaptive control of plants with parameterized uncertainty, i.e., control of $\hat{P}_{\theta^*}$. Specifically, we choose $\hat{P}_{\theta^*}$ of the form $k_p \hat{n}_p / \hat{d}_p$, where $\hat{n}_p$, $\hat{d}_p$ are monic, coprime polynomials of degrees m, n respectively. We assumed that

(a) The number of poles of $\hat{P}_{\theta^*}$, that is, n, is known.

(b) The number of zeros of $\hat{P}_{\theta^*}$, that is, $m \leq n$, is known.

(c) The sign of the high-frequency gain k_p is known (a bound may also be required) .

(d) $\hat{P}_{\theta^*}$ is minimum phase, that is, the zeros of $\hat{n}_p$ lie in the open left half plane (LHP).

It is important to note that the assumptions apply to the *nominal* plant $\hat{P}_{\theta^*}$. In particular, $\hat{P}$ may have many more stable poles and zeros than $\hat{P}_{\theta^*}$. Further, the sign of the high-frequency gain of $\hat{P}$ is usually indeterminate as shown in Figure 5.1.

The question is, of course,: how will the adaptive algorithms described in previous chapters behave with the true plant $\hat{P}$? A basic desirable property of the control algorithm is to maintain stability in the presence of uncertainties. This property is usually referred to as the *robustness* of the control algorithm.

A major difficulty in the definition of robustness is that it is very problem dependent. Clearly, an algorithm which could not tolerate *any* uncertainty (that is, no matter how small) would be called non robust. However, it would also be considered non robust in practice, if the range of tolerable uncertainties were smaller than the actual uncertainties present in the system. Similarly, an algorithm may be sufficiently robust for one application, and not for another. A key set of observations made by Rohrs, Athans, Valavani & Stein [1982, 1985] is that adaptive control algorithms which are proved stable by the techniques of previous chapters can become unstable in the presence of mild unmodeled dynamics or arbitrarily small output disturbances. We start by reviewing their examples.

5.2 THE ROHRS EXAMPLES

Despite the existence of stability proofs for adaptive control systems (cf. Chapter 3), Rohrs *et al* [1982], [1985] showed that several algorithms can become unstable when some of the assumptions required by the stability proofs are not satisfied. While Rohrs (we drop the *et al* for compactness) considered several continuous and discrete time algorithms, the results are qualitatively similar for the various schemes. We consider one of these schemes here, which is the output error direct adaptive control scheme of Section 3.3.2, assuming that the degree and the relative degree of the plant are 1.

The adaptive control scheme of Rohrs examples is designed assuming a first order plant with transfer function

$$\hat{P}_{\theta^*}(s) = \frac{k_p}{s + a_p} \tag{5.2.1}$$

and the strictly positive real (SPR) reference model

$$\hat{M}(s) = \frac{k_m}{s + a_m} = \frac{3}{s + 3} \tag{5.2.2}$$

The output error adaptive control scheme (cf. Section 3.3.2) is described by

$$u = c_0 r + d_0 y_p \tag{5.2.3}$$

$$e_0 = y_p - y_m \tag{5.2.4}$$

$$\dot{c}_0 = -g\, e_0 r \tag{5.2.5}$$

$$\dot{d}_0 = -g\, e_0 y_p \tag{5.2.6}$$

As a first step, we assume that the plant transfer function is given by (5.2.1), with $k_p = 2$, $a_p = 1$. The nominal values of the controller parameters are then

$$c_0^* = \frac{k_m}{k_p} = 1.5 \tag{5.2.7}$$

$$d_0^* = \frac{a_p - a_m}{k_p} = -1 \tag{5.2.8}$$

The behavior of the adaptive system is then studied, assuming that the *actual* plant does not satisfy exactly the assumptions on which the adaptive control system is based. The actual plant is only *approximately* a first order plant and has the third order transfer function

$$\hat{P}(s) = \frac{2}{s+1} \cdot \frac{229}{s^2 + 30s + 229} \tag{5.2.9}$$

In analogy with nonadaptive control terminology, the second term is called the *unmodeled dynamics*. The poles of the unmodeled dynamics are located at $-15 \pm j2$, and, at low frequencies, this term is approximately equal to 1.

In Rohrs examples, the measured output $y_p(t)$ is also affected by a measurement noise $n(t)$. The actual plant with the reference model and the controller are shown in Figure 5.6.

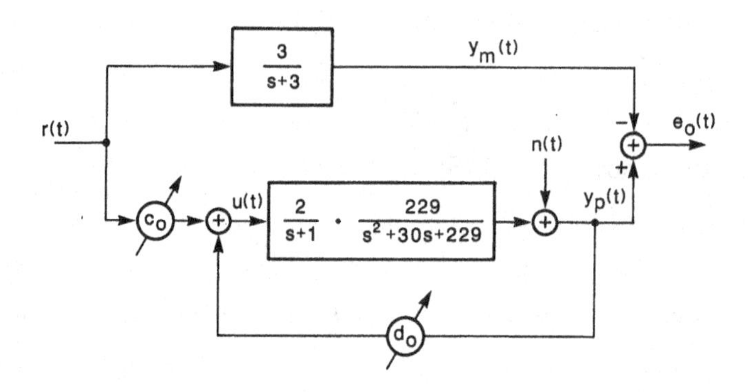

Figure 5.6: Rohrs Example—Plant, Reference Model, and Controller

An important aspect of Rohrs examples is that the modes of the actual plant and those of the model are well within the stability region. Moreover, the unmodeled dynamics are well-damped, stable modes. From a traditional control design standpoint, they would be considered rather innocuous.

At the outset, Rohrs showed through simulations that, without measurement noise or unmodeled dynamics, the adaptive scheme is stable and the output error converges to zero, as predicted by the stability analysis.

However, *with unmodeled dynamics*, three different mechanisms of instability appear:

(R1) With a *large, constant* reference input and no measurement noise, the output error initially converges to zero, but eventually diverges to infinity, along with the controller parameters c_0 and d_0.

Figures 5.7 and 5.8 show a simulation with $r(t) = 4.3$, $n(t) = 0$, that illustrates this behavior ($c_0(0) = 1.14$, $d_0(0) = -0.65$ and other initial conditions are zero).

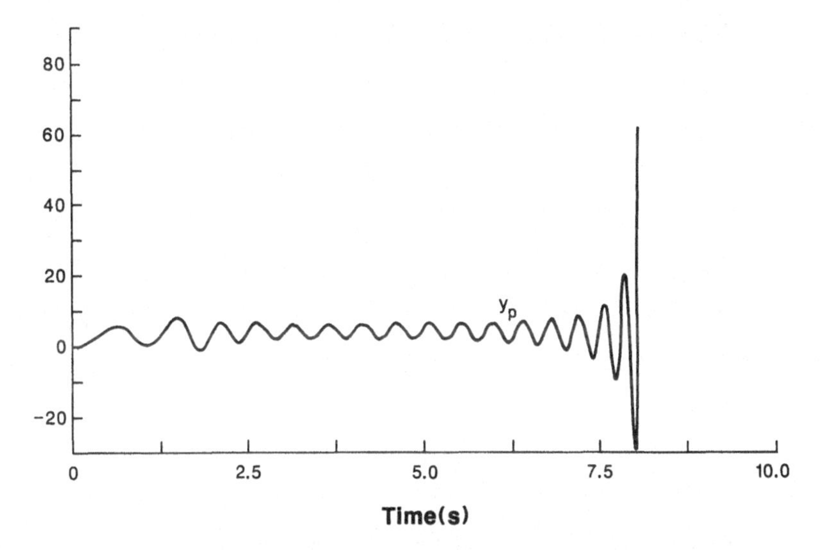

Figure 5.7 Plant Output ($r = 4.3$, $n = 0$)

(R2) With a reference input having a *small constant* component and a *large high frequency* component, the output error diverges at first slowly, and then more rapidly to infinity, along with the controller parameters c_0 and d_0.

Figures 5.9 and 5.10 show a simulation with $r(t) = 0.3 + 1.85 \sin 16.1t$, $n(t) = 0$ ($c_0(0) = 1.14$, $d_0(0) = -0.65$, and other initial conditions are zero).

(R3) With a moderate *constant input* and a small *output disturbance*, the output error initially converges to zero. After staying in the neighborhood of zero for an extended period of time, it diverges to infinity. On the other hand, the controller parameters c_0 and d_0 drift apparently at a constant rate, until they suddenly diverge to infinity.

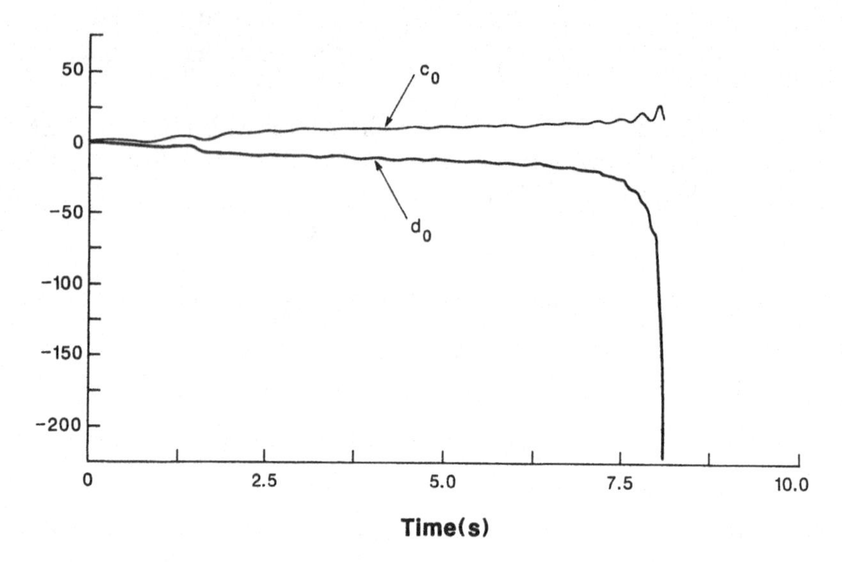

Figure 5.8 Controller Parameters ($r = 4.3$, $n = 0$)

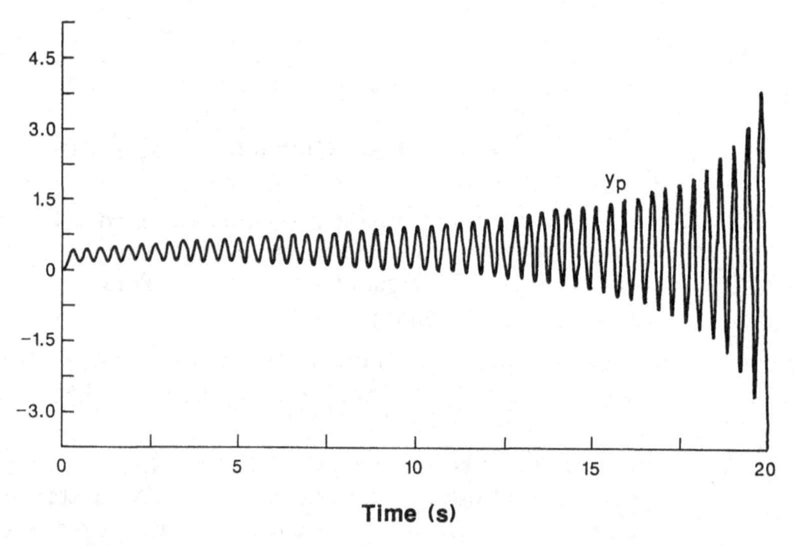

Figure 5.9 Plant Output ($r = 0.3 + 1.85 \sin 16.1t$, $n = 0$)

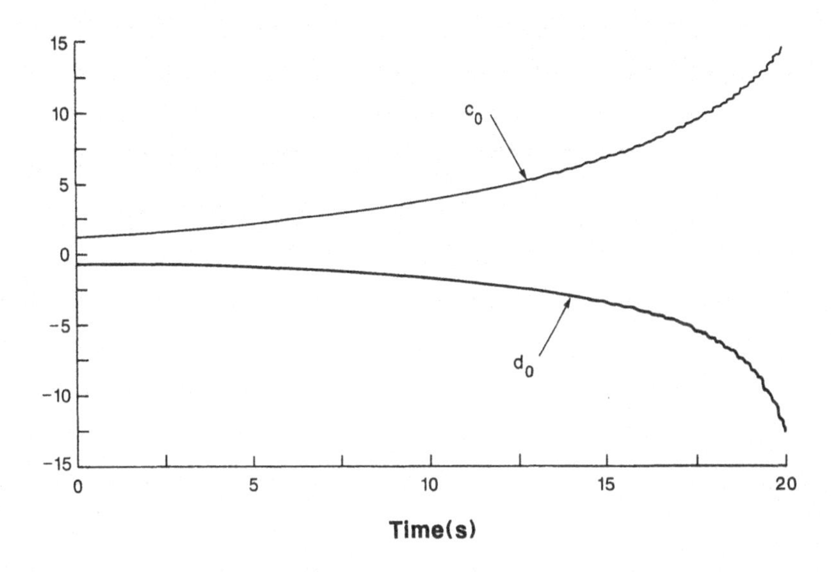

Figure 5.10 Controller Parameters ($r = 0.3 + 1.85 \sin 16.1t$, $n = 0$)

Figures 5.11 and 5.12 show a simulation with $r(t) = 2$, $n(t) = 0.5 \sin 16.1t$ ($c_0(0) = 1.14$, $d_0(0) = -0.65$, and other initial conditions are zero).

Although this simulation corresponds to a comparatively high value of $n(t)$, simulations show that when smaller values of the output disturbance $n(t)$ are present, instability still appears, but after a longer period of time. The controller parameters simply drift at a slower rate. Instability is also observed with other frequencies of the disturbance, including a constant $n(t)$.

Rohrs examples stimulated much research about the robustness of adaptive systems. Examination of the mechanisms of instability in Rohrs examples show that the instabilities are related to the identifier. In identification, such instabilities involve computed signals, while in adaptive control, variables associated with the plant are also involved. This justifies a more careful consideration of robustness issues in the context of adaptive control.

5.3 ROBUSTNESS OF ADAPTIVE ALGORITHMS WITH PERSISTENCY OF EXCITATION

Rohrs examples show that the bounded-input bounded-state (BIBS) stability property obtained in Chapter 3 is not robust to uncertainties. In

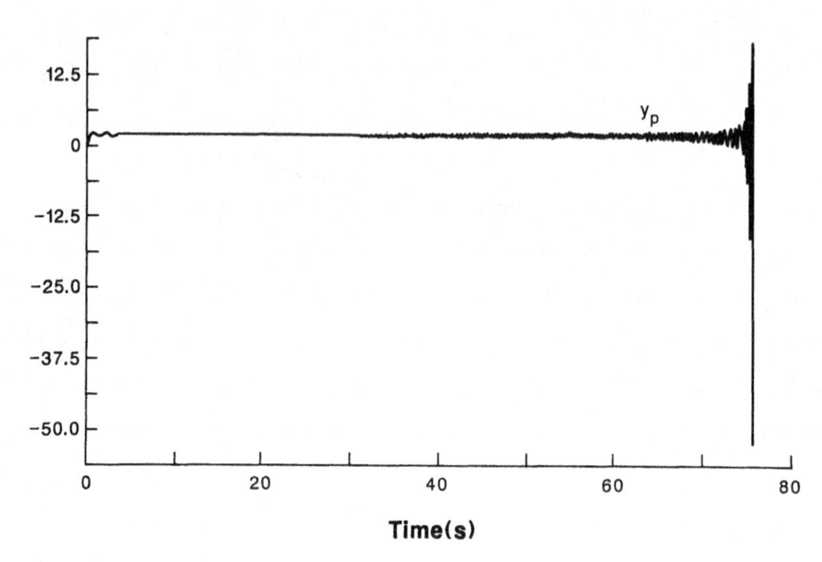

Figure 5.11 Plant Output ($r = 2$, $n = 0.5 \sin 16.1t$)

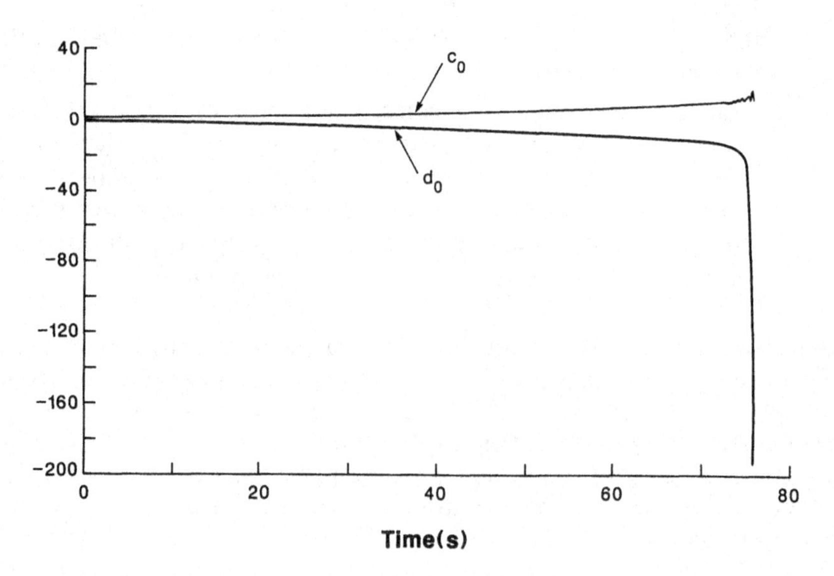

Figure 5.12 Controller Parameters ($r = 2$, $n = 0.5 \sin 16.1t$)

some cases, an arbitrary small disturbance can destabilize an adaptive system, which is otherwise proved to be BIBS stable. In this section, we will show that the property of *exponential stability* is robust, in the sense that exponentially stable systems can tolerate a certain amount of disturbance. Thus, provided that the nominal adaptive system is exponentially stable (guaranteed by a persistency of excitation (PE) condition), we will obtain robustness margins, that is, bounds on disturbances and unmodeled dynamics that do not destroy the stability of the adaptive system. Our presentation follows the lines of Bodson & Sastry [1984].

Of course, the practical notion of robustness is that stability should be preserved in the presence of actual disturbances present in the system. Robustness margins must include actual disturbances for the adaptive system to be robust in that sense. The main difference from classical linear time-invariant (LTI) control system robustness margins is that *robustness does not depend only on the plant and control system, but also on the reference input*, which must guarantee persistent excitation of the nominal adaptive system (that is, without disturbances or unmodeled dynamics).

5.3.1 Exponential Convergence and Robustness

In this section, we consider properties of a so-called *perturbed* system

$$\dot{x} = f(t, x, u) \qquad x(0) = x_0 \tag{5.3.1}$$

and relate its properties to those of the *unperturbed* system

$$\dot{x} = f(t, x, 0) \qquad x(0) = x_0 \tag{5.3.2}$$

where $t \geq 0$, $x \in \mathbb{R}^n$, $u \in \mathbb{R}^m$. Depending on the interpretation, the signal u will be considered either a disturbance or an input.

We restrict our attention to solutions x and inputs u belonging to some arbitrary balls $B_h \in \mathbb{R}^n$ and $B_c \in \mathbb{R}^m$.

Theorem 5.3.1 Small Signal I/O Stability

Consider the perturbed system (5.3.1) and the unperturbed system (5.3.2). Let $x = 0$ be an equilibrium point of (5.3.2), i.e., $f(t, 0, 0) = 0$, for all $t \geq 0$. Let f be piecewise continuous in t and have continuous and bounded first partial derivatives in x, for all $t \geq 0$, $x \in B_h$, $u \in B_c$. Let f be Lipschitz in u, with Lipschitz constant l_u, for all $t \geq 0$, $x \in B_h$, $u \in B_c$. Let $u \in L_\infty$.

If $x = 0$ is an exponentially stable equilibrium point of the unperturbed system

Then

(a) The perturbed system is *small-signal L_∞ - stable,* that is, there exist γ_∞ , $c_\infty > 0$, such that $\| u\|_\infty < c_\infty$ implies that

$$\| x\|_\infty \leq \gamma_\infty \| u\|_\infty < h \tag{5.3.3}$$

where x is the solution of (5.3.1) starting at $x_0 = 0$;

(b) There exists $m \geq 1$ such that, for all $|x_0| < h/m$, $0 < \| u\|_\infty < c_\infty$ implies that $x(t)$ converges to a B_δ ball of radius $\delta = \gamma_\infty \| u\|_\infty < h$, that is: for all $\epsilon > 0$, there exists $T \geq 0$ such that

$$|x(t)| \leq (1 + \epsilon)\delta \tag{5.3.4}$$

for all $t \geq T$, along the solutions of (5.3.1) starting at x_0. Also, for all $t \geq 0$, $|x(t)| < h$.

Comments

Part (a) of theorem 5.3.1 is a direct extension of theorem 1 of Vidyasagar & Vannelli [1982] (see also Hill & Moylan [1980]) to the non autonomous case. Part (b) further extends it to non zero initial conditions.

Theorem 5.3.1 relates *internal* exponential stability to *external* input/output stability (the output is here identified with the state). In contrast with the definition of BIBS stability of Section 3.4, we require a linear relationship between the norms in (5.3.3) for L_∞ stability.

Although lack of exponential stability does not imply input/output instability, it is known that simple stability and even (non uniform) asymptotic stability are *not* sufficient conditions to guarantee I/O stability (see e.g., Kalman & Bertram [1960], Ex. 5, p. 379).

Proof of Theorem 5.3.1

The differential equation (5.3.2) satisfies the conditions of theorem 1.5.1, so that there exists a Lyapunov function $v(t, x)$ satisfying the following inequalities

$$\alpha_1 |x|^2 \leq v(t,x) \leq \alpha_2 |x|^2 \tag{5.3.5}$$

$$\left.\frac{dv(t,x)}{dt}\right|_{(5.3.2)} \leq -\alpha_3 |x|^2 \tag{5.3.6}$$

$$\left|\frac{\partial v(t,x)}{\partial x}\right| \leq \alpha_4 |x| \tag{5.3.7}$$

for some strictly positive constants $\alpha_1 \cdots \alpha_4$, and for all $t \geq 0$, $x \in B_h$.

If we consider the same function to study the perturbed differential equation (5.3.1), inequalities (5.3.5) and (5.3.7) still hold, while (5.3.6) is modified, since the derivative is now to be taken along the trajectories of (5.3.1) instead of (5.3.2). The two derivatives are related through

$$\left.\frac{dv(t,x)}{dt}\right|_{(5.3.1)} = \frac{\partial v(t,x)}{\partial t} + \sum_{i=1}^{n} \frac{\partial v(t,x)}{\partial x_i} f_i(t,x,u)$$

$$= \left.\frac{dv(t,x)}{dt}\right|_{(5.3.2)} + \sum_{i=1}^{n} \frac{\partial v(t,x)}{\partial x_i} \cdot \left[f_i(t,x,u) - f_i(t,x,0) \right] \tag{5.3.8}$$

Using (5.3.5)–(5.3.7), and the Lipschitz condition on f

$$\left.\frac{dv(t,x)}{dt}\right|_{(5.3.1)} \leq -\alpha_3 |x|^2 + \alpha_4 |x| \; l_u \| u \|_\infty \tag{5.3.9}$$

Define

$$\gamma_\infty := \frac{\alpha_4}{\alpha_3} l_u \left[\frac{\alpha_2}{\alpha_1} \right]^{\frac{1}{2}} \tag{5.3.10}$$

$$\delta := \gamma_\infty \| u \|_\infty \tag{5.3.11}$$

$$m := \left[\frac{\alpha_2}{\alpha_1} \right]^{\frac{1}{2}} \geq 1 \tag{5.3.12}$$

Inequality (5.3.9) can now be written

$$\left.\frac{dv(t,x)}{dt}\right|_{(5.3.1)} \leq -\alpha_3 |x| \left[|x| - \frac{\delta}{m} \right] \tag{5.3.13}$$

This inequality is the basis of the proof.

Part (a) Consider the situation when $|x_0| \leq \delta / m$ (this is true in particular if $x_0 = 0$). We show that this implies that $x(t) \in B_\delta$ for all $t \geq 0$ (note that $\delta / m \leq \delta$, since $m \geq 1$).

Suppose, for the sake of contradiction, that it were not true. Then, by continuity of the solutions, there would exist T_0, $T_1 (T_1 > T_0 \geq 0)$, such that

$$|x(T_0)| = \delta / m \quad \text{and} \quad |x(T_1)| > \delta$$

and for all $t \in [T_0, T_1] : |x(t)| \geq \delta / m$. Consequently, inequality

(5.3.13) shows that, in $[T_0, T_1]$, $\dot{v} \le 0$. However, this contradicts the fact that

$$v(T_0, x(T_0)) \le \alpha_2 (\delta / m)^2 = \alpha_1 \delta^2$$

and

$$v(T_1, x(T_1)) > \alpha_1 \delta^2$$

Part (b) Assume now that $|x_0| > \delta / m$. We show the result in two steps.
(b1) for all $\epsilon > 0$, there exists $T \ge 0$ such that $|x(T)| \le (\delta / m)(1 + \epsilon)$.
Suppose it was not true. Then, for some $\epsilon > 0$ and for all $t \ge 0$

$$|x(t)| > (\delta / m)(1 + \epsilon)$$

and, from (5.3.13)

$$\dot{v} < -\alpha_3 (\delta / m)^2 (1 + \epsilon) \epsilon$$

which is a strictly negative constant. However, this contradicts the fact that

$$v(0, x_0) \le \alpha_2 |x_0|^2 < \alpha_2 \frac{h^2}{m^2}$$

and $v(t, x(t)) \ge 0$ for all $t \ge 0$. Note that an upper bound on T is

$$T \le \frac{\alpha_2 h^2}{\alpha_3 \delta^2 \epsilon}$$

(b2) for all $t \ge T$, $|x(t)| \le \delta(1 + \epsilon)$. This follows directly from (b1), using an argument identical to the one used to prove (a).

Finally, recall that the assumptions require that $x(t) \in B_h$, $u(t) \in B_c$, for all $t \ge 0$. This is also guaranteed, using an argument similar to (a), provided that $|x_0| < h/m$ and $\| u \|_\infty < c_\infty$, where m is defined in (5.3.12), and

$$c_\infty := \min(c, h / \gamma_\infty) \tag{5.3.14}$$

(5.3.14) implies that $\delta < h$, and $|x_0| < h/m \le h$ implies that $|x(t)| \le m |x_0| < h$ for all $t \ge 0$.

Note that although part (a) of the proof is, in itself, a result for non zero initial conditions, the size of the ball $B_{\delta / m}$ involved decreases when the amplitude of the input decreases, while the size of $B_{h / m}$ is independent of it. $\square$

Additional Comments

a) The proof of the theorem gives an interesting interpretation of the interaction between the exponential convergence of the original system and the effect of the disturbances on the perturbed system. To see this, consider (5.3.9): the term $-\alpha_3 |x|^2$ acts like a restoring force bringing the state vector back to the origin. This term originates from the exponential stability of the unperturbed system. The term $\alpha_4 |x| \, l_u \| u \|_\infty$ acts like a disturbing force, pulling the state *away* from the origin. This term is caused by the input u (i.e. by the disturbance acting on the system). While the first term is proportional to the norm squared, the second is only proportional to the norm, so that when $|x|$ is sufficiently large, the restoring force equilibrates the disturbing force. In the form (5.3.13), we see that this happens when $|x| = \delta / m = \gamma_\infty / m \| u \|_\infty$.

b) If the assumptions are valid *globally,* then the results are valid globally too. The system remains stable and has finite I/O gain, independent of the size of the input. In the example of Section 5.3.2, and for a wide category of nonlinear systems (bilinear systems for example), the Lipschitz condition is not verified globally. Yet, given *any* balls B_h, B_c, the system satisfies a Lipschitz condition with constant l_u depending on the size of the balls (actually increasing with it). The balls B_h, B_c are consequently arbitrary in that case, but the values of γ_∞ (the L_∞ gain) and c_∞ (the stability margin) will vary with them. In general, it can be expected that c_∞ will remain bounded despite the freedom left in the choice of h and c, so that the I/O stability will only be local.

c) *Explicit* values of γ_∞ and c_∞ can be obtained from parameters of the differential equation, using equations (5.3.10) and (5.3.14). Note that if we used the Lyapunov function satisfying (5.3.5)–(5.3.7) to obtain a convergence rate for the unperturbed system, this rate would be $\alpha_3 / 2\alpha_1$. Therefore, it can be verified that, with other parameters remaining identical, the L_∞ gain is decreased and the stability margin c_∞ is increased, *when the rate of exponential convergence is increased.*

5.3.2 Robustness of an Adaptive Control Scheme

For the purpose of illustration, we consider the output error direct adaptive control algorithm of Section 3.3.2, when the relative degree of the plant is 1. This example contains the specific cases of the Rohrs examples.

In Section 3.5, we showed that the overall output error adaptive scheme for the relative degree 1 case is described by (cf. (3.5.28))

$$\dot{e}(t) = A_m e(t) + b_m \phi^T(t) w_m(t) + b_m \phi^T(t) Q\, e(t)$$

$$\dot{\phi}(t) = -g c_m^T e(t) w_m(t) - g c_m^T e(t) Q\, e(t) \qquad (5.3.15)$$

where $e(t) \in \mathbb{R}^{3n-2}$, and $\phi(t) \in \mathbb{R}^{2n}$. A_m is a stable matrix, and $w_m(t) \in \mathbb{R}^{2n}$ is bounded for all $t \geq 0$. (5.3.15) is a nonlinear ordinary differential equation (actually it is bilinear) of the form

$$\dot{x} = f(t, x) \qquad x(0) = x_0 \qquad (5.3.16)$$

which is of the form (5.3.2), where

$$x := \begin{bmatrix} e \\ \phi \end{bmatrix} \in \mathbb{R}^{5n-2} \qquad (5.3.17)$$

Recall that we also found, in Section 3.8, that (5.3.15) (i.e. (5.3.16)) is exponentially stable in any closed ball, provided that w_m is PE.

Robustness to Output Disturbances

Consider the case when the measured output is affected by a measurement noise $n(t)$, as in Figure 5.6. Denote by y_p^* the output of the plant $\hat{P}_{\theta^*}(s)$ (that is the output without measurement noise) and by $y_p(t)$, the measured output, affected by noise, so that

$$y_p(t) = y_p^*(t) + n(t) = \hat{P}_{\theta^*}(u) + n(t) \qquad (5.3.18)$$

To find a description of the adaptive system in the presence of the measurement noise $n(t)$, we return to the derivation of (5.3.15) (that is (3.5.28)) in Section 3.5. The plant $\hat{P}_{\theta^*}$ has a minimal state-space representation $[A_p, b_p, c_p^T]$ such that

$$\dot{x}_p = A_p x_p + b_p u$$

$$y_p^* = c_p^T x_p \qquad (5.3.19)$$

The observers are described by

$$\dot{w}^{(1)} = \Lambda w^{(1)} + b_\lambda u$$

$$\dot{w}^{(2)} = \Lambda w^{(2)} + b_\lambda y_p$$

$$= \Lambda w^{(2)} + b_\lambda c_p^T x_p + b_\lambda n \qquad (5.3.20)$$

and the control input is given by $u = \theta^T w = \phi^T w + \theta^{*^T} w$.

As previously, we let $x_{pw}^T = (x_p^T, w^{(1)^T}, w^{(2)^T})$. Using the definition of A_m, b_m and c_m in (3.5.18)–(3.5.19), the description of the plant with controller is now

$$\begin{aligned} \dot{x}_{pw} &= A_m x_{pw} + b_m \phi^T w + b_m c_0^* r + b_n n \\ y_p^* &= c_m^T x_{pw} \end{aligned} \tag{5.3.21}$$

where we define $b_n^T = (0, 0, b_\lambda^T) \in (\mathbb{R}^n, \mathbb{R}^{n-1}, \mathbb{R}^{n-1}) = \mathbb{R}^{3n-2}$.

As previously, we represent the model and its output by

$$\begin{aligned} \dot{x}_m &= A_m x_m + b_m c_0^* r \\ y_m &= c_m^T x_m \end{aligned} \tag{5.3.22}$$

and we let $e = x_{pw} - x_m$.

The update law is given by

$$\begin{aligned} \dot{\phi} &= -g(y_p - y_m) w \\ &= -g c_m^T e w - g n w \end{aligned} \tag{5.3.23}$$

and the regressor is now related to the state e by

$$\begin{aligned} w &= \begin{bmatrix} r \\ w^{(1)} \\ y_p \\ w^{(2)} \end{bmatrix} = w_m + \begin{bmatrix} 0 \\ w^{(1)} - w_m^{(1)} \\ y_p^* - y_m \\ w^{(2)} - w_m^{(2)} \end{bmatrix} + \begin{bmatrix} 0 \\ 0 \\ n \\ 0 \end{bmatrix} \\ &= w_m + Qe + q_n n \end{aligned} \tag{5.3.24}$$

where we define $q_n^T = (0, 0, 1, 0) \in (\mathbb{R}, \mathbb{R}^{n-1}, \mathbb{R}, \mathbb{R}^{n+1}) = \mathbb{R}^{2n}$.

Using these results, the adaptive system with measurement noise is described by

$$\begin{aligned} \dot{e}(t) &= A_m e(t) + b_m \phi^T(t) w_m(t) + b_m \phi^T(t) Q e(t) \\ &\quad + b_m \phi^T(t) q_n n(t) + b_n n(t) \\ \dot{\phi}(t) &= -g c_m^T e(t) w_m(t) - g c_m^T e(t) Q e(t) - g c_m^T e(t) q_n n(t) \\ &\quad - g n(t) w_m(t) - g n(t) Q e(t) - g n^2(t) q_n \end{aligned} \tag{5.3.25}$$

which, with the definition of x in (5.3.17) and the definition of f in (5.3.15)–(5.3.16) can be written

$$\dot{x} = f(t, x) + p_1(t) + P_2(t)x(t) \tag{5.3.26}$$

where $p_1(t) \in \mathbb{R}^{5n-2}$ and $P_2(t) \in \mathbb{R}^{5n-2 \times 5n-2}$ are given by

$$p_1(t) = \begin{bmatrix} b_n\, n(t) \\ -gn(t)\, w_m(t) - gn^2(t)\, q_n \end{bmatrix}$$

$$P_2(t) = \begin{bmatrix} 0 & b_m\, n(t)\, q_n^T \\ -gn(t)\, q_n\, c_m^T - gn(t)\, Q & 0 \end{bmatrix} \tag{5.3.27}$$

Note that if $n \in L_\infty$, then p_1 and $P_2 \in L_\infty$. Therefore, the perturbed system (5.3.26) is a special form of system (5.3.1), where u contains the components of p_1 and P_2. Although $p_1(t)$ depends quadratically on n, given a bound on n, there exists $k_n \geq 0$ such that

$$\| p_1 \|_\infty + \| P_2 \|_\infty \leq k_n \| n \|_\infty \tag{5.3.28}$$

From these derivations, we deduce the following theorem.

Theorem 5.3.2 Robustness to Disturbances

Consider the output error direct adaptive control scheme of Section 3.2.2, assuming that the relative degree of the plant is 1. Assume that the measured output y_p of the plant is given by (5.3.18), where $n \in L_\infty$. Let $h > 0$.

If w_m is PE

Then there exists γ_n, $c_n > 0$ and $m \geq 1$, such that $\| n \|_\infty < c_n$ and $|x(0)| < h/m$ implies that $x(t)$ converges to a B_δ ball of radius $\delta = \gamma_n \| n \|_\infty$, with $|x(t)| \leq m\, |x_0| < h$ for all $t \geq 0$.

Proof of Theorem 5.3.2

Since w_m is PE, the unperturbed system (5.3.15) (i.e. (5.3.16)) is exponentially stable in any B_h by theorem 3.8.2. The perturbed system (5.3.25) (i.e. (5.3.26)) is a special case of the general form (5.3.1), so that theorem 5.3.1 can be applied with u containing the components of $p_1(t), P_2(t)$. The results on $p_1(t), P_2(t)$ can be translated into similar results involving $n(t)$, using (5.3.28). □

Comments

a) A specific bound c_n on $\| n \|_\infty$ can be obtained such that, within this bound, and provided the initial error is sufficiently small, *the stability of the adaptive system will be preserved.* For this reason, c_n is called a robustness margin of the adaptive system to output disturbances.

b) The deviations from equilibrium are locally *at most proportional* to the disturbances (in terms of L_∞ norms), and their bounds can be made arbitrarily small by reducing the bounds on the disturbances.

c) The L_∞ gain from the disturbances to the deviations from equilibrium can be reduced by *increasing the rate of exponential convergence of the unperturbed system* (provided that other constants remain identical).

d) Rohrs example (R3) of instability of an adaptive scheme with output disturbances on a non persistently excited system, is an example of instability when the persistency of excitation condition of the nominal system is not satisfied.

Robustness to Unmodeled Dynamics

We assume again that there exists a nominal plant $\hat{P}_{\theta^*}(s)$, satisfying the assumptions on which the adaptive control scheme is based, and we define the *output of the nominal plant* to be

$$y_p^* = \hat{P}_{\theta^*}(u) \tag{5.3.29}$$

The actual output is modeled as the output of the nominal plant, plus some additive uncertainty represented by a bounded operator H_a

$$y_p(t) = y_p^*(t) + H_a(u)(t) \tag{5.3.30}$$

The operator H_a represents the difference between the real plant, and the idealized plant $\hat{P}(s)$. We refer to it as an *additive unstructured uncertainty,* and it constitutes all the uncertainty, since it is the purpose of the adaptive scheme to reduce to zero the *structured* or *parametric* uncertainty.

We assume that $H_a : L_{\infty e} \to L_{\infty e}$ is a causal operator satisfying

$$\| H_a(u)_t \|_\infty \le \gamma_a \| u_t \|_\infty + \beta_a \tag{5.3.31}$$

for all $t \ge 0$. β_a may include the effect of initial conditions in the unmodeled dynamics and the possible presence of bounded output disturbances.

The following theorem guarantees the stability of the adaptive system in the presence of unmodeled dynamics satisfying (5.3.31).

Theorem 5.3.3 Robustness to Unmodeled Dynamics

Consider the output error direct adaptive control scheme of Section 3.3.2, assuming that the relative degree of the plant is 1. Assume that the nominal plant output and actual measured plant output satisfy (5.3.29)–(5.3.30), where $\hat{P}_{\theta^*}$ satisfies the assumptions of Section 3.3.2. H_a satisfies (5.3.31) and is such that trajectories of the adaptive system are continuous with respect to t.

If w_m is PE

Then for x_0, γ_a, β_a sufficiently small, the state trajectories of the adaptive system remain bounded.

Proof of Theorem 5.3.3

Let $T > 0$ such that $x(t) \leq h$ for all $t \in [0, T]$. Define $n = H_a(u)$, so that, by assumption

$$\| n_t \|_\infty \leq \gamma_a \| u_t \|_\infty + \beta_a \tag{5.3.32}$$

for all $t \in [0, T]$. Using (5.3.24), the input u is given by

$$\begin{aligned} u &= \theta^T w = \theta^{*^T} w + \phi^T w \\ &= \theta^{*^T} w_m + \theta^{*^T} Qe + \theta^{*^T} q_n n + \phi^T w_m + \phi^T Qe + + \phi^T q_n n \end{aligned} \tag{5.3.33}$$

Since $x \in B_h$, there exist γ_u, $\beta_u \geq 0$ such that

$$\| u_t \|_\infty \leq \gamma_u \| n_t \|_\infty + \beta_u \tag{5.3.34}$$

for all $t \in [0, T]$. Let γ_a, β_a sufficiently small that

$$\gamma_a \gamma_u < 1 \tag{5.3.35}$$

$$\frac{\beta_a + \gamma_a \beta_u}{1 - \gamma_a \gamma_u} < c_n \tag{5.3.36}$$

where c_n is the constant found in theorem 5.3.2. Applying the small gain theorem (lemma 3.6.6), and using (5.3.32), (5.3.35) and (5.3.36), it follows that $\| n_t \|_\infty < c_n$. By theorem 5.3.2, this implies that $|x(t)| < h$ for all $t \in [0, T]$. Since none of the constants γ_a, β_a, γ_n and β_n is dependent on T, $|x(t)| < h$ for all $t \geq 0$. Indeed, suppose it was not true. Then, by continuity of the solutions, there would exist a $T > 0$ such that $|x(t)| \leq h$ for all $t \in [0, T]$, and $x(T) = h$. The theorem would then apply, resulting in a contradiction since $|x(T)| < h$. □

Comments

Condition (5.3.24) is very general, since it includes possible nonlinearities, unmodeled dynamics, and so on, provided that they can be represented by additive, bounded-input bounded-output operators.

If the operator H_a is linear time invariant, the stability condition is a condition on the L_∞ gain of H_a. One can use

$$\gamma_a = \| h_a \|_1 = \int_0^\infty | h_a(\tau) | \, d\tau \tag{5.3.37}$$

where $h_a(\tau)$ is the impulse response of $\hat{H}_a$. The constant β_a depends on the initial conditions in the unmodeled dynamics.

The proof of theorem 5.3.3 gives some margins of unmodeled dynamics that can be tolerated without loss of stability of the adaptive system. Given γ_a, β_a it is actually possible to compute these values. The most difficult parameter to determine is possibly the rate of convergence of the unperturbed system, but we saw in Chapter 4 how some estimate could be obtained, under the conditions of averaging. Needless to say the expression for these robustness margins depends in a complex way on unknown parameters, and it is likely that the estimates would be conservative. The importance of the result is to show that if the unperturbed system is persistently excited, it will tolerate *some* amount of disturbance, or conversely that an arbitrary small disturbance *cannot* destabilize the system, such as in example (R3).

5.4 HEURISTIC ANALYSIS OF THE ROHRS EXAMPLES

By considering the overall adaptive system, including the plant states, observer states, and the adaptive parameters, we showed in Section 5.3 the importance of the exponential convergence to guarantee some robustness of the adaptive system. This convergence depends especially on the *parameter* convergence, and therefore on conditions on the input signal $r(t)$.

A heuristic analysis of the Rohrs examples gives additional insight into the mechanisms leading to instability, and suggest practical methods to improve robustness. Such an analysis can be found in Astrom [1983], and its success relies mainly on the separation of time scales between the evolution of the plant/observer states, and the evolution of the adaptive parameters. This separation of time scales is especially suited for the application of averaging methods (cf. Chapter 4).

Following Astrom [1983], we will show that instability in the Rohrs examples are due to one or more of the following factors

a) the lack of sufficiently rich inputs to

- allow for parameter convergence in the nominal system,
- prevent the drift of the parameters due to unmodeled dynamics or output disturbances.

b) the presence of significant excitation at high frequencies, originating either from the reference input, or from output disturbances. These signals cause the adaptive loop to try to get the plant loop to match the model at high frequencies, resulting in a closed-loop unstable plant.

c) a large reference input with a non-normalized identification (adaptation) algorithm and unmodeled dynamics, resulting in the instability of the identification algorithm.

Analysis

Consider now the mechanisms of instability corresponding to these three cases.

a) Consider first the case when the input is not sufficiently rich (example (R3)).

In the nominal case, the output error tends to zero. When the PE condition is not satisfied, the controller parameter does not necessarily converge to its nominal value, but to a value such that the closed-loop transfer function matches the model transfer function at the frequencies of the reference input. Consider Rohrs example, without unmodeled dynamics. The closed-loop transfer function from $r \rightarrow y_p$, assuming that c_0 and d_0 are fixed, is

$$\frac{\hat{y}_p}{\hat{r}} = \frac{2c_0}{s+1-2d_0} \tag{5.4.1}$$

If a constant reference input is used, only the DC gain of this transfer function must be matched with the DC gain of the reference model. This implies the condition that

$$\frac{2c_0}{1-2d_0} = 1 \tag{5.4.2}$$

Any value of c_0, d_0 satisfying (5.4.2) will lead to $y_p - y_m \rightarrow 0$ as $t \rightarrow \infty$ for a constant reference input. Conversely, when $e_0 \rightarrow 0$, so do $\dot{c}_0$, and $\dot{d}_0$, so that the assumption that c_0, d_0 are fixed is justified.

If an output disturbance $n(t)$ enters the adaptive system, it can cause the parameters c_0, d_0 to move along the line (more generally the surface) defined by (5.4.2), leaving $e_0 = y_p - y_m$ at zero. In particular, note that when output disturbances are present, the actual update law for d_0 is not (5.2.6) anymore, but

$$\dot{d}_0 = -gy_p^*(y_p^* - y_m) - gy_m n - gn^2 \tag{5.4.3}$$

where we find the presence of the term $-gn^2$, which will tend to make d_0 slowly drift toward the negative direction.

In example (R3), unmodeled dynamics are present, so that the transfer function from $r \rightarrow y_p$ is in fact given by

$$\frac{\hat{y}_p}{\hat{r}} = \frac{458c_0}{(s+1)(s^2+30s+229) - 458d_0} \tag{5.4.4}$$

which is identical to (5.4.1) for DC signals, but which is unstable for $d_0 \geq 0.5$ and $d_0 \leq -17.03$.

The result is observed in Figures 5.11 and 5.12, where d_0 slowly drifts in the negative direction, until it reaches the limit of stability of the closed-loop plant with unmodeled dynamics. This instability is called the *slow drift instability.* The error converges to a neighborhood of zero, and the signal available for parameter update is very small and unreliable, since it is indistinguishable from the output noise $n(t)$. It is the accumulation of updates based on incorrect information that leads to parameter drift, and eventually to instability.

In terms of the discussion of Section 5.3, we see that the constant disturbance $-gn^2$ is not counteracted by any restoring force, as would be the case if the original system was exponentially stable. For example, consider the case where $n = 0.1 \sin 16.1\, t$. Figure 5.13 shows the evolution of the parameter d_0 in a simulation where $r(t) = 2$ and where $r(t) = 2 \sin t$. In the first case, the parameter slowly drifts, leading eventually to instability. When $r(t) = 2 \sin t$, so that PE conditions are satisfied, the parameter d_0 deviates from d_0^* but remains close to the nominal value.

Finally, note that instabilities of this type can be obtained for systems of relative degree greater than two even without unmodeled dynamics and can lead to the so-called *bursting phenomenon* (cf. Anderson [1985]). The presence of noise in the update law leads to drift in the feedback coefficients to a region where they are large, resulting in a closed loop unstable system and a large increase in e_0. The output error e_0 eventually converges back to zero, but a large 'blip' is observed. This repeats at random instants, and is referred to as bursting. As before a safeguard, against bursting is persistent excitation.

b) Consider now the case when the reference input, or the output disturbance, contain a large component at a frequency where unmodeled dynamics are significant (example (R2)).

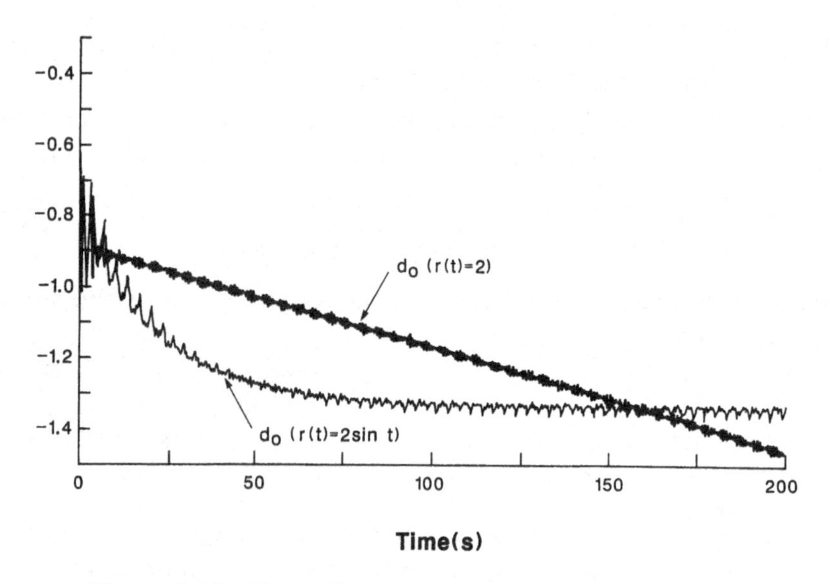

Figure 5.13 Controller Parameter $d_0(n = 0.1 \sin 16.1t)$

Let us return to Rohrs example, with a sinusoidal reference input $r(t) = r_0 \sin(\omega_0 t)$. With unmodeled dynamics, there are still *unique* values of c_0, d_0 such that the transfer function from $r \to y_p$ matches $\hat{M}$ at the frequency of the reference input ω_0. Without unmodeled dynamics, these would be the nominal c_0^*, d_0^*, but now they are the values c_0^+, d_0^+ given by

$$\left. \frac{458 c_0^+}{(s+1)(s^2+30s+229) - 458 d_0^+} \right|_{j\omega_0} = \left. \frac{3}{s+3} \right|_{j\omega_0} \tag{5.4.5}$$

where ω_0 is the frequency of the reference input. Note that the values of c_0^+, d_0^+ depend on $\hat{M}$, $\hat{P}$, the unmodeled dynamics, and also *on the reference input r*.

On the other hand, it may be verified through simulations that the output error tends to zero and that the controller parameters converge to the following values $c_{0_{ss}}$ and $d_{0_{ss}}$ (cf. Astrom [1983]).

ω_0	$c_{0_{ss}}$	$d_{0_{ss}}$
1	1.69	-1.26
2	1.67	-1.44
5	1.53	-2.72
10	1.04	-7.31

It may be verified that these values are identical to c_0^+, d_0^+ defined earlier. Therefore, the adaptive control system updates the parameters, trying to match the closed-loop transfer function—including the unmodeled dynamics—to the model reference transfer function. Note that the parameter $d_{0_{ss}} = d_0^+$ quickly decreases for $\omega_0 > 5$. On the other hand, the closed-loop system is unstable when $d_0 \leq -17.03$ and $d_0^+ \leq -17.03$, when $\omega_0 \geq 16.09$. Therefore, by attempting to match the reference model at a high frequency, the adaptive system leads to an unstable closed-loop system, and thereby to an unstable overall system.

This is the instability observed in example (R2). In contrast, Figure 5.14 shows a simulation where $r = 0.3 + 1.85 \sin t$, that is where the sinusoidal component of the input is at a frequency where model matching is possible.

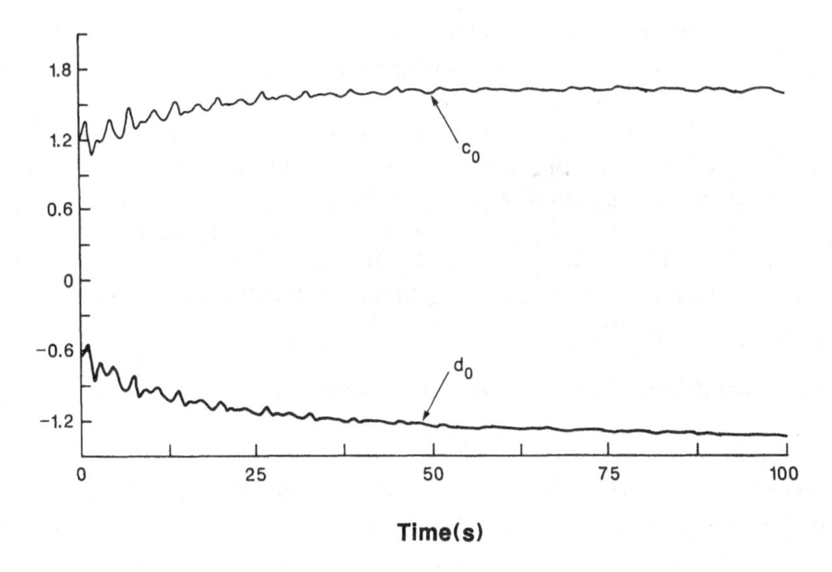

Figure 5.14 Controller Parameters ($r = 0.3 + 1.85 \sin t$, $n = 0$)

Then, the parameters converge to values c_0^+, d_0^+ close to c_0^*, d_0^*, and the adaptive system remains stable, despite the unmodeled dynamics.

c) Consider finally the mechanism of instability observed with a large reference input (example (R1)).

This mechanism will be called the *high-gain identifier instability.* Although we do not have explicitly a high adaptation gain g, recall that the adaptation law is given by

$$\dot{c}_0 = -g\, e_0\, r \tag{5.4.6}$$

$$\dot{d}_0 = -g\, e_0\, y_p \tag{5.4.7}$$

Roughly speaking, multiplying r by 2 means multiplying y_m, y_p and e_0 by 2 and therefore is equivalent to multiplying the adaptation gain by 4.

The instabilities obtained for high values of the adaptation gain are comparable to instabilities caused by high gain feedback in LTI systems with relative degree greater than 2 (cf. Astrom [1983] for a simple root-locus argument). A simple fix to these problems is to replace the identification algorithm by a *normalized* algorithm.

5.5 AVERAGING ANALYSIS OF SLOW DRIFT INSTABILITY

As was pointed out in Section 5.4, Astrom [1983] introduced an analysis of instability based on slow parameter adaptation, to separate the evolution of the plant/observer states and the adaptive parameters. The phenomenon under study is the so-called *slow drift instability* and is caused by either a lack of sufficiently rich inputs, or the presence of significant excitation at high frequencies, originating either from the reference input or output disturbances.

A heuristic analysis of this phenomenon was already given in the preceding section. In this section, we make the analysis more rigorous using the averaging framework of Chapter 4. In Section 5.5.1, we develop general instability theorems for averaging of one and two time scale systems. In Section 5.5.2, we apply these results to an output error adaptive scheme. Our treatment is based on Riedle & Kokotovic [1985] and Fu & Sastry [1986].

5.5.1 Instability Theorems Using Averaging

One Time Scale Systems

Recall the setup of Section 4.2, where we considered differential equations of the form

$$\dot{x} = \epsilon f(t, x, \epsilon) \tag{5.5.1}$$

and their *averaged* versions

$$\dot{x}_{av} = \epsilon f_{av}(x_{av}) \tag{5.5.2}$$

where

$$f_{av}(x) = \lim_{T \to \infty} \frac{1}{T} \int_{t_0}^{t_0+T} f(\tau, x, 0)\, d\tau \tag{5.5.3}$$

assuming that the limit exists uniformly in t_0 and x. We will not repeat the definitions and the assumptions (A1)–(A5) of Section 4.2, but we will assume that the systems (5.5.1), (5.5.2) satisfy those identical assumptions. The reader may wish to review those assumptions before proceeding with the proof of the following theorem.

Theorem 5.5.1 Instability Theorem for One Time Scale Systems

If the original system (5.5.1) and the averaged system (5.5.2) satisfy assumptions (A1)–(A5), the function $f_{av}(x)$ has continuous and bounded first partial derivatives in x, and there exists a continuously differentiable, decrescent function $v(t, x)$ such that

(i) $v(t, 0) = 0$

(ii) $v(t, x) > 0$ for some x arbitrarily close to 0

(iii) $\left| \frac{\partial v(t, x)}{\partial x} \right| \le k_1 |x|$ for some $k_1 > 0$

(iv) the derivative of $v(t, x)$ along the trajectories of (5.5.2) satisfies

$$\dot{v}(t, x) \Big|_{(5.5.2)} \ge \epsilon k_2 |x|^2 \tag{5.5.4}$$

for some $k_2 > 0$.

Then the original system (5.5.1) is unstable for ϵ sufficiently small.

Remark

By an instability theorem of Lyapunov (see for example Vidyasagar [1978]), the additional assumptions (i)–(iv) of the theorem guarantee that the averaged system (5.5.2) is unstable. By definition, a system is *unstable* if it is not stable, meaning that there exists a neighborhood of the origin and arbitrarily small initial conditions so that the state vectors originating from them are expelled from the neighborhood of the origin.

Proof of Theorem 5.5.1

As in Chapter 4, the first step is to use the transformation of lemma 4.2.3 to transform the original system. Thus, we use

$$x = z + \epsilon w_\epsilon(t, z) \tag{5.5.5}$$

with $w_\epsilon(t, z)$ satisfying, for some $\xi(\epsilon) \in K$

$$|\epsilon w_\epsilon(t, z)| \leq \xi(\epsilon)|z| \tag{5.5.6}$$

$$\left|\epsilon \frac{\partial w_\epsilon(t, z)}{\partial z}\right| \leq \xi(\epsilon) \tag{5.5.7}$$

to transform (5.5.1) into

$$\dot{z} = \epsilon f_{av}(z) + \epsilon p(t, z, \epsilon) \tag{5.5.8}$$

where $p(t, z, \epsilon)$ satisfies

$$|p(t, z, \epsilon)| \leq \psi(\epsilon)|z| \tag{5.5.9}$$

for some $\psi(\epsilon) \in K$.

Now, consider the derivative of $v(t, z)$ along the trajectories of (5.5.8), namely

$$\dot{v}(t, z)\Big|_{(5.5.8)} = \dot{v}(t, z)\Big|_{(5.5.2)} + \frac{\partial v(t, z)}{\partial z}\, \epsilon p(t, z, \epsilon) \tag{5.5.10}$$

Using the inequalities (5.5.4) and (5.5.9), we have that

$$\dot{v}(t, z)\Big|_{(5.5.8)} \geq \epsilon k_2 |z|^2 - \epsilon\psi(\epsilon) k_1 |z|^2 \tag{5.5.11}$$

If ϵ_0 is chosen so that $k_2 - \psi(\epsilon_0) k_1 > 0$, then it is clear that $\dot{v}(t, z)|_{(5.5.8)}$ is positive definite. By the same Lyapunov instability theorem as was mentioned in the remark preceding the theorem, it follows from (5.5.11) that for $\epsilon \leq \epsilon_0$, the system (5.5.8) is unstable, and consequently so is the original system (5.5.1). □

Comments

The continuously differentiable, decrescent function required by theorem 5.5.1 can be found prescriptively, if the averaged system is linear, that is

$$\dot{x}_{av} = \epsilon A x_{av} \tag{5.5.12}$$

and if A has at least one eigenvalue in the open right-half plane, but no eigenvalue on the $j\omega$-axis. In this case, the function v can be chosen to be

$$v(x) = x^T P x$$

where P satisfies the Lyapunov equation

$$A^T P + PA = I \tag{5.5.13}$$

The *Taussky lemma generalized* (see Vidyasagar [1978]) says that P has at least one positive eigenvalue, so that $v(x)$ takes on positive values in some directions (and arbitrarily close to the origin). It is also easy to verify that the conditions (iii), (iv) of theorem 5.5.1 are also satisfied by $v(x)$.

Two Time Scale Systems

We now consider the system of Section 4.4 namely

$$\dot{x} = \epsilon f(t, x, y) \tag{5.5.14}$$

$$\dot{y} = A\, y + \epsilon g(t, x, y) \tag{5.5.15}$$

with $x \in \mathbb{R}^n$, $y \in \mathbb{R}^m$. The only difference between (5.5.14), (5.5.15) and the system (4.4.1), (4.4.2) of Chapter 4 is that the A matrix of (5.5.15) is now assumed to be constant and stable rather than a function of x which is uniformly stable. The averaged system is

$$\dot{x}_{av} = \epsilon f_{av}(x_{av}) \tag{5.5.16}$$

where $f_{av}(x)$ is defined to be

$$f_{av}(x) = \lim_{T \to \infty} \frac{1}{T} \int_{t_0}^{t_0+T} f(\tau, x, 0)\, d\tau \tag{5.5.17}$$

The functions f, g satisfy assumptions (B1), (B2), (B3) and (B5) (only assumption (B4) is not necessary). As in the case of theorem 5.5.1, we advise the reader to review the results of Section 4.4 before following the next theorem.

Theorem 5.5.2 Instability Theorem for Two Time Scale Systems

If the original system (5.5.14), (5.5.15) and the averaged system (5.5.16) satisfy assumptions (B1), (B2), (B3) and (B5), along with the assumption that there exists a continuously differentiable decrescent function $v(t, x)$ such that

(i) $v(t, 0) = 0$

(ii) $v(t, x) > 0$ for some x arbitrarily close to 0

(iii) $\left| \dfrac{\partial v(t, x)}{\partial x} \right| \le k_1 |x|$ for some $k_1 > 0$

(iv) the derivative of $v(t, x)$ along the trajectory of (5.5.16) satisfies

$$\dot{v}(t, x)\Big|_{(5.5.16)} \ge \epsilon k_2 |x|^2 \tag{5.5.18}$$

for some $k_2 > 0$.

Then the original system (5.5.14), (5.5.15) is unstable for ϵ sufficiently small.

Proof of Theorem 5.5.2

To study the instability of (5.5.14), (5.5.15), we consider another decrescent function v_1,

$$v_1(t, x, y) = v(t, x) - k_3 y^T P y \tag{5.5.19}$$

where P is the symmetric positive definite matrix satisfying the Lyapunov equation

$$A^T P + P A = -I$$

Using the transformation of lemma 4.4.1, we may transform (5.5.14), (5.5.15)—as in the proof of theorems 4.4.2 and 4.4.3 —into

$$\dot{z} = \epsilon f_{av}(z) + \epsilon p_1(t, z, \epsilon) + \epsilon p_2(t, z, y, \epsilon) \tag{5.5.20}$$

$$\dot{y} = Ay + \epsilon g(t, x(z), y) \tag{5.5.21}$$

where $p_1(t, z, \epsilon)$ and $p_2(t, z, y, \epsilon)$ satisfy

$$|p_1(t, z, \epsilon)| \leq \xi(\epsilon) k_4 |z| \tag{5.5.22}$$

$$|p_2(t, z, y, \epsilon)| \leq k_5 |y| \tag{5.5.23}$$

and $\xi(\epsilon) \in K$. Clearly, $v_1(t, z, y) > 0$ for some (x, y) values arbitrarily close to the origin (let $y = 0$ and use assumption (ii)). Now, consider

$$\dot{v}_1(t, z, y)\Big|_{(5.5.20,\ 21)} = \dot{v}(t, z)\Big|_{(5.5.20)} + k_3 |y|^2 - 2\epsilon k_3 y^T P g(t, z, y)$$

Using exactly the same techniques as in the proof of theorem 4.4.3 (the reader may wish to follow through the details), it may be verified that

$$\dot{v}_1(t, z, y)\Big|_{(5.5.20,\ 21)} \geq \epsilon \alpha(\epsilon) |z|^2 + q(\epsilon) |y|^2$$

for some $\alpha(\epsilon) \to k_2$ and $q(\epsilon) \to k_3$ as $\epsilon \to 0$. Thus $\dot{v}_1(t, z, y)$ is a positive definite function along the trajectories of (5.5.20, 21). Hence, the system (5.5.20), (5.5.21) and consequently the original system (5.5.14), (5.5.15) is unstable for ϵ sufficiently small. □

Mixed Time Scales

As was noted in Chapter 4, a more general class of two-time scale systems arises in adaptive control, having the form

$$\dot{x} = \epsilon f'(t, x, y') \tag{5.5.24}$$

$$\dot{y}' = Ay' + h(t, x) + \epsilon g'(t, x, y') \tag{5.5.25}$$

Again, for simplicity, we let the matrix A be a constant matrix (we will only consider linearized adaptive control schemes in the next section).

In (5.5.24), (5.5.25), x is the slow variable but y' has both a fast and a slow component. As we saw in Section 4.4, the system (5.5.24), (5.5.25) can be transformed into the system (5.5.20), (5.5.21) through the use of the coordinate change

$$y = y' - v(t, x) \tag{5.5.26}$$

where $v(t, x)$ is defined to be

$$v(t, x) := \int_0^t e^{A(t-\tau)} h(\tau, x)\, d\tau \tag{5.5.27}$$

The averaged system of (5.5.24), (5.5.25) of the form of (5.5.16) will exist if the following limit exists uniformly in t_0 and x

$$f_{av}(x) = \lim_{T \to \infty} \frac{1}{T} \int_{t_0}^{t_0+T} f'(\tau, x, v(\tau, x))\, d\tau$$

The instability theorem of 5.5.2 is applicable with the additional assumption (B6) of Section 4.4.

5.5.2 Application to the Output Error Scheme

Tuned Error Formulation with Unmodeled Dynamics

We will apply the results of the previous section to an output error adaptive control scheme (see Section 3.3.2) designed for a plant of order n and relative degree one. The controller is, however, applied to a plant of order $n + n_u$, where the extra n_u states represent the unmodeled dynamics. In analogy to (3.5.16), the total state of the plant and observers $x_{pw} \in \mathbb{R}^{3n-2+n_u}$ satisfies the equations

$$\begin{bmatrix} \dot{x}_p \\ \dot{w}^{(1)} \\ \dot{w}^{(2)} \end{bmatrix} = \begin{bmatrix} A_p & 0 & 0 \\ 0 & \Lambda & 0 \\ b_\lambda c_p^T & 0 & \Lambda \end{bmatrix} \begin{bmatrix} x_p \\ w^{(1)} \\ w^{(2)} \end{bmatrix} + \begin{bmatrix} b_p \\ b_\lambda \\ 0 \end{bmatrix} u$$

$$y_p = [c_p^T \; 0 \; 0] \begin{bmatrix} x_p \\ w^{(1)} \\ w^{(2)} \end{bmatrix} \tag{5.5.28}$$

(5.5.28) is precisely (3.5.16) with the difference that $x_p \in \mathbb{R}^{n+n_u}$ rather to $\mathbb{R}^n$.

Now, it may no longer be possible to find a $\theta^* \in \mathbb{R}^{2n}$ such that the closed loop plant transfer function equals the model transfer function. Instead, we will assume that there is a value of θ which is at least *stabilizes* the closed loop system, and refer to it as the *tuned value* θ_*. We define

$$A_* = \begin{bmatrix} A_p + b_p d_{0*} c_p^T & b_\lambda c_*^T & b_p d_*^T \\ b_\lambda d_{0*} c_p^T & \Lambda + b_\lambda c_*^T & b_\lambda d_*^T \\ b_\lambda c_p^T & 0 & \Lambda \end{bmatrix} . \tag{5.5.29}$$

We will call A_* the *tuned closed-loop matrix* (cf. (3.5.18)), and we define the *tuned plant* as

$$\begin{aligned} \dot{x}_{pw*} &= A_* x_{pw*} + b_* c_{0*} r \\ y_{p*} &= c_*^T x_{pw*} \end{aligned} \tag{5.5.30}$$

where

$$b_* = \begin{bmatrix} b_p \\ b_\lambda \\ 0 \end{bmatrix} \in \mathbb{R}^{3n + n_u - 2} \quad \text{and} \quad c_* = \begin{bmatrix} c_p \\ 0 \\ 0 \end{bmatrix} \in \mathbb{R}^{3n + n_u - 2}$$

Note the analogy between (5.5.30) and (3.5.20). Now, the transfer function of the tuned plant is not exactly equal to the transfer function of the model, and the error between the tuned plant output and the model output is referred to as the *tuned error*

$$e_* = y_{p*} - y_m \tag{5.5.31}$$

Typically, the values θ_* which are chosen as tuned values correspond to those values of θ_* for which the tuned plant transfer function approximately matches the model transfer function at low frequencies (at those frequencies, the effect of unmodeled dynamics is small).

An error formulation may now be derived, with respect to the tuned system instead of the model system of Section 3.5. Let $\tilde{\theta} := \theta - \theta_*$ represent the parameter error with respect to the tuned parameter value, and rewrite (5.5.28) as

$$\begin{aligned} \dot{x}_{pw} &= A_* x_{pw} + b_* \tilde{\theta}^T w + b_* c_{0*} r \\ y_p &= c_*^T x_{pw} \end{aligned} \tag{5.5.32}$$

This is similar to (3.5.19). The output error parameter update law is

$$\dot{\theta} = \dot{\tilde{\theta}} = -g e_0 w \tag{5.5.33}$$

In turn the output error $e_0 := y_p - y_m$ can be decomposed as

$$e_0 = y_p - y_m = y_p - y_{p*} + e_* \tag{5.5.34}$$

Now, defining $\tilde{e} = x_{pw} - x_{pw*}$ and $\tilde{e}_0 = y_p - y_{p*}$ we may subtract equation (5.5.30) from equation (5.5.31) to get

$$\begin{aligned} \dot{\tilde{e}} &= A_* \tilde{e} + b_* \tilde{\theta}^T w \\ \tilde{e}_0 &= c_*^T \tilde{e} \end{aligned} \tag{5.5.35}$$

along with the update law

$$\dot{\tilde{\theta}} = -g(c_*^T \tilde{e} + e_*) w \tag{5.5.36}$$

As in the ideal case, w can be written as $w = w_* + Q\tilde{e}$ (with $w_* \in \mathbb{R}^{2n}$ having the obvious interpretation), so that (5.5.35), (5.5.36) may be combined to yield

$$\begin{aligned} \dot{\tilde{e}} &= A_* \tilde{e} + b_* \tilde{\theta}^T w_* + b_* \tilde{\theta}^T Q \tilde{e} \\ \dot{\tilde{\theta}} &= -g c_*^T \tilde{e} w_* - g c_*^T \tilde{e} Q \tilde{e} - g e_* w_* - g e_* Q \tilde{e} \end{aligned} \tag{5.5.37}$$

Comparing the equations (5.5.37) with the corresponding equation (3.5.28) for the adaptive system, one sees the presence of two new terms in the second equation. If the tuned error $e_* = 0$, the terms disappear and the equations (5.5.37) reduce to (3.5.28). The first of the new terms is an exogenous forcing term and the second a term which is linear in the error state variables. Without the term $e_* w_*$, the origin $\tilde{e} = 0, \tilde{\theta} = 0$ is an equilibrium of the system. Consequently, we will drop this term for the sake of our local stability/instability analysis. We will also treat the second term $g e_* Q \tilde{e}$ as a small perturbation term (which it is if e_* is small) and focus attention on the linearized and simplified system

$$\begin{aligned} \dot{\tilde{e}} &= A_* \tilde{e} + b_* w_*^T \tilde{\theta} \\ \dot{\tilde{\theta}} &= -g c_*^T \tilde{e} w_* \end{aligned} \tag{5.5.38}$$

Averaging Analysis

To apply the averaging theory of the previous section, we set the gain $g = \epsilon$, a small parameter. Since A_* is stable, it is easy to see that the system (5.5.38) is of the form of the mixed time scale system (5.5.24), (5.5.25) so that averaging may be applied. The averaged parameter error

$\tilde{\theta}_{av}$ satisfies

$$\dot{\tilde{\theta}}_{av} = \epsilon f_{av}(\tilde{\theta}_{av}) \tag{5.5.39}$$

where $f_{av}(\tilde{\theta})$ is defined as

$$f_{av}(\tilde{\theta}) = -\lim_{T \to \infty} \frac{1}{T} \int_{t_0}^{t_0+T} w_*(t)\, c_*^T \left[\int_0^t e^{A_*(t-\tau)} b_* w_*^T(\tau) d\tau \right] dt\, \tilde{\theta} \tag{5.5.40}$$

Note that $f_{av}(\tilde{\theta})$ is a linear function of $\tilde{\theta}$ so that the stability/instability of (5.5.38) for ϵ small is easily determined from the eigenvalues of the matrix in (5.5.36). As previously, the matrix in (5.5.40) may be written as the cross-correlation at 0 between $w_*(t)$ and

$$w_{*f}(t) := \int_0^t c_*^T e^{A_*(t-\tau)} b_* w_*(\tau)\, d\tau \tag{5.5.41}$$

Thus (5.5.40) may be written as

$$f_{av}(\tilde{\theta}) = -R_{w_* w_{*f}}(0)\, \tilde{\theta} \tag{5.5.42}$$

Frequency Domain Analysis

To derive a frequency domain interpretation, we assume that r is stationary. The spectral measure of w_* is related to that of r by

$$S_{w_*}(d\omega) = \hat{H}_{w_*r}^*(j\omega)\, \hat{H}_{w_*r}^T(j\omega)\, S_r(d\omega) \tag{5.5.43}$$

where the transfer function from r to w_* is $\hat{H}_{w_*r}(s)$. This transfer function is obtained by denoting the transfer function of the tuned plant

$$c_{0*}\, c_*^T(sI - A_*)^{-1} b_* = \hat{M}_*(s) \tag{5.5.44}$$

so that

$$\hat{H}_{w_*r}(s) = \begin{bmatrix} 1 \\ (sI - \Lambda)^{-1} b_\lambda \hat{P}^{-1} \hat{M}_* \\ \hat{M}_* \\ (sI - \Lambda)^{-1} b_\lambda \hat{M}_* \end{bmatrix} \tag{5.5.45}$$

The cross-correlation between w_* and w_{*f} is then given by

$$R_{w_* w_{*f}}(0) = \frac{1}{2\pi c_{0*}} \int_{-\infty}^{\infty} \hat{H}^*_{w_* r}(j\omega)\, \hat{H}^T_{w_* r}(j\omega)\, \hat{M}_*(j\omega)\, S_r(d\omega) \qquad (5.5.46)$$

Note the similarity between (5.5.45), (5.5.46) and (4.5.6), (4.5.9) for averaging in the ideal case. The chief difference is the presence of the tuned plant transfer function $\hat{M}_*(s)$ in place of the model transfer function $\hat{M}(s)$. $\hat{M}_*(s)$ may not be strictly positive real, even if $\hat{M}(s)$ is so. Consequently $R_{w_* w_{*f}}(0)$ *may not be a positive semi-definite* matrix. Heuristically speaking, if a large part of the frequency support of the reference signal lies in a region where the real part of $\hat{M}_*(j\omega)$ is negative, then $R_{w_* w_{*f}}(0)$ may fail to be positive semi-definite.

It is easy to see that if all the eigenvalues of $R_{w_* w_{*f}}(0)$ are in the right half plane, the (simplified) overall system (5.5.38) is *globally* asymptotically stable for ϵ small enough. Also, if even one of the eigenvalues of $R_{w_* w_{*f}}(0)$ lies in the left half plane, then the system (5.5.38) is unstable (in the sense of Lyapunov). From the form of the integral in (5.5.46), one may deduce that a necessary condition for $R_{w_* w_{*f}}(0)$ to have no zero eigenvalues is for the reference input to have at least $2n$ points of support. In fact, heuristically speaking, for $R_{w_* w_{*f}}(0)$ to have no negative eigenvalues, the reference input is required to have at least $2n$ points of support in the frequency range where $\mathrm{Re}\, \hat{M}_*(j\omega) > 0$ (the reason that this is heuristic rather than precise is because the columns of $\hat{H}_{w_* r}(j\omega)$ may not be linearly independent at every set of $2n$ frequencies). Since the tuned plant transfer function $\hat{M}_*(s)$ is close to the model transfer function $\hat{M}(s)$ at least for low frequencies (where there are no unmodeled dynamics), it follows that to keep the adaptive system stable, sufficient excitation at lower frequencies is required. It is also important to see that the stability/instability criterion is both *signal-dependent* as well as dependent on the *tuned plant transfer function* $\hat{M}_*(s)$.

It is important at this point to note that all of the analysis is being performed on the averaged version of the simplified linearized system (5.5.38). As far as the original system (5.5.31) is concerned, we can make the following observations

a) If the simplified, linearized system (5.5.38) is exponentially stable, then the original system (5.5.37) is locally stable in the sense that if the tuned error and the initial conditions on $\tilde{e}_*$, $\tilde{\theta}$ are sufficiently small, trajectories will eventually be confined to a neighborhood of the origin (the size of the neighborhood depends on e_* and goes to zero as $|e_*(t)|$ goes to zero). The proof of this follows from theorem 5.3.1.

b) If the simplified linearized system is unstable, then the original system (5.5.37) is also unstable, using arguments from theorem 5.5.1.

Of course, the averaging analysis may be inconclusive if the averaged system $R_{w_* w_{*f}}(0)$ has some zero eigenvalues. In this instance, if $R_{w_* w_{*f}}(0)$ has at least one eigenvalue in the open left half plane, then the original system is unstable. However, if $R_{w_* w_{*f}}(0)$ has all its eigenvalues in the closed right half plane, including some at zero, the averaging is inconclusive for (5.5.38) and for (5.5.37). Simulations seem to suggest that, in this case, the parameter error vector $\tilde{\theta}$ driven by e_* drifts away from the origin in the presence of noise. This is what happens in Rohrs example (R3), where the reference input is only a DC input: e_* corresponding to the tuned error is small, since the closed loop plant matches the model at low frequencies, but its place is taken by the output disturbance which causes the parameters to drift away from their tuned values.

The result of this section also makes rigorous the heuristic explanation for the instability mechanism of the example in (R2) where a significant high frequency signal is present in a range where the tuned plant transfer function is not strictly positive real. A tuned plant is easily obtained by removing the unmodeled poles at $-15 \pm j2$ to get the tuned values $c_{0*} = 1.5$, $d_{0*} = 1$ identical to θ^*.

Example

In this section, we discuss an example from Riedle & Kokotovic [1985]. We consider the plant

$$\hat{P}(s) = \frac{k_p}{\mu s^2 + (1+\mu)s + 1} \tag{5.5.47}$$

where $\mu > 0$ is a small parameter. The adaptive controller is designed assuming a first order plant with relative degree 1. Thus, we assume that the 'nominal' plant is of the form

$$P_{\theta^*} = \frac{k_p}{s+1} \tag{5.5.48}$$

with k_p unknown. The model is of the form $1/(s+1)$ and we set the tuned value of c_0, namely c_{0*} to be $1/k_p$ for the analysis. For the example, k_p is chosen to be 1. The error system is

$$\begin{bmatrix} \dot{\tilde{e}} \\ \dot{\tilde{\theta}} \end{bmatrix} = \begin{bmatrix} -1 & 1 & 0 \\ 0 & \dfrac{-1}{\mu} & \dfrac{1}{\mu} r(t) \\ -\epsilon r(t) & 0 & 0 \end{bmatrix} \begin{bmatrix} \tilde{e} \\ \tilde{\theta} \end{bmatrix} + \begin{bmatrix} 0 \\ 0 \\ -g \end{bmatrix} r(t) e_*(t) \qquad (5.5.49)$$

Note that (5.5.49) is simpler than (5.5.37) since there are no adaptive parameters in the feedback loop. $R_{w_* w_{*f}}(0)$ is a scalar, and is easily computed to be

$$R_{w_* w_{*f}}(0) = \int_{-\infty}^{\infty} \frac{1 - \mu\omega^2}{(1 - \mu\omega^2)^2 + (1 + \mu)^2 \omega^2} S_r(d\omega) \qquad (5.5.50)$$

Note that the integrand is positive for $|\omega| < 1/\sqrt{\mu}$ and negative for $|\omega| > 1/\sqrt{\mu}$. For example, if $\mu = 0.1$ and

$$r_1(t) = \sin 5t \qquad R_{w_* w_{*f}}(0) = -0.046$$

$$r_2(t) = 0.4 \sin t + \sin 5t \qquad R_{w_* w_{*f}}(0) = 0.026$$

Thus, the first input results in an unstable system and the second in a stable one. These results are borne out in the simulations of Figure 5.15.

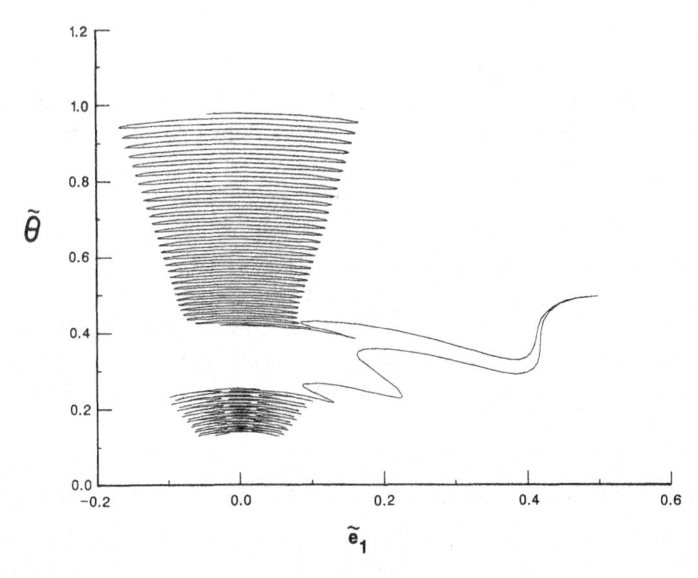

Figure 5.15 Stability–Instability Boundary for $r_1(t)$ and $r_2(t) = r_1(t) + 0.4 \sin t$.

5.6 METHODS FOR IMPROVING ROBUSTNESS–QUALITATIVE DISCUSSION

Adaptive systems are not magically robust: several choices need to be carefully made in the course of design, and they need to explicitly take into account the limitations and flexibilities of the rather general algorithms presented in earlier chapters. We begin with a qualitative discussion of methods to improve the robustness of adaptive systems. A review of a few specific update law modifications is given in the next section.

5.6.1 Robust Identification Schemes

An important part of the adaptive control scheme is the identifier, or adaptation algorithm. When only parametric uncertainty is present, adaptive schemes are proved to be stable, with asymptotic tracking. Parameter convergence is not guaranteed in general, but is not necessary to achieve stability. In the presence of unmodeled dynamics and measurement noise, drift instabilities may occur, so that the spectral content of the input becomes important. The robustness of the identifier is fundamental to the robustness the adaptive system, and may be influenced by a careful design.

An initial choice of the designer is the frequency range of interest. In an adaptive control context, it is the frequency range over which accurate tracking is desired, and is usually limited by actuators' bandwith and sensor noise.

The order of the plant model must then be selected. The order should be sufficient to allow for modeling of the plant dynamics in the frequency range of interest. On the other hand, if the plant is of high order, a great deal of excitation (a number of independent frequencies) will be required. The presence of a large parameter vector in the identifier may also cause problems of numerical conditioning in the identification procedure. Then, the covariance matrix $R_w(0)$ (see Chapters 2 and 3) measuring the extent of persistent excitation, is liable to be ill-conditioned, resulting in slow parameter convergence along certain directions in parameter space. In summary, it is important to choose a low enough order plant model capable of representing all the plant dynamics in the frequency range of interest.

Filtering of the plant input and output signals is achieved by the observer, with a bandwith determined by the filter polynomial (denoted earlier $\hat{\lambda}(s)$). To reduce the effect of noise, it may be reasonable to further filter the regression vectors in the identification algorithm, so as to exclude the contribution of data from frequency regions lying outside the range of frequencies of importance to the controller (i.e. low pass filtering with a cut-off somewhat higher than the control bandwith).

The spectrum of the reference input is another parameter, partially left to the designer. Recall that the identifier identifies the portion of the plant dynamics in the frequency range of the input spectrum. Thus, it is important that the input signal: a) be rich enough to guarantee parameter convergence, and b) have energy content in the frequency range where the plant model is of sufficient order to represent the actual plant. The examples of Rohrs consisted of scenarios in which a) the input was not rich enough (only a DC signal), and b) the output had energy in the frequencies of the unmodeled dynamics (a DC signal and a high-frequency sinusoid). In the first case, noise caused parameter drift, consistent with a good low frequency model of the plant, into a region of instability. In the second, an incorrect plant model resulted in an unstable loop.

From a practical viewpoint, it is important to monitor the signal excitation in the identifier loop and to turn off the adaptation when the excitation is poor. This includes the case when the level of excitation is so low as to make it difficult to distinguish between the excitation and the noise. It is also clear that if the excitation is poor over periods of time where parameters vary, the parameter identification will be ineffectual. In such an event, the only cure is to inject extra perturbation signals into the reference input so as to provide excitation for the identification algorithm.

We summarize this discussion in the form of the following table for a robust identification scheme.

Steps of Robust Identification

	Step	Considerations
1.	Choice of the frequency range of interest	Frequency range over which tracking is desired
2.	Plant Order Determination	Modeling of the plant dynamics in the frequency range of interest Low
3.	Regressor Filter Selection	Filter high frequency components (unmodeled dynamics range)
4.	Reference Input Selection	Sufficient richness Spectrum within frequency range of interest
	If not, check step 5	
5.	Turn off parameter update when not rich enough	

If the excitation is not rich over periods of time where parameters vary, add perturbation signal	Limit perturbation to plant

5.6.2 Specification of the Closed Loop Control Objective—Choice of the Reference Model and of Reference Input

The reference model must be chosen to reflect a desirable response of the closed-loop plant. From a robust control standpoint, however, control should only be attempted over a frequency range where a satisfactory plant model and controller parameterization exists. Therefore, the control objective (or reference model choice) should have a bandwidth no greater than that of the identifier. In particular, the reference model should not have large gain in those frequency regions in which the unmodeled dynamics are significant.

The choice of reference input is also one of the choices in the overall control objective. We indicated above how important the choice is for the identification algorithm. However, persistent excitation in the correct frequency range for identification may require added reference inputs not intended for tuned controller performance. In some applications (such as aircraft flight control), the insertion of perturbation signals into the reference input can result in undesirable dithering of the plant output. The reference input in adaptive systems plays a dual role, since the input is required both for generating the reference output required for tracking, as well as furnishing the excitation needed for parameter convergence (this dual role is sometimes referred to as the *dual control concept*).

5.6.3 The Usage of Prior Information

The schemes and stability proofs thus far have involved very little *a priori* knowledge about the plant under control. In practice, one is often confronted with systems which are fairly well modeled, except for a few unknown and uncertain components which need to be identified. In order to use the schemes in the form presented so far, all of the prior knowledge needs to be completely discounted. This, however, increases the order of complexity of the controller, resulting in extra requirements on the amount of excitation needed. In certain instances, the problem of incorporating prior information can be solved in a neat and consistent fashion—for this we refer the reader to Section 6.1 in the next chapter.

5.6.4 Time Variation of the Parameters

The adaptive control algorithms so far have been derived and analyzed for the case of unknown but fixed parameter values. In practice, adaptive control is most useful in scenarios involving slowly changing plant parameters. In these instances the estimator needs to converge much faster than the rate of plant parameter variation. Further, the estimator needs to discount old input-output data: old data should be discounted quickly enough to allow for the estimator parameters to track the time-varying ones. The discounting should not, however, be too fast since this would involve an inconsistency of parameter values and sensitivity to noise.

We conclude this section with Figure 5.16, inspired from Johnson [1988], indicating the desired ranges of the different dynamics in an adaptive system.

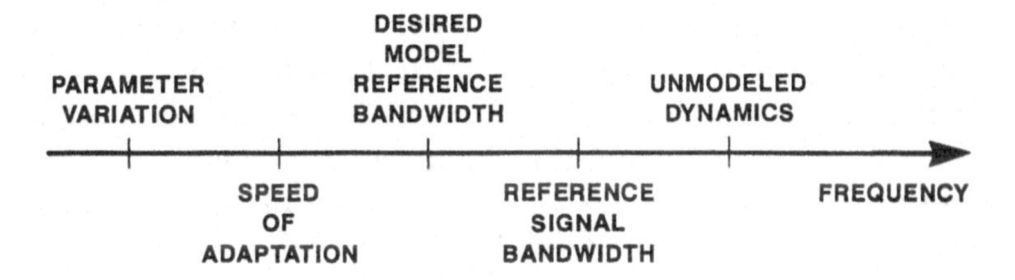

Figure 5.16 Desirable Bandwidths of Operation of an Adaptive Control System.

5.7 ROBUSTNESS VIA UPDATE LAW MODIFICATIONS

In the previous sections, we reviewed some of the reasons for the loss of robustness in adaptive schemes and qualitatively discussed how to remedy them. In this section, we present modifications of the parameter update laws which were recently proposed as robustness enhancement techniques.

5.7.1 Deadzone and Relative Deadzone

The general idea of a *deadzone* is to stop updating the parameters when the excitation is insufficient to distinguish between the regressor signal and the noise. Thus, the adaptation is turned off when the identifier error is smaller than some threshold.

More specifically, consider the input error direct adaptive control algorithm with the generalized gradient algorithm and projection. The update law with deadzone is given by

$$\dot{\theta} = -g \frac{e_2 v}{1 + \gamma v^T v} \qquad \text{if } |e_2| > \Delta \tag{5.7.1}$$

$$\dot{\theta} = 0 \qquad \text{if } |e_2| \le \Delta \tag{5.7.2}$$

and as before, if $c_0 = c_{\min}$ and $\dot{c}_0 < 0$, then set $\dot{c}_0 = 0$. The parameter Δ in equations (5.7.1) and (5.7.2) represents the size of the deadzone. Similarly, the output error direct adaptive control algorithm with gradient algorithm is modified to

$$\dot{\bar{\theta}} = -g e_1 \bar{v} \qquad \text{if } |e_1| > \Delta \tag{5.7.3}$$

$$\dot{\bar{\theta}} = 0 \qquad \text{if } |e_1| \le \Delta \tag{5.7.4}$$

where Δ is, as before, the deadzone threshold. It is easy to see how the other schemes (including the least-squares update laws) are modified.

The most critical part in the application of these schemes is the selection of the width of the deadzone Δ. If the deadzone Δ is too large, e_2 in equations (5.7.1), (5.7.2) and e_1 in (5.7.3), (5.7.4) will not tend to zero, but will only be asymptotically bounded by a large Δ, resulting in undesirable closed-loop performance. A number of recent papers (for example, Peterson & Narendra [1982], Samson [1983], Praly [1983], Sastry [1984], Ortega, Praly & Landau [1985], Kreisselmeier & Anderson [1986], Narendra & Annaswamy [1986]) have suggested different techniques for the choice of the deadzone Δ. The approach taken by Peterson & Narendra [1982]—for the case when the plant output is corrupted by additive noise—and by Praly [1983] and Sastry [1984]—for the case of both output noise and unmodeled dynamics—is to use some prior bounds on the disturbance magnitude and some prior knowledge about the plant to find a (conservative) bound on Δ and establish that the tracking error eventually converges to the region $|e_1| \le \Delta$. The bounds on Δ which follow from their calculations are, however, extremely conservative. From a practical standpoint, these results are to be interpreted as mere existence results. Practically, one would choose Δ from observing the noise floor of the parameter update variable e_1 (with no exogenous reference input present). It is also possible to modify it on-line depending on the quality of the data.

The approach of Samson [1983], Ortega, Praly & Landau [1985] and Kreisselmeier & Anderson [1986] is somewhat different in that it involves a deadzone size Δ which is not determined by e_1 or e_2 alone, but by how large the regressor signal in the adaptive loop is (the deadzone acts on a suitably normalized, relative identification error). The logic behind this so-called *relative deadzone* is that if the regressor vector is large, then the identification error may be large even for a small transfer function error due to unmodeled dynamics. The details of the relative deadzone are somewhat involved (the three papers referenced above are also for discrete time algorithms). However, the adaptive law can only guarantee that the *relative* (or normalized) identification error becomes smaller than the deadzone eventually. Thus, if the closed-loop system were unstable, the absolute identification error could be unbounded. To complete the proofs of stability with relative deadzones, it is then important to prove that the regressor vector is bounded. It is claimed (cf. Kreisselmeier & Anderson [1986]) that the relative deadzone approach will not suffer from "bursting," unlike the absolute deadzone approach.

5.7.2 Leakage Term (σ–Modification)

Ioannou & Kokotovic [1983] suggested modifying the parameter update law to counteract the drift of parameter values into regions of instability in the absence of persistent excitation. The original form of the modification is, for the direct output error scheme

$$\dot{\bar{\theta}} = -g e_1 \bar{v} - \sigma \bar{\theta} \tag{5.7.5}$$

where σ is chosen small but positive to keep $\bar{\theta}$ from growing unbounded. Two other interesting modifications in the spirit of (5.7.5) are

$$\dot{\bar{\theta}} = -g e_1 \bar{v} - \sigma (\bar{\theta} - \bar{\theta}_0) \tag{5.7.6}$$

where $\bar{\theta}_0$ is a prior estimate of $\bar{\theta}$ (for this and other modifications see Ioannou [1986] and Ioannou and Tsakalis [1986]), and one suggested by Narendra and Annaswamy [1987]

$$\dot{\bar{\theta}} = -g e_1 \bar{v} - \sigma |e_1| \bar{\theta} \tag{5.7.7}$$

Both (5.7.6) and (5.7.7) attempt to capture the spirit of (5.7.5) without its drawback of causing $\bar{\theta} \to 0$ if e_1 is small. Equation (5.7.6) tries to bias the direction of the drift towards $\bar{\theta}_0$ rather than 0 and (5.7.7) tries to turn off the drift towards 0 when $|e_1|$ is small. The chief advantage of the update law (5.7.7) is that it retains features of the algorithm without leakage (such as convergence of the parameters to their true values when the excitation is persistent). Also, the algorithm (5.7.7) may be less susceptible to bursting than (5.7.5), though this claim has not been fully substantiated.

5.7.3 Regressor Vector Filtering

The concept of low pass filtering or pre-conditioning the regressor vector in the parameter update law was discussed in Section 5.6. It is usually accomplished by low pass filtering of the process input and output in the identification algorithm and is widely prevalent (cf. the remarks in Wittenmark & Astrom [1984]). Some formalization of the concept and its analysis is in the work of Johnson, Anderson & Bitmead [1984]. The logic is that low pass filtering tends to remove noise and contributions of high frequency unmodeled dynamics.

5.7.4 Slow Adaptation, Averaging and Hybrid Update Laws

A key characteristic of the parameter update laws even with the addition of deadzones, leakage and regressor vector filtering is their "impatience." Thus, if the identification error momentarily becomes large, perhaps for reasons of spurious, transient noise, the parameter update operates instantaneously. A possible cure for this impatience is to slow down the adaptation. In Chapter 4, we studied in great detail the averaging effects of using a small adaptation gain on the parameter trajectories. In fact, a reduction of the effect of additive noise (by averaging) is also observed.

Another modification of the parameter update law in the same spirit is the so-called hybrid update law involving discrete updates of continuous time schemes. One such modification of the gradient update law (due to Narendra, Khalifa & Annaswamy [1985]) is

$$\theta(t_{k+1}) = \theta(t_k) - \int_{t_k}^{t_{k+1}} g\, e_1 v\, dt \qquad (5.7.8)$$

In (5.7.8), the t_k refer to parameter update times, and the controller parameters are held constant on $[t_k, t_{k+1}]$. The law (5.7.8) relies on the averaging inherent in the integral to remove noise.

Slow adaptation and hybrid adaptation laws suffer from two drawbacks. First, they result in undesirable transient behavior if the initial parameter estimates result in an unstable closed loop (since stabilization is slow). Second, they are incapable of tracking fast parameter variations. Consequently, the best way to use them is after the initial part of the transient in the adaptation algorithm or a short while after a parameter change, which the "impatient" algorithms are better equipped to handle.

5.8 CONCLUSIONS

In this chapter, we studied the problem of the robustness of adaptive systems, that is, their ability to maintain stability despite modeling errors and measurement noise.

We first reviewed the Rohrs examples, illustrating several mechanisms of instability. Then, we derived a general result relating exponential stability to robustness. The result indicated that the property of exponential stability is robust, while examples show that the BIBS stability property is not (that is, BIBS stable systems can become unstable in the presence of arbitrarily small disturbances). In practice, the amplitude of the disturbances should be checked against robustness margins to determine if stability is guaranteed. The complexity of the relationship between the robustness margins and known parameters, and the dependence of these margins on external signals unfortunately made the result more conceptual than practical.

The mechanisms of instability found in the Rohrs examples were discussed in view of the relationship between exponential stability and robustness, and a heuristic analysis gave additional insight. Further explanations of the mechanisms of instability were presented, using an averaging analysis. Finally, various methods to improve robustness were reviewed, together with recently proposed update law modifications.

We have attempted to sketch a sampling of what is a very new and active area of research in adaptive systems. We did not give a formal statement of the convergence results for all the adaptation law modifications. The results are not yet in final form in the literature and estimates accruing from systematic calculations are conservative and not very insightful. A great deal of the preceding discussion should serve as design guidelines: the exact design trade-offs will vary from application to application. The general message is that it is perhaps not a good idea to treat adaptive control design as a "black box" problem, but rather to use as much process knowledge as is available in a given application.

A guideline for design might run as follows

a) Determine the frequency range beyond which one chooses not to model the plant (where unmodeled dynamics appear) and find a parameterization which is likely to yield a good model of the plant in this frequency range, yet without excessive parameterization. If prior information is available, use it (see Section 6.1 for more on this).

b) Choose a reference model (performance objective) whose bandwidth does not extend into the range of unmodeled dynamics.

c) In the course of adaptation, implement the adaptive law with a deadzone whose size is determined by observing the amount of noise in the absence of exogenous input. Also, *monitor the excitation* and turn off the adaptation when the excitation is not rich over a large interval of time. If necessary, *inject extra excitation* into the exogenous reference input as a perturbation signal. If it appears that the plant parameters are not varying very rapidly, slow down the rate of adaptation or use a hybrid update algorithm (this is rather like a variable time step feature in numerical integration routines). Other modifications, such as leakage, may be added as desired.

d) Implement the appropriate start-up features for the algorithm using prior knowledge about the plant to choose initial parameter values and include "safety nets" to cover start-up, shut-down and transitioning between various modes of operation of the overall controller.

The guidelines given in this chapter are for the most part conceptual: in applications, questions of numerical conditioning of signals, sampling intervals (for digital implementations), anti-aliasing filters (for digital implementations), controller-architecture featuring several levels of interruptability, resetting, and so on are important. Even with a considerable wealth of theory and analysis of the algorithms, the difference an adaptive controller makes in a given application is chiefly due to the art of the designer!

CHAPTER 6

ADVANCED TOPICS IN IDENTIFICATION AND ADAPTIVE CONTROL

6.1 USE OF PRIOR INFORMATION

6.1.1 Identification of Partially Known Systems

We consider in this section the problem of identifying partially known single-input single-output (SISO) transfer functions of the form

$$\hat{P}(s) = \frac{\hat{N}_0(s) + \sum_{i=1}^{m} \alpha_i \hat{N}_i(s)}{\hat{D}_0(s) - \sum_{j=1}^{n} \beta_j \hat{D}_j(s)} \tag{6.1.1}$$

where $\hat{N}_i$ and $\hat{D}_j$ are known, proper, stable rational transfer functions and α_i, β_j are unknown, real parameters. The identification problem is to identify α_i, β_j from input-output measurements of the system. The problem was recently addressed by Clary [1984], Dasgupta [1984], and Bai and Sastry [1986].

The representation (6.1.1) is general enough to model several kinds of "partially known" systems.

Examples

a) *Network functions of RLC circuits with some elements unknown.* Consider for example the circuit of Figure 6.1, with the resistor R unknown (the circuit is drawn as a two port to exhibit the unknown resistance).

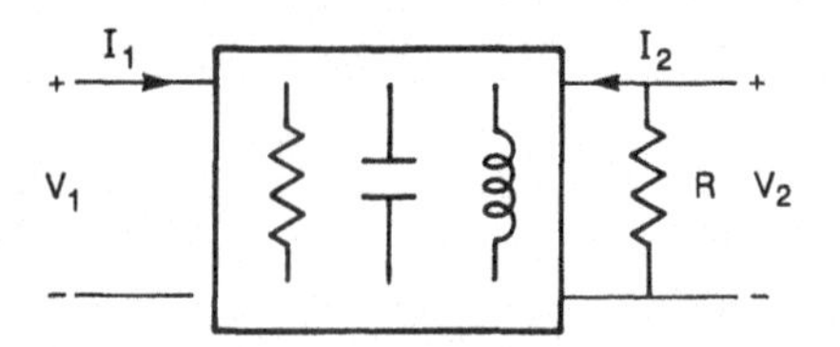

Figure 6.1 Two Port with Unknown Resistance R

If the short circuit admittance matrix of the two port in Figure 6.1 is

$$\begin{bmatrix} i_1 \\ i_2 \end{bmatrix} = \begin{bmatrix} y_{11}(s) & y_{12}(s) \\ y_{21}(s) & y_{22}(s) \end{bmatrix} \cdot \begin{bmatrix} v_1 \\ v_2 \end{bmatrix} \tag{6.1.2}$$

then a simple calculation yields the admittance function

$$\frac{i_1}{v_1} = \frac{y_{11} + R(y_{11}y_{22} - y_{12}y_{21})}{1 + Ry_{22}} \tag{6.1.3}$$

which is of the form of (6.1.1). Circuits with more than one unknown element can be drawn as multiports to show that the admittance function is of the form of (6.1.1).

b) *Interconnections of several known systems with unknown interconnection gains.* A simple example of this is shown in Figure 6.2, with a plant $\hat{P}(s)$ known, and a feedback gain k unknown.

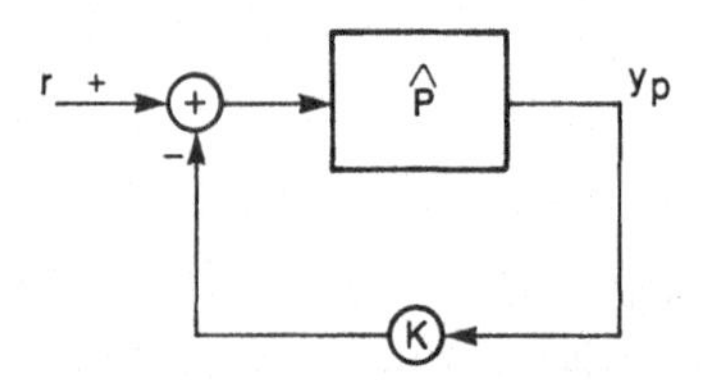

Figure 6.2 Plant with Unknown Feedback Gain

The closed-loop transfer function, namely $\hat{P} / 1 + k\hat{P}$ is of the form of (6.1.1) if $\hat{P}$ is stable. If $\hat{P}$ is unstable, then by writing $\hat{P} = \hat{N}_p / \hat{D}_p$ as the ratio of two proper stable rational transfer functions, the closed-loop transfer function is $\hat{N}_p / \hat{D}_p + k\hat{N}_p$, which is of the form (6.1.1).

c) *Classical transfer function models,* that is, plants of the form studied in Chapter 2

$$\hat{P}(s) = \frac{\alpha_m s^{m-1} + \cdots + \alpha_1}{s^n + \beta_n s^{n-1} + \cdots + \beta_1} \tag{6.1.4}$$

with $m \le n$ and α_i, β_j unknown, can be stated in terms of the set up of (6.1.1) by choosing

$$\begin{aligned} \hat{N}_0(s) &= 0 \\ \hat{D}_0(s) &= s^n / \hat{\lambda}(s) \\ \hat{N}_i(s) &= s^{i-1} / \hat{\lambda}(s) \qquad i = 1, \ldots, m \\ \hat{D}_j(s) &= -s^{j-1} / \hat{\lambda}(s) \quad j = 1, \ldots, n \end{aligned} \tag{6.1.5}$$

where $\hat{\lambda}(s)$ is (any) Hurwitz polynomial of order n.

d) Systems with some known poles and zeros. Consider the system of Figure 6.3, with unknown plant, but known actuator and sensor dynamics (with transfer functions $\hat{P}_a(s)$ and $\hat{P}_s(s)$ respectively).

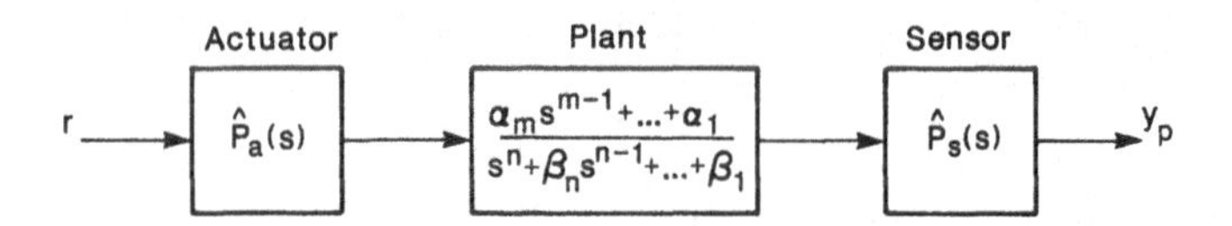

Figure 6.3 Unknown Plant with Known Actuator and Sensor Dynamics

The overall transfer function is written as

$$\hat{P}(s) = \hat{P}_s(s) \cdot \frac{\sum_{i=1}^{m} \alpha_i s^{i-1}}{s^n + \sum_{j=1}^{n} \beta_j s^{j-1}} \cdot \hat{P}_a(s) \tag{6.1.6}$$

which is of the form (6.1.1) by choosing, as above

$$\begin{aligned} \hat{N}_0(s) &= 0 \\ \hat{D}_0(s) &= s^n / \hat{\lambda}(s) \\ \hat{N}_i(s) &= s^{i-1} \hat{P}_a(s) \hat{P}_s(s) / \hat{\lambda}(s) \qquad i = 1, \ldots, m \\ \hat{D}_j(s) &= -s^{j-1} / \hat{\lambda}(s) \qquad j = 1, \ldots, n \end{aligned} \tag{6.1.7}$$

where $\hat{\lambda}(s)$ is (any) Hurwitz polynomial of order n.

Identification Scheme

We now return to the general system described by (6.1.1). The identification problem is to determine the unknown parameters α_i, β_j from input-output measurements. One could, of course, neglect the prior information embedded in the form of the transfer function (6.1.1) and identify the whole transfer function using one of the procedures of Chapter 2. However, usage of the particular structure embodied in (6.1.1) will result in the identification of fewer unknown parameters, with a reduction of computational requirements and usually faster convergence properties.

Let $\hat{r}(s)$, $\hat{y}_p(s)$ denote the input and output of the plant. Using (6.1.1)

$$\hat{D}_0 \hat{y}_p - \hat{N}_0 \hat{r} = \sum_{j=1}^{n} \beta_j \hat{D}_j \hat{y}_p + \sum_{i=1}^{m} \alpha_i \hat{N}_i \hat{r} \tag{6.1.8}$$

Defining the signals

$$\begin{aligned} \hat{z}_p &:= \hat{D}_0 \hat{y}_p - \hat{N}_0 \hat{r} \\ \hat{w}_i &:= \hat{N}_i \hat{r} \qquad i = 1, \ldots, m \\ \hat{w}_{m+j} &:= \hat{D}_j \hat{y}_p \qquad j = 1, \ldots, n \end{aligned} \tag{6.1.9}$$

and the nominal parameter vector θ^*

$$\theta^{*^T} := (\alpha_1, \ldots, \alpha_m, \beta_1, \ldots, \beta_n) \in \mathbb{R}^{n+m} \tag{6.1.10}$$

we may rewrite (6.1.8) as

$$\hat{z}_p = \theta^{*^T} \begin{bmatrix} \hat{w}_1 \\ \cdot \\ \cdot \\ \cdot \\ \hat{w}_{n+m} \end{bmatrix} \tag{6.1.11}$$

or, in the time domain

$$z_p(t) = \theta^{*^T} w(t) \tag{6.1.12}$$

where

$$w^T(t) := (w_1(t), \ldots, w_{n+m}(t)) \in \mathbb{R}^{n+m}$$

Note the close resemblance between (6.1.12) and the plant parameterization of (2.2.14) in Chapter 2. It is easy to verify that in the instance in which no prior information about the plant is available (case c) above), equation (6.1.12) above is a reformulation of equation

(2.2.14).

The purpose of the identifier is to produce a recursive estimate $\hat{\theta}(t)$ of the parameter vector θ^*. Since r and y_p are available, the signals $z_p(t)$, $w(t)$ are obtainable through stable filtering of r and y_p, up to some exponentially decaying terms (since $\hat{N}_i$ and $\hat{D}_j$ are stable, proper, rational functions). These decaying terms will be neglected for simplicity.

In analogy to the expression of the plant equation (6.1.12), we define the output of the identifier as

$$z_i(t) := \theta^T(t)\,w(t) \tag{6.1.13}$$

We also define the parameter error

$$\phi(t) := \theta(t) - \theta^* \tag{6.1.14}$$

and the identifier error

$$e_1(t) := z_i(t) - z_p(t) \tag{6.1.15}$$

so that, for the analysis

$$e_1(t) = \phi^T(t)\,w(t) \tag{6.1.16}$$

Equation (6.1.16) is now exactly the same as (2.3.2) of Chapter 2, so that all the update algorithms and properties of Chapter 2 can be used verbatim in this context. Thus, for example, the *gradient algorithm*

$$\dot{\theta} = -g e_1 w \qquad g > 0 \tag{6.1.17}$$

or the *least-squares algorithm*

$$\dot{\theta} = -gPe_1 w$$

$$\dot{P} = -gPww^T P \qquad g > 0 \tag{6.1.18}$$

along with covariance resetting, are appropriate parameter update laws. As in Chapter 2, the parameter error will converge (exponentially) to zero for the gradient or the least-squares algorithm with resetting if the vector w is persistently exciting, i.e. if there exist α_1, α_2, $\delta > 0$, such that

$$\alpha_2 I \geq \int_{t_0}^{t_0+\delta} w(\tau)\,w^T(\tau)\,d\tau \geq \alpha_1 I \qquad \text{for all } t_0 \geq 0 \tag{6.1.19}$$

Frequency Domain Conditions for Parameter Convergence

The techniques of Chapter 2 can be used to give frequency domain conditions on $r(t)$ to guarantee (6.1.19). To guarantee the upper bound, we simply assume:

(A1) Boundedness of the Regressor

The plant $\hat{P}(s)$ is stable, and the reference signal r is piecewise continuous and bounded.

As usual, the nontrivial condition comes from the lower bound in (6.1.19). To relate this condition on w to the frequency content of r, we will need a new *identifiability* condition on the transfer function from r to w, which we will denote

$$\hat{H}_{wr}^T := (\hat{N}_1(s), \ldots, \hat{N}_m(s), \hat{D}_1(s)\hat{P}(s), \ldots, \hat{D}_n(s)\hat{P}(s)) \qquad (6.1.20)$$

(A2) Identifiability

The system is assumed to be *identifiable*, meaning that for every choice of $n+m$ distinct frequencies $\omega_1, \ldots, \omega_{n+m}$, the vectors $\hat{H}_{wr}(j\omega_i) \in \mathbb{C}^{n+m}$ $(i = 1, \ldots, n+m)$ are linearly independent.

Proposition 6.1.1

Under assumptions (A1) and (A2), w is persistently exciting *if and only if* r is sufficiently rich of order $n+m$.

Proof of Proposition 6.1.1 similar to the proof of theorem 2.7.2.

Comments

a) From (6.1.20) and (6.1.11), it follows that if an input with $(n+m)$ spectral lines is applied to the system, we would have

$$(\hat{z}_p(j\omega_1), \ldots, \hat{z}_p(j\omega_{n+m})) = \theta^{*^T}(\hat{H}_{wr}(j\omega_1), \ldots, \hat{H}_{wr}(j\omega_{n+m}))$$

$$\operatorname{diag}(\hat{r}(j\omega_1), \ldots, \hat{r}(j\omega_{n+m})) \qquad (6.1.21)$$

The identifiability condition implies that (6.1.21) has a unique solution for θ^*, while proposition 6.1.1 shows that the identifier parameter will converge to this value.

b) It is difficult to give a more concrete characterization of identifiability, since the components of $\hat{H}_{wr}(s)$ are proper stable rational functions of different orders. An exception is the case of example c) and discussed in Chapter 2. In that case, it was shown that the identifiability condition holds if the numerator and denominator of the plant transfer function are coprime polynomials.

6.1.2 Effect of Unmodeled Dynamics

The foregoing set up used transfer functions of the form (6.1.1) with $\hat{N}_i$, and $\hat{D}_j$ known exactly. In practice, the $\hat{N}_i$ and $\hat{D}_j$ are only known approximately. In particular, the transfer functions used to approximate the $\hat{N}_i$ and $\hat{D}_j$ will generally be low order, proper stable transfer functions (neglecting high-frequency dynamics and replacing near pole-zero cancellations by exact pole-zero cancellations). Thus, the identifier model of the plant is of the form

$$\hat{P}_a(s) = \frac{\hat{N}_{a0}(s) + \sum_{i=1}^{m} \alpha_i \hat{N}_{ai}(s)}{\hat{D}_{a0}(s) - \sum_{j=1}^{n} \beta_j \hat{D}_{aj}(s)} \tag{6.1.22}$$

where $\hat{P}_a(s)$ is a proper stable transfer function, and $\hat{N}_{ao}$, $\hat{N}_{ai}$, $\hat{D}_{ao}$, $\hat{D}_{aj}$ are approximations of the actual transfer functions $\hat{N}_0$, $\hat{N}_i$, $\hat{D}_0$, $\hat{D}_j$. We will assume that

$$|\Delta \hat{N}_i(j\omega)| := |(\hat{N}_{ai} - \hat{N}_i)(j\omega)| < \epsilon \qquad \text{for all } \omega, \quad i = 0, \ldots, m \tag{6.1.23}$$

$$|\Delta \hat{D}_j(j\omega)| := |(\hat{D}_{aj} - \hat{D}_j)(j\omega)| < \epsilon \qquad \text{for all } \omega, \quad j = 0, \ldots, n \tag{6.1.24}$$

The identifier uses the form (6.1.22), while the true plant $\hat{P}(s)$ is accurately described by (6.1.1). Consequently, the signals of (6.1.9) are now replaced by

$$\begin{aligned}
\hat{z}_{ap} &:= \hat{D}_{ao}\,\hat{y}_p - \hat{N}_{ao}\,\hat{r} \\
\hat{w}_{ai} &:= \hat{N}_{ai}\,\hat{r} \qquad i = 1, \ldots, m \\
\hat{w}_{am+j} &:= \hat{D}_{aj}\,\hat{y}_p \qquad j = 1, \ldots, n
\end{aligned} \tag{6.1.25}$$

It is important to note that (6.1.12) is still valid, since the signals w, z_p pertain to the true plant. The identifier, however, uses the signals $w_{ai}(t)$ so that the identifier output

$$z_i(t) := \theta^T(t)\, w_a(t) \tag{6.1.26}$$

where $\theta(t)$ is the parameter estimate at time t. The parameter update laws of (6.1.17), (6.1.18) are modified by replacing w by w_a, while

$$e_1(t) = z_i(t) - z_p(t)$$

$$= \theta^T(t)\, w_a(t) - \theta^{*^T} w(t)$$

$$= \phi^T(t)\, w_a(t) + \theta^{*^T} (w_a(t) - w(t)) \qquad (6.1.27)$$

Define

$$\Delta w(t) \;=\; w_a(t) - w(t)$$

Consequently, the gradient algorithm is described by

$$\dot{\phi} \;=\; \dot{\theta} \;=\; -g e_1 w_a$$

$$= -g w_a\, w_a^T \phi - g \theta^{*^T} \Delta w\, w_a \qquad (6.1.28)$$

and the least-squares algorithm by

$$\dot{\phi} \;=\; \dot{\theta} \;=\; -gPe_1\, w_a$$

$$= -gPw_a\, w_a^T \phi - gP\theta^{*^T} \Delta w\, w_a$$

$$\dot{P} \;=\; -gPw_a\, w_a^T P \qquad (6.1.29)$$

Equations (6.1.28), (6.1.29) have a similar form as without unmodeled dynamics with the exception of the extra forcing terms

$$-g\theta^{*^T} \Delta w\, w_a \qquad \text{in (6.1.28)} \qquad (6.1.30)$$

and

$$-gP\theta^{*^T} \Delta w\, w_a \qquad \text{in (6.1.29)} \qquad (6.1.31)$$

Note that Δw_p is bounded, since it is the difference between the outputs of two proper transfer functions with a bounded reference input r. Consequently, the terms in (6.1.30) and (6.1.31) are bounded. Thus, if the systems of (6.1.28) and (6.1.29) are exponentially stable in the absence of the driving terms, the robustness results of Chapter 5 (specifically theorem 5.3.1) can be used to guarantee the convergence of the parameter error to a ball around the origin. It is further readily obvious from the estimates in the statement of the theorem that the size of the ball goes to zero as Δw shrinks (or, equivalently, the inaccuracy of modeling decreases).

It therefore remains to give conditions under which the undriven systems of (6.1.28), (6.1.29) with resetting are exponentially stable. It is easy to see that this is guaranteed if w_a is persistently exciting, i.e., condition (6.1.19) holds with w replaced by w_a. It is plausible that if ϵ (the extent of mismodeling of the $\hat{N}_i$, $\hat{D}_j$) is small enough and w is persistently exciting, then w_a is also persistently exciting. This is established in the following two lemmas.

Lemma 6.1.2 Persistency of Excitation under Perturbation

If the signal $w(t) \in \mathbb{R}^{n+m}$ is persistently exciting, i.e. there exist $\alpha_1, \alpha_2, \delta > 0$ such that

$$\alpha_2 I \geq \int_{t_0}^{t_0+\delta} w(\tau) w^T(\tau) d\tau \geq \alpha_1 I \quad \text{for all } t_0 \geq 0 \tag{6.1.32}$$

and the signal $\Delta w(t) \in \mathbb{R}^{n+m}$ satisfies

$$\sup_t |\Delta w(t)| < \left[\frac{\alpha_1}{\delta}\right]^{\frac{1}{2}} \tag{6.1.33}$$

Then $w + \Delta w$ is also persistently exciting.

Proof of Lemma 6.1.2

$w + \Delta w$ is persistently exciting if there exist $\alpha'_1, \alpha'_2, \delta' > 0$, such that for all $x \in \mathbb{R}^{n+m}$ of unit norm,

$$\alpha'_2 \geq \int_{t_0}^{t_0+\delta'} \left[x^T(w(\tau) + \Delta w(\tau))\right]^2 d\tau \geq \alpha'_1 \tag{6.1.34}$$

Let $\delta' = \delta$. The upper bound of the integral in (6.1.34) is automatically verified, simply because Δw is bounded and w satisfied a similar inequality. For the lower bound, we use the triangle inequality to get

$$\left[\int_{t_0}^{t_0+\delta} \left[x^T(w(\tau) + \Delta w(\tau))\right]^2 d\tau\right]^{\frac{1}{2}}$$

$$\geq \left[\int_{t_0}^{t_0+\delta} (x^T w(\tau))^2 d\tau\right]^{\frac{1}{2}} - \left[\int_{t_0}^{t_0+\delta} (x^T \Delta w(\tau))^2 d\tau\right]^{\frac{1}{2}}$$

$$\geq \alpha_1^{1/2} - \delta^{1/2} \sup_\tau |\Delta w(\tau)| \tag{6.1.35}$$

The conclusion now follows readily from (6.1.33). □

Thus, we see that w_a is guaranteed to be persistently exciting, when w persistently exciting and Δw is sufficiently small. The claim that Δw is small, when ϵ in (6.1.23), (6.1.24) is small enough, follows from the next lemma.

Lemma 6.1.3

If $\hat{g}(s)$ is a proper, stable, nth order rational function with corresponding impulse response $g(t)$

Then the L_1 norm of $g(t)$ can be bounded by

$$\|g\|_1 = \int_0^\infty |g(\tau)|\, d\tau \le 2n \sup_\omega |\hat{g}(j\omega)| \tag{6.1.36}$$

Proof of Lemma 6.1.3 see Doyle [1984].

From lemma 6.1.3, and the definition of Δw, it is easy to verify that

$$|\Delta w(t)| \le 2\epsilon N \sup_\tau |r(\tau)| \tag{6.1.37}$$

where N is the maximum order of the ΔN_i, ΔD_j. Thus, $\Delta w(t)$ is small enough for ϵ small enough, and the persistency of excitation of w guarantees that of w_a. Using the estimate (6.1.37) and applying theorem 5.3.1, we see that the parameter error will converge to a ball of radius of order ϵ.

6.2 GLOBAL STABILITY OF INDIRECT ADAPTIVE CONTROL SCHEMES

The indirect approach is a popular technique of adaptive control. First, a non-adaptive controller is designed parametrically, that is, the controller parameters are written as functions of the plant parameters. Then, the scheme is made adaptive by replacing the plant parameters in the design calculation by their estimates at time t obtained from an on-line identifier. A reason for the popularity of the indirect approach is the considerable flexibility in the choice of both the controller and the identifier. Global stability of indirect schemes was shown by several authors in the discrete time case (Goodwin & Sin [1984], Anderson & Johnstone [1985], and others). In a continuous time context, Elliott, Cristi & Das [1985] used random parameter update times for proving convergence, and Kreisselmeier [1985, 1986] assumed that the parameters lie in a convex set in which no unstable pole-zero cancellations occur.

In Section 3.3.3, we considered the specific case of a model reference *indirect* adaptive control algorithm. We also indicated how it could be replaced by a pole placement algorithm for nonminimum phase systems in Section 3.3.5. In this section, we consider more general controller designs, following the approach of Bai & Sastry [1987]. We discuss a general, indirect adaptive control scheme for a SISO continuous time system using an identifier in conjunction with an *arbitrary* stabilizing controller. We will show that when the reference input is sufficiently rich, the input to the identifier is also sufficiently rich to cause parameter convergence in the identifier. The controller is updated only when adequate information has been obtained for a "meaningful" update. Thus, roughly speaking, the adaptive system consists of a fast parameter identification loop and a slow controller update loop. We will not need any conditions calling for the parameters to lie in a convex set or calling for lack of unstable pole-zero cancellations in the identifier. However, we will need sufficient richness conditions on the input that we were not assumed previously to establish global stability.

For application of the results, we will specialize our scheme to a pole placement adaptive controller, as well as to a factorization based adaptive stabilizer (of a kind that has attracted a great deal of interest in the literature on non-adaptive, robust control—see, for example, Vidyasagar [1985]).

6.2.1 Indirect Adaptive Control Scheme

The basic structure of an indirect adaptive controller is shown in Figure 6.4.

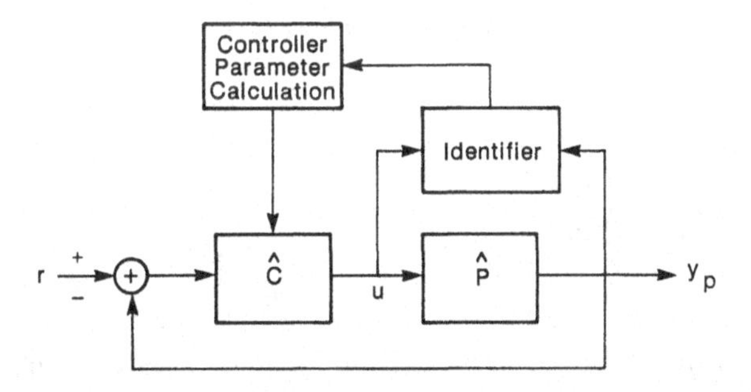

Figure 6.4 Basic Structure of an Indirect Adaptive Controller

The unknown, strictly proper plant is assumed to be described by

$$\hat{P}(s) = \frac{k_p \hat{n}_p(s)}{\hat{d}_p(s)} = \frac{\alpha_n s^{n-1} + \cdots + \alpha_1}{s^n + \beta_n s^{n-1} + \cdots + \beta_1} \tag{6.2.1}$$

with $\hat{n}_p(s)$, $\hat{d}_p(s)$ monic, coprime polynomials. The degree of $\hat{d}_p$ is n, and that of $\hat{n}_p$ is less than or equal to $n-1$ (consequently, some of the α_i's may be zero).

The compensator is a proper, m th order compensator of the form

$$\hat{C}(s) = \frac{\hat{n}_c(s)}{\hat{d}_c(s)} = \frac{a_{m+1}s^m + \cdots + a_1}{b_{m+1}s^m + \cdots + b_1} \tag{6.2.2}$$

The adaptive scheme proceeds as follows: the identifier obtains an estimate of the plant parameters. The compensator design (pole placement, model reference, etc.) is performed, assuming that the plant parameter estimates are the true parameter values (*certainty equivalence principle*). We will assume that there exists a unique compensator of the form (6.2.2) for every value of the plant parameter estimates. The hope is that, as $t \to \infty$, the identifier identifies the plant correctly, and therefore the compensator converges asymptotically to the desired one.

We first design an identifier as in Chapter 2. Define $w(t) \in \mathbb{R}^{2n}$ with Laplace transform

$$\hat{w}^T := \left[\frac{\hat{u}}{\hat{\lambda}}, \frac{s\,\hat{u}}{\hat{\lambda}}, \ldots, \frac{s^{n-1}\hat{u}}{\hat{\lambda}}, \frac{\hat{y}_p}{\hat{\lambda}}, \ldots, \frac{s^{n-1}\hat{y}_p}{\hat{\lambda}} \right] \tag{6.2.3}$$

with $\hat{\lambda}(s)$, a monic Hurwitz polynomial of the form $s^n + \lambda_n s^{n-1} + \cdots + \lambda_1$. Then

$$y_p(t) = \theta^{*^T} w(t) \tag{6.2.4}$$

where

$$\theta^{*^T} = \left[\alpha_1, \ldots, \alpha_n, \lambda_1 - \beta_1, \ldots, \lambda_n - \beta_n \right] \tag{6.2.5}$$

The identifier output is

$$y_i(t) = \theta^T(t) w(t) \tag{6.2.6}$$

where $\theta(t)$ is the estimate of θ^* at time t. If $\phi(t)$ is the parameter error $\phi(t) = \theta(t) - \theta^*$, then the identifier error $e_1(t) = y_i(t) - y_p(t)$ has the form

$$e_1(t) = \phi^T(t) w(t) \tag{6.2.7}$$

For the identification algorithm, we will use the least-squares with resetting

$$\dot{\phi}(t) = -P(t) w(t) e_1(t) \tag{6.2.8}$$

$$\dot{P}(t) = -P(t) w(t) w^T(t) P(t) \qquad t \neq t_i \tag{6.2.9}$$

with $P(t_i^+) = k_0 I > 0$, where t_i is the sequence of resetting times, *to be specified hereafter*. It is shown in the Appendix (lemma A6.2.1) that the parameter error $\phi(t)$ is bounded, even if $y_p(t)$ may not be. Further,

$\phi(t) \to 0$ asymptotically, if $w(t)$ is persistently exciting, i.e., if there exist $\alpha_1, \alpha_2, \delta > 0$, such that

$$\alpha_2 I \ \geq \ \int_{t_0}^{t_0+\delta} w(\tau) w^T(\tau) d\tau \ \geq \ \alpha_1 I \qquad \text{for all } t_0 \geq 0 \qquad (6.2.10)$$

Note however that the upper bound in (6.2.10) is not needed for the specific algorithm chosen here. Further, it has been shown in Chapter 2 that under the condition that $\hat{n}_p$, $\hat{d}_p$ are coprime polynomials, w satisfies the lower bound condition (6.2.10) if u is sufficiently rich, i.e., if the support of the spectrum of u has at least $2n$ points (assuming that $u(t)$ is stationary).

The design of the compensator is based on the plant parameter estimates $\theta(t)$. It would appear intuitive that if $\theta(t) \to \theta^*$ as $t \to \infty$, then the time-varying compensator will converge to the nominal compensator and the closed-loop system will be asymptotically stable. Therefore, the system of Figure 6.4 can be understood as a time-varying linear system which is asymptotically time-invariant and stable. The following lemma guarantees the asymptotic stability of the linear time varying system.

Lemma 6.2.1

Consider the time-varying system

$$\dot{x} \ = \ (A + \Delta A(t))x \qquad (6.2.11)$$

where A is a constant matrix with eigenvalues in the open LHP and $\| \Delta A(t) \|$ is a bounded function of t converging to zero as $t \to \infty$.

Then (6.2.11) is asymptotically stable, i.e. there exist m, $\alpha > 0$ such that the state transition matrix $\Phi(t, t_0)$ of $A + \Delta A(t)$ satisfies

$$\| \Phi(t, t_0) \| \ \leq \ m e^{-\alpha(t - t_0)} \qquad \text{for all } t > t_0$$

Update Sequence for the Controller

Although the update law (6.2.8), (6.2.9) may be shown to be asymptotically stable when w satisfies (6.2.10) (in fact only the lower bound), it is of practical importance to limit the update of the controller to instants when sufficient new information has been obtained. This is measured through the "information matrix"

$$\int_{t_i}^{t_i+\delta} w(\tau) w^T(\tau) d\tau$$

Thus, given $\gamma > 0$, we choose the update sequence t_i by $t_0 = 0$, and $t_{i+1} = t_i + \delta_i$, where δ_i satisfies

$$\delta_i \ := \ \underset{\Delta}{\text{argmin}} \int_{t_i}^{t_i + \Delta} w \, w^T d\tau \ \geq \ \gamma I \tag{6.2.12}$$

Then, the compensator parameters are held constant between t_i and t_{i+1}. We will assume that the compensator parameters are continuous functions of θ. We may now relate the richness of the reference signal $r(t)$ in Figure 6.4 to the convergence of the identifier.

Lemma 6.2.2 Convergence of the Identifier

Consider the system of Figure 6.4 with the least-squares update law (6.2.8), (6.2.9) and resetting times identical to the controller update times given in (6.2.12).

If the input r is stationary and its spectral support contains at least $3n + m$ points,

Then the identifier parameter error converges to zero exponentially as $t \to \infty$. More precisely, there exists $0 < \xi < 1$ such that

$$| \phi(t_i) | \ \leq \ \xi^i | \phi(0) | \tag{6.2.13}$$

and $\delta_i := t_{i+1} - t_i$ is a bounded sequence.

Proof of Lemma 6.2.2

The proof uses lemmas which are collected in the Appendix. In particular, lemma A6.2.1 shows the conclusion (6.2.13) if the sequence δ_i is bounded. Thus, we will establish this fact. We proceed by contradiction. If δ_i is an unbounded sequence, then one of two following possibilities occurs

(a) There exists an $i < \infty$ such that $\delta_i = \infty$ or

(b) $\delta_i \to \infty$ as $i \to \infty$.

Consider the scenario (a) first. If this happens the system becomes time invariant after t_i, since the controller is not updated. Consequently, one can find the transfer function from r to u to be

$$\hat{H}_{ur} \ = \ \frac{k_p \hat{n}_c(t_i) \hat{n}_p}{k_p \hat{n}_p \hat{n}_c(t_i) + \hat{d}_p \hat{d}_c(t_i)} \ := \ \frac{\hat{n}}{\hat{d}} \tag{6.2.14}$$

where $\hat{d}_c(t_i)$ and $\hat{n}_c(t_i)$ are the denominator and numerator of the controller at time t_i. Using (6.2.14), we may write the transfer function from r to w to be

$$\hat{H}_{wr}(s) = \frac{\hat{n}}{\hat{\lambda}\hat{d}_p\hat{d}}\left[\hat{d}_p, \ldots, s^{n-1}\hat{d}_p, k_p\hat{n}_p, \ldots, k_p\hat{n}_p s^{n-1}\right]^T \qquad (6.2.15)$$

Since the degree of $\hat{n}$ is $(n+m)$, no more than $(n+m)$ of the spectral lines of the input can correspond to zeros of the numerator polynomial. Even in this (worst-case) situation, we see from the arguments of Chapter 2 that w is persistently exciting. This fact, however, contradicts the assumption that $\delta_i = \infty$.

Now, consider scenario (b). First, notice that when the plant parameters are known, the closed loop system is time-invariant and stable so that we may write the equation relating $r(t)$ to the signal $w_m(t)$ ($w_m(t)$ corresponds to $w(t)$ in the case when $\phi(t) = 0$, as in Chapter 3)

$$\dot{z}_m = Az_m + br$$
$$w_m = Cz_m$$

where A, C are constant matrices, b is a constant vector and A is stable. For the adaptive control situation, $\phi(t) \neq 0$, but we may still write the following equation relating $r(t)$ to $w(t)$

$$\dot{z} = (A + \Delta A(t))z + (b + \Delta b(t))r$$
$$w = (C + \Delta C(t))z$$

where $\Delta A(t), \Delta C(t), \Delta b(t)$ are continuous functions of $\phi(t)$ and $\Delta A(t), \Delta C(t)$ and $\Delta b(t) \to 0$ as $\phi(t) \to 0$. If scenario (b) happens, we still have that $\phi(t) \to 0$ as $t \to 0$ from lemma A6.2.1. Further, from lemma A6.2.2 it follows that $w_m(t)$ approaches $w(t)$ for t large enough. Then the persistency of excitation of $w(t)$ follows as a consequence of the result of lemma A6.2.2 and the fact that $w_m(t)$ is persistently exciting. This, however, contradicts the hypothesis that $\delta_i \to \infty$ as $i \to \infty$.
□

Theorem 6.2.3 Asymptotic Stability of the Indirect Adaptive System
Consider the system of Figure 6.4, with the plant and compensator as in lemma 6.2.2.

If the input r is stationary and its spectral support contains at least $3n + m$ points

Then the adaptive system is asymptotically stable.

Proof of Theorem 6.2.3 follows readily from lemmas 6.2.1 and 6.2.2.

6.2.2 Indirect Adaptive Pole Placement

Pole placement is easily described in the context of Figure 6.4. The compensator $\hat{C}$ is chosen so that the closed-loop poles lie at the zeros of a given characteristic polynomial $\hat{d}_{\text{cl}}(s)$, typically of degree $(2n - 1)$. In other words, $\hat{n}_c$, $\hat{d}_c$ need to be found to satisfy

$$k_p \hat{n}_c \hat{n}_p + \hat{d}_c \hat{d}_p = \hat{d}_{\text{cl}} \tag{6.2.16}$$

When $\hat{n}_p$, $\hat{d}_p$ are coprime, equation (6.2.16) may be solved (see lemma A6.2.3 in the Appendix) for any arbitrary $\hat{d}_{\text{cl}}$ of degree $(2n - 1)$, with $\hat{n}_c$, $\hat{d}_c$ each of order $n - 1$.

When the plant is unknown, the "adaptive" pole placement scheme is mechanized by using the estimates $\hat{n}_p(t_i)$, $\hat{d}_p(t_i)$ of the numerator and denominator polynomials. Using lemma A6.2.3 again, it is easy to verify that if $k_p(t_i)\, \hat{n}_p(t_i)$, $\hat{d}_p(t_i)$ are coprime, there exist unique $\hat{n}_c(t_i)$ and $\hat{d}_c(t_i)$ of order $n - 1$ such that

$$k_p(t_i)\, \hat{n}_c(t_i) \hat{n}_p(t_i) + \hat{d}_c(t_i) \hat{d}_p(t_i) = \hat{d}_{\text{cl}} \tag{6.2.17}$$

The estimates for $k_p(t_i)\, \hat{n}_p(t_i)$ and $\hat{d}_p(t_i)$ follow from the plant parameter estimates $\theta(t)$ of the identifier. In analogy to the plant parameter vector θ, we have the parameter vector of the compensator (cf. (6.2.2) with $m = n - 1$)

$$\theta_c(t) = \left[b_0(t), \ldots, b_n(t), a_0(t), \ldots, a_n(t) \right] \tag{6.2.18}$$

As usual, θ_c^* has the interpretation of being the nominal compensator parameter vector. Further, to guarantee that $k_p(t_i)\, \hat{n}_p(t_i)$ and $\hat{d}_p(t_i)$ are coprime at the instants t_i, we need to modify the definitions of the update times. Let

$$t_{i+1} = t_i + \delta_i \tag{6.2.19}$$

where δ_i is the smallest real number satisfying

$$\int_{t_i}^{t_i + \delta_i} w w^T dt \geq \gamma I \tag{6.2.20}$$

$$k_p(t_i)\, \hat{n}_p(t_i + \delta_i) \text{ and } \hat{d}_p(t_i + \delta_i) \text{ are coprime} \tag{6.2.21}$$

More precisely (6.2.21) is satisfied by guaranteeing that the smallest singular value of the matrix in lemma A6.2.3 of the Appendix—

measuring the coprimeness of $k_p(t_i + \delta_i)\, \hat{n}_p(t_i + \delta_i)$, $\hat{d}_p(t_i + \delta_i)$—exceeds a number $\sigma > 0$. □

Theorem 6.2.4 Asymptotic Stability of the Adaptive Pole Placement Scheme

Consider the adaptive pole placement scheme, with the least squares identifier of (6.2.8)–(6.2.9) and the update sequence t_i defined by (6.2.19)–(6.2.21).

If the input $r(t)$ is stationary and its spectral support contains at least $4n - 1$ points

Then all signals in the loop are bounded and the characteristic polynomial of the closed loop system tends to $\hat{d}_{\text{cl}}(s)$. Moreover, $|\,\theta_c(t) - \theta_c^*\,| \to 0$ exponentially.

Proof of Theorem 6.2.4

The first half of the theorem follows from lemmas 6.2.1, 6.2.2—a slight modification of the arguments of lemma 6.2.2 is needed to account for the new condition (6.2.21) in the update time, but this is easy because of the convergence of the identifier and the coprimeness of the *true* $\hat{n}_p$, $\hat{d}_p$. For the second half, we see from lemma A6.2.3 in the Appendix that

$$A(\theta(t_i))\, \theta_c(t_i) \;=\; d \tag{6.2.22}$$

with d the vector of coefficients of the polynomial $\hat{d}_{\text{cl}}$. It follows from lemma A6.2.3 in the Appendix that there is an $m > 0$ such that

$$\| A(\theta(t_i)) - A(\theta^*) \| \;\le\; m |\, \theta(t_i) - \theta^* |$$

Further, subtracting (6.2.22) from

$$A(\theta^*)\theta_c^* \;=\; d \tag{6.2.23}$$

we see that

$$\Big[A(\theta(t_i)) - A(\theta^*)\Big]\, \theta_c(t_i) \;=\; -A(\theta^*)(\theta_c(t_i) - \theta_c^*)$$

Thus

$$|\, \theta_c(t_i) - \theta_c^*\,| \;\le\; m \| A^{-1}(\theta^*) \|\; |\, \theta(t_i) - \theta^*\,|\; |\, \theta_c(t_i) | \tag{6.2.24}$$

Since $|\, \theta_c(t_i) |$ is bounded (by (6.2.21)), we get that

$$|\, \theta_c(t_i) - \theta_c^* | \;\le\; m_1 |\, \theta(t_i) - \theta^* | \tag{6.2.25}$$

for some m_1. Since $\theta(t_i)$ converges to θ^* exponentially, so does $\theta_c(t_i)$ to

θ_c^*. □

6.2.3 Indirect Adaptive Stabilization—The Factorization Approach

The Factorization Approach to Controller Design

We will briefly review the factorization approach to controller design (the non-adaptive version). Consider the controller structure of Figure 6.5.

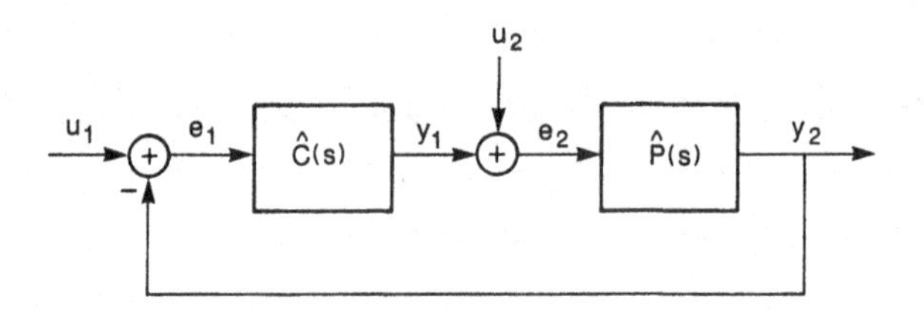

Figure 6.5 Standard One Degree of Freedom Controller

The plant $\hat{P}$ is as defined in (6.2.1) and the compensator $\hat{C}$ as in (6.2.2). The transfer function relating e_1, e_2 to u_1, u_2 is given by

$$\hat{H}_{eu} = \frac{1}{1+\hat{P}\hat{C}} \begin{bmatrix} 1 & -\hat{P} \\ \hat{C} & 1 \end{bmatrix} \tag{6.2.26}$$

The system of Figure 6.5 is BIBO stable if and only if each of the four elements of (6.2.26) is stable, i.e., belongs to $\boldsymbol{R}$, the ring of proper, stable, rational functions. The ring $\boldsymbol{R}$ is a more convenient ring than is the ring of polynomials for several reasons (see for example, Vidyasagar [1985])—including the study of the robustness properties of the closed loop systems. We assumed that $\hat{P}$ and $\hat{C}$ are factored coprimely in $\boldsymbol{R}$ (not uniquely!) as

$$\hat{P}(s) = \hat{N}_p(s)\hat{D}_p^{-1}(s)$$

$$\hat{C}(s) = \hat{D}_c^{-1}(s)\hat{N}_c(s) \tag{6.2.27}$$

From Vidyasagar [1985], it follows that the system of Figure 6.5 is BIBO stable if and only if $(\hat{N}_p\hat{N}_c + \hat{D}_p\hat{D}_c)^{-1}$ belongs to $\boldsymbol{R}$, or, equivalently, $\hat{N}_p\hat{N}_c + \hat{D}_p\hat{D}_c$ is a unimodular element of $\boldsymbol{R}$. Without loss of generality, we can state that a compensator stabilizes the system of Figure 6.5 if and only if

$$\hat{N}_p \hat{N}_c + \hat{D}_p \hat{D}_c = 1 \qquad (6.2.28)$$

Equation (6.2.28) parameterizes all stabilizing compensators. We will be interested in a parameterization of all solutions of (6.2.28) in terms of the coefficients of $\hat{N}_p$, $\hat{D}_p$. For this purpose, let (A_p, b_p, c_p^T) be a minimal realization of $\hat{P}(s)$. If $f \in \mathbb{R}^n$ and $l \in \mathbb{R}^n$ are chosen to stabilize $A_{pf} := A_p - b_p f^T$ and $A_{pl} := A_p - l c_p^T$ (such a choice is possible by the minimality of the realization of A_p, b_p, c_p^T), then, it may be shown (see Vidyasagar [1985]) that a right coprime fraction of $\hat{P}$ is given by

$$\hat{N}_p = c_p^T (sI - A_{pf})^{-1} b_p \qquad (6.2.29)$$

$$\hat{D}_p = 1 - c_p^T (sI - A_{pl})^{-1} b_p \qquad (6.2.30)$$

and further that all solutions of (6.2.28) may be written as

$$\hat{D}_c = 1 + c_p^T (sI - A_{pf})^{-1} l - \hat{Q}(s) c_p^T (sI - A_{pf})^{-1} b_p \qquad (6.2.31)$$

$$\hat{N}_c = f^T (sI - A_{pf})^{-1} l + \hat{Q}(s) \left[1 - f^T (sI - A_{pf})^{-1} b_p \right] \qquad (6.2.32)$$

with $\hat{Q}(s) \in \boldsymbol{R}$ an arbitrary element chosen to meet other performance criteria (such as minimization of the disturbance to output map, obtaining the desired closed loop transfer function, optimal desensitization to unmodeled dynamics, etc.).

The optimal choice of $\hat{Q}(s)$ depends on the plant parameters. However, such a choice of $\hat{Q}(s)$ may not be unique or depend continuously on plant parameters. The optimal choice of $\hat{Q}(s)$ is the topic of the so-called H^∞ *optimal control* systems design methodology. In this chapter, we will not concern ourselves with anything more than stabilization, and use a fixed $\hat{Q}(s)$ rather than one whose calculation depends on the current estimate of plant parameters. For simplicity we will, in fact, fix $\hat{Q}(s) = 0$ in the adaptive stabilization which follows.

Adaptive Stabilization Using the Factorization Approach

The objective is to design the compensator $\hat{C}(s)$ adaptively based on the estimate θ of the plant parameters so that the closed loop system is asymptotically stable with all signals bounded. Set $u_2(t) \equiv 0$.

The identifier and compensator update sequence $\{ t_i \}$ are specified in (6.2.19)–(6.2.21). The only difficulty in mechanizing the n th order compensator of (6.2.31) and (6.2.32) (with $\hat{Q}(s) = 0$) lies in calculating

$f(t_i), l(t_i) \in \mathbb{R}^n$ to stabilize $A_p(t_i)$, the estimate of A_p based on the current plant parameter estimate. To that effect, we consider the controllable form and observable form realizations of the plant estimate

$$A_p(t_i) = \begin{bmatrix} 0 & 1 & & 0 \\ 0 & 0 & & \cdot \\ \cdot & \cdot & \cdots & \cdot \\ \cdot & \cdot & & 1 \\ -\beta_1(t_i) & -\beta_2(t_i) & & -\beta_n(t_i) \end{bmatrix} \qquad b_p = \begin{bmatrix} 0 \\ \cdot \\ \cdot \\ 0 \\ 1 \end{bmatrix} \tag{6.2.33}$$

$$c_p^T(t_i) = [\alpha_1(t_i)\, \alpha_2(t_i) \cdots \alpha_n(t_i)]$$

and

$$\tilde{A}_p(t_i) = \begin{bmatrix} 0 & 0 & 0 & -\beta_1(t_i) \\ 1 & 0 & \cdot & \cdot \\ \cdot & \cdot & \cdot & \cdot \\ \cdot & \cdot & \cdot & \cdot \\ 0 & \cdot & 1 & -\beta_n(t_i) \end{bmatrix} \qquad \tilde{b}_p(t_i) = \begin{bmatrix} \alpha_1(t_i) \\ \cdot \\ \cdot \\ \cdot \\ \alpha_n(t_i) \end{bmatrix} \tag{6.2.34}$$

$$\tilde{c}_p^T(t_i) = [0 \cdots 0\, 1]$$

Let the transformation matrix $M(t_i)$ relate the realization (6.2.33) to (6.2.34)

$$\tilde{A}_p(t_i) = M(t_i)\, A_p M(t_i)^{-1}$$
$$\tilde{b}_p(t_i) = M(t_i)\, b_p(t_i)$$
$$\tilde{c}_p^T(t_i) = c_p^T(t_i) M(t_i)^{-1}$$

Note that $M(t_i)$ is the only calculation that needs to be performed. Now $f(t_i)$ and $l(t_i)$ are easily read off. Indeed, consider any stable polynomial $s^n + p_n s^{n-1} + \cdots + p_1$. Then it is easy to see that

$$f^T(t_i) = [-p_1 + \beta_1(t_i), \ldots, -p_n + \beta_n(t_i)]^T$$

and

$$l(t_i) = M^{-1}(t_i)[-p_1 + \beta_1(t_i), \ldots, -p_n + \beta_n(t_i)]^T \tag{6.2.35}$$

Therefore $A_p(t_i) - b_p f^T(t_i)$ and $A_p(t_i) - l(t_i) c_p^T(t_i)$ have their eigenvalues at the zeros of $s^n + p_n s^{n-1} + \cdots + p_1$. The compensator of (6.2.31), (6.2.32) with $\hat{Q}(s) = 0$ can be made adaptive by choosing $\hat{C}(t_i) = \hat{D}_c^{-1}(t_i)\, \hat{N}_c(t_i)$ with

$$\hat{N}_c(t_i) = f^T(t_i)(sI - A_{pf}(t_i))^{-1} l(t_i) \tag{6.2.36}$$

$$\hat{D}_c(t_i) = 1 + c_p^T(t_i)(sI - A_{pf}(t_i))^{-1} l(t_i) \tag{6.2.37}$$

Then, as before we have the following theorem:

Theorem 6.2.5 Asymptotic Stability of the Adaptive Identifier Using the Factorization Approach

Consider the set up of Figure 6.5, with the least squares identifier of (6.2.8), (6.2.9), the update sequence $\{t_i\}$ of (6.2.19)–(6.2.21), and the compensator of (6.2.36), (6.2.37).

If the input is stationary and its spectral support is not concentrated on $k \leq 4n$ points,

Then the adaptive system is asymptotically stable.

Proof of Theorem 6.2.5 follows as the proof of theorem 6.2.4.

6.3 MULTIVARIABLE ADAPTIVE CONTROL

6.3.1 Introduction

The extension of adaptive control algorithms for single-input single-output systems (SISO) to multi-input multi-output systems (MIMO) is far from trivial. Indeed, transfer function properties which are easily established for SISO systems are much more complex for MIMO systems. The issue of the parameterization of the adaptive controllers becomes a dominant problem. Several MIMO adaptive control algorithms were proposed based on the model reference approach (Singh & Narendra [1982], Elliott & Wolovich [1982], Goodwin & Long [1980], Johansson [1987]), the pole placement approach (Prager & Wellstead [1981], Elliot, Wolovich & Das [1984]), and quadratic optimization approaches (Borisson [1979], Koivo [1980]). See also the review/survey papers by Dugard & Dion [1985] and Elliott & Wolovich [1984].

The understanding of parameterization issues benefited significantly from progress in the theory of nonadaptive MIMO control theory. In Section 6.3.2, we briefly review some basic results. More details may be found in Kailath [1980], Callier & Desoer [1984], and Vidyasagar [1985]. These results will help us to establish the parameterization of MIMO adaptive controllers in Section 6.3.3. Once the parameterization is established, the design of adaptive schemes will follow in a more straightforward manner from SISO theory (Section 6.3.4).

As previously, we will concentrate our discussion on a model reference adaptive control scheme and follow an approach parallel to that of Chapter 3. Alternate adaptive control schemes may be found in the references just mentioned. We will say very little about the dynamic properties of MIMO adaptive control systems. Indeed, this topic is not well understood so far (even less than for SISO systems!).

6.3.2 Preliminaries

6.3.2.1 Factorization of Transfer Function Matrices

Right and Left Fractions

Consider a square transfer function matrix $\hat{P}(s)$, with p rows and p columns, whose elements are rational functions of s with real coefficients. The set of such matrices is denoted $\hat{P}(s) \in \mathbb{R}^{p \times p}(s)$. A rational transfer function is expressed as the ratio of two polynomials in s. Similarly, a rational transfer function matrix may be represented as the ratio of two polynomial matrices. The set of polynomial matrices of dimension $p \times p$ is denoted $\mathbb{R}^{p \times p}[s]$ (note the slight difference in notation). A pair of polynomial matrices $(\hat{N}_R, \hat{D}_R)$ is called a *right fraction (r.f.)* of $\hat{P}(s) \in \mathbb{R}^{p \times p}(s)$ if

- $\hat{N}_R, \hat{D}_R \in \mathbb{R}^{p \times p}[s]$
- $\hat{D}_R$ nonsingular, i.e., $\det(\hat{D}_R(s)) \neq 0$ for almost all s.
- $\hat{P} = \hat{N}_R \hat{D}_R^{-1}$

Similarly, a pair $(\hat{D}_L, \hat{N}_L)$ is called a *left fraction (l.f.)* of $\hat{P}(s) \in \mathbb{R}^{p \times p}(s)$ *if*

- $\hat{N}_L, \hat{D}_L \in \mathbb{R}^{p \times p}[s]$
- $\hat{D}_L$ nonsingular, i.e. $\det(\hat{D}_L(s)) \neq 0$ for almost all s
- $\hat{P} = \hat{D}_L^{-1} \hat{N}_L$

Given a right fraction $(\hat{N}_R, \hat{D}_R)$, another right fraction may be obtained by multiplying $\hat{N}_R$ and $\hat{D}_R$ on the right by a nonsingular polynomial matrix $\hat{R}(s) \in \mathbb{R}^{p \times p}[s]$, i.e.

$$\begin{array}{lll} \hat{N}_R, \hat{D}_R \text{ r.f. of } \hat{P} & \Rightarrow & \hat{N}_{R_1} = \hat{N}_R \hat{R} \quad \text{r.f. of } \hat{P} \\ \hat{R} \text{ nonsingular} & & \hat{D}_{R_1} = \hat{D}_R \hat{R} \end{array}$$

The matrix $\hat{R}$ is called a *common right divisor* of $\hat{N}_{R_1}$ and $\hat{D}_{R_1}$. It is called the *greatest common right divisor (gcrd)* of $\hat{N}_{R_1}$ and $\hat{D}_{R_1}$, if any other common right divisor of $\hat{N}_{R_1}$ and $\hat{D}_{R_1}$ is also a common right divisor of $\hat{R}$. In fact, "the" gcrd is not unique, but all gcrd's are equivalent in a sense to be defined hereafter.

Similar definitions follow for a left fraction $(\hat{D}_L, \hat{N}_L)$. Given $(\hat{D}_L, \hat{N}_L)$ a left fraction of $\hat{P}(s)$ and a nonsingular matrix $\hat{L}$, $(\hat{L}\hat{D}_L, \hat{L}\hat{N}_L)$ is also a left fraction of $\hat{P}(s)$. Greatest common left divisors are defined by making the appropriate transpositions.

Right and Left Coprime Fractions—Poles and Zeros

A polynomial matrix $\hat{D}_R(s)$ is called *unimodular* if its inverse is a polynomial matrix. A necessary and sufficient condition is that $\det \hat{D}_R(s)$ is a real number different from 0. Clearly, a unimodular matrix is nonsingular. Further, multiplying a polynomial matrix by a unimodular matrix does not affect its rank or the degree of its determinant.

Two matrices $\hat{R}_1$ and $\hat{R}_2$ are said to be *right equivalent* if there exists a unimodular matrix $\hat{R}$ such that $\hat{R}_2 = \hat{R}_1 R$. Given a pair $(\hat{N}_R, \hat{D}_R)$ and a gcrd $\hat{R}_1$, the matrix $\hat{R}_2 = \hat{R}_1\hat{R}$ is also a gcrd if $\hat{R}$ is unimodular. In fact, it may be shown that all gcrd's are so related, that is, *all gcrd's are right equivalent.* Extracting the gcrd $\hat{R}$ of a right fraction $(\hat{N}_{R_1}, \hat{D}_{R_1})$ is pretty much like extracting common factors in a scalar transfer function. The new pair $(\hat{N}_R = \hat{N}_{R_1}\hat{R}^{-1}, \hat{D}_R = \hat{D}_{R_1}\hat{R}^{-1})$ is also a right fraction so that $\hat{P} = \hat{N}_{R_1} D_{R_1}^{-1} = \hat{N}_R \hat{D}_R^{-1}$. Note that the gcrd's of $(\hat{N}_R, \hat{D}_R)$ are unimodular. In general, two matrices $(\hat{N}_R, \hat{D}_R)$ are called *right coprime* if their gcrd is unimodular. It may be shown that for all $\hat{P}(s) \in \mathbb{R}^{p \times p}(s)$, there exists a *right coprime fraction* of $\hat{P}(s)$; that is,

- $\hat{N}_R, \hat{D}_R \in \mathbb{R}^{p \times p}[s]$ right coprime
- $\hat{D}_R$ nonsingular
- $\hat{P} = \hat{N}_R \hat{D}_R^{-1}$

In analogy to the SISO case, the poles and zeros of $\hat{P}$ are defined as

- p is a *pole* of $\hat{P}$ if $\det(\hat{D}_R(p)) = 0$
- z is a *zero* of $\hat{P}$ if $\det(\hat{N}_R(z)) = 0$

where $(\hat{N}_R, \hat{D}_R)$ is a right coprime fraction of $\hat{P}$. Similarly, n = order of the system $= \partial \det(\hat{D}_R(s))$. It may be shown that these definitions correspond to the similar definitions for a minimal state-space realization of a proper $\hat{P}(s)$.

Similar definitions are found for left coprime fractions. It follows that rcf and lcf of a plant $\hat{P}$ must satisfy $\det(\hat{D}_R(s)) = \det(\hat{D}_L(s))$ and $\det(\hat{N}_R(s)) = \det(\hat{N}_L(s))$, except for a constant real factor.

Properness, Column Reduced Matrices

We restrict our attention to proper and strictly proper transfer function matrices, defined as follows:

$$\hat{P}(s) \in \mathbb{R}_p^{p \times p}(s) \quad \text{if} \quad \lim_{s \to \infty} \hat{P}(s) \text{ exists}$$

$$\hat{P}(s) \in \mathbb{R}_{p,o}^{p \times p}(s) \quad \text{if} \quad \lim_{s \to \infty} \hat{P}(s) = 0$$

Consider a right fraction $(\hat{N}_R, \hat{D}_R)$, and define the *column degrees* as

$$\partial_{cj}(\hat{D}_R) = \max_i [\partial(\hat{D}_R)_{ij}]$$

The following fact is easily established:

$$\hat{P}(s) \in \mathbb{R}_p^{p \times p}(s) \Rightarrow \partial_{cj}(\hat{N}_R) \leq \partial_{cj}(\hat{D}_R)$$

$$\hat{P}(s) \in \mathbb{R}_{p,o}^{p \times p}(s) \Rightarrow \partial_{cj}(\hat{N}_R) < \partial_{cj}(\hat{D}_R)$$

The converse is true if we introduce the concept of column reduced matrices. First define the *highest column degree coefficient matrix* D_{hc}

$$(D_{hc})_{ij} = \text{coefficient of the term of degree } \partial_{cj}(\hat{D}_R) \text{ in } (D_R)_{ij}$$

A matrix is called *column reduced* (also *column proper*) if D_{hc} is non-singular. If $(\hat{N}_R, \hat{D}_R)$ is a right fraction of $\hat{P}(s)$ and $\hat{D}_R$ is column reduced, then

$$\partial_{cj}(\hat{N}_R) \leq \partial_{cj}(\hat{D}_R) \Leftrightarrow \hat{P}(s) \in \mathbb{R}_p^{p \times p}(s)$$

$$\partial_{cj}(\hat{N}_R) < \partial_{cj}(\hat{D}_R) \Leftrightarrow \hat{P}(s) \in \mathbb{R}_{p,o}^{p \times p}(s)$$

If $(\hat{N}_R, \hat{D}_R)$ is a right *coprime* fraction of $\hat{P}(s)$, and $\hat{D}_R$ is column reduced, we call

$$\mu_j = \partial_{cj}(\hat{D}_R) := \textit{controllability indices of } \hat{P}$$

$$\mu = \max_j (\mu_j) := \textit{controllability index of } \hat{P}$$

It is a remarkable fact that $\{ \mu_j \}$ is invariant (see Kailath [1980]). In other words, the controllability indices are the same (modulo permutations) for all r.c.f. $(\hat{N}_R, \hat{D}_R)$ with $\hat{D}_R$ column reduced. Note also that

$$n = \text{order of the system} = \sum_{j=1}^{p} \mu_j$$

It may also be shown that the definition of controllability indices correspond to the alternate definition for a minimal state-space realization.

Given a matrix $\hat{P}(s) \in \mathbb{R}_p^{p \times p}(s)$ or $\hat{P}(s) \in \mathbb{R}_{p,o}^{p \times p}(s)$, it is always possible to find a rcf $(\hat{N}_R, \hat{D}_R)$ such that $\hat{D}_R$ is column reduced. The procedure is somewhat lengthy and is discussed in Kailath [1980]. The first step is to obtain a right coprime fraction (using a Hermite row form decomposition), and the second step is to reduce the matrix to a column reduced form, multiplying further on the right by some appropriately chosen unimodular matrix.

Properness, Row Reduced Matrices

Similar facts and definitions hold for left fractions and are briefly summarized. The *highest row degree coefficient matrix* D_{hr} is defined as:

$$(D_{hr})_{ij} = \text{coefficient of the term of degree } \partial_{ri}(\hat{D}_L) \text{ in } (\hat{D}_L)_{ij}$$

$$\text{where} \quad \partial_{ri}(\hat{D}_L) = \max_j (\partial(\hat{D}_L)_{ij}) \text{ are the } \textit{row degrees} \text{ of } \hat{D}_L .$$

The matrix $\hat{D}_L$ is called *row reduced* if D_{hr} is nonsingular. If $(\hat{D}_L, \hat{N}_L)$ is a left fraction of $\hat{P}(s)$, and $\hat{D}_L$ is row reduced, then

$$\partial_{ri}(\hat{N}_L) \leq \partial_{ri}(\hat{D}_L) \iff \hat{P}(s) \in \mathbb{R}_p^{p \times p}(s)$$

$$\partial_{ri}(\hat{N}_L) < \partial_{ri}(\hat{D}_L) \iff \hat{P}(s) \in \mathbb{R}_{p,o}^{p \times p}(s)$$

When $(\hat{D}_L, \hat{N}_L)$ is a left *coprime* fraction of $\hat{P}(s)$, with $\hat{D}_L$ row reduced, we define

$$\nu_i = \partial_{ri}(\hat{D}_L) := \textit{observability indices} \text{ of } \hat{P}$$

$$\nu = \max_i (\nu_i) := \textit{observability index} \text{ of } \hat{P}$$

The set of observability indices is invariant. They are different from the controllability indices, although related by

$$n = \text{system order} = \sum_{i=1}^{p} \nu_i = \sum_{j=1}^{p} \mu_j$$

Polynomial Matrix Division

Consider, first, scalar polynomials. Given two polynomials $\hat{n}$ and $\hat{d}$, the standard division algorithm provides $\hat{q}$ and $\hat{r}$ such that

$$\hat{n} = \hat{q}\,\hat{d} + r \qquad \partial\hat{r} < \partial\hat{d}$$

This procedure is equivalent to separating the strictly proper and not strictly proper parts of a transfer function:

$$\frac{\hat{n}}{\hat{d}} = \hat{q} + \frac{\hat{r}}{\hat{d}} \qquad \hat{q} \text{ polynomial },\ \frac{\hat{r}}{\hat{d}} \text{ strictly proper}$$

From this observation, the following proposition is obtained in the multivariable case.

Proposition 6.3.1 Polynomial Matrix Division

Let $\hat{N}_R, \hat{D}_R, \hat{N}_L, \hat{D}_L \in \mathbb{R}^{p \times p}[s]$ with $\hat{D}_R, \hat{D}_L$ nonsingular. Let $\hat{D}_R$ be column reduced and $\hat{D}_L$ be row reduced.

Then There exists $\hat{Q}_R, \hat{R}_R, \hat{Q}_L, \hat{R}_L \in \mathbb{R}^{p \times p}[s]$ such that

$$\hat{N}_R = \hat{Q}_R \hat{D}_R + \hat{R}_R \qquad \partial_{cj}(\hat{R}_R) < \partial_{cj}(\hat{D}_R)$$

$$\hat{N}_L = \hat{D}_L \hat{Q}_L + \hat{R}_L \qquad \partial_{ri}(\hat{R}_L) < \partial_{ri}(\hat{D}_L)$$

Proof of Proposition 6.3.1

The elements of $\hat{N}_R \hat{D}_R^{-1}$ are rational functions of s. Divide each numerator by its denominator and call the matrix of quotients $\hat{Q}_R$. Therefore $\hat{N}_R \hat{D}_R^{-1} = \hat{Q}_R + \hat{S}_R$ where $\hat{S}_R \in \mathbb{R}_{p,o}^{p \times p}(s)$. Let $\hat{R}_R = \hat{S}_R \hat{D}_R$, i.e. $\hat{S}_R = \hat{R}_R \hat{D}_R^{-1}$.

Since $\hat{S}_R \hat{D}_R = \hat{N}_R - \hat{Q}_R \hat{D}_R$, $\hat{R}_R$ is a polynomial matrix. Further, $\hat{S}_R$ being strictly proper and $\hat{D}_R$ column reduced, the column degrees of $\hat{R}_R$ must be strictly less than those of $\hat{D}_R$. A similar proof establishes the fact for left fractions. □

6.3.2.2 Interactor Matrix and Hermite Form

Interactor Matrix

In Chapter 3, we observed that SISO model reference adaptive control requires the knowledge of the relative degree of the transfer function $\hat{P}(s)$. The extension of the concept of relative degree to transfer function matrices is not trivial and must take into account the high-frequency interactions between different inputs and outputs. The concept of an *interactor matrix* follows in a natural way, by taking the following approach. Note that the knowledge of the relative degree of a scalar transfer function $\hat{P}(s) \in \mathbb{R}_{p,o}(s)$ is equivalent to the knowledge of a monic polynomial $\hat{\xi}(s)$ such that

$$\lim_{s \to \infty} \hat{\xi}(s) \hat{P}(s) = k_p \neq 0$$

Then, the relative degree of $\hat{P}(s)$ is equal to the degree of $\hat{\xi}(s)$. The scalar k_p was earlier called the high-frequency gain of the plant $\hat{P}$.

The high-frequency behavior of MIMO systems is similarly determined by a polynomial matrix such that

$$\lim_{s \to \infty} \hat{\xi}(s)\hat{P}(s) = K_p \quad \text{nonsingular}$$

We must assume here that $\hat{P}(s)$ is itself nonsingular. The matrix $\hat{\xi}(s)$ is not unique, unless its structure is somewhat restricted. It is shown in Wolovich & Falb [1976] that there exists a unique matrix $\hat{\xi}(s)$ satisfying the following conditions.

Definition Interactor Matrix

The *interactor matrix* of a nonsingular plant $\hat{P}(s) \in \mathbb{R}_p^{p \times p}(s)$ is the unique matrix $\hat{\xi} \in \mathbb{R}^{p \times p}(s)$ such that

$$\lim_{s \to \infty} \hat{\xi}(s)\hat{P}(s) = K_p \quad \text{nonsingular}$$

$$\hat{\xi}(s) = \hat{\Sigma}(s)\hat{\Delta}(s)$$

where

- $\hat{\Delta}(s) = \text{diag}\,(s^{r_i})$

- $\hat{\Sigma}(s) = \begin{bmatrix} 1 & 0 & 0 & \cdot & 0 \\ \hat{\sigma}_{21}(s) & 1 & 0 & \cdot & 0 \\ \hat{\sigma}_{31}(s) & \hat{\sigma}_{32}(s) & 1 & \cdot & 0 \\ \cdot & \cdot & \cdot & \cdot & \cdot \\ \hat{\sigma}_{p1}(s) & \cdot & \cdot & \cdot & 1 \end{bmatrix}$

- any polynomial $\hat{\sigma}_{ij}(s)$ is divisible by s (or is zero)

The matrix K_p is called the *high frequency gain* of the plant $\hat{P}(s)$. The integers r_i extend the notion of the relative degree r of an SISO transfer function. The matrix $\hat{\Sigma}(s)$, which becomes 1 in the SISO case, describes the high-frequency interconnections between different inputs and outputs.

Hermite Normal Form

Another approach, leading to an equivalent definition, is found in Morse [1976]. For this, one notes that proper rational functions of s form a *ring*. A division algorithm may also be defined, where the *gauge* of an element is its relative degree. Therefore, $\hat{P}(s)$ is a matrix whose elements belong to a principal ideal domain, and may be factored as

$$\hat{P}(s) = \hat{H}(s) \cdot \hat{U}(s) \qquad \hat{H}, \hat{U} \in \mathbb{R}_p^{p \times p}(s)$$

where $\hat{H}$ is the *Hermite column form* of $\hat{P}$. The Hermite column form of $\hat{H}$ is a lower triangular matrix, such that the elements below the diagonal are either zero, or have relative degree strictly less than the diagonal element on the same row. The matrix $\hat{U}$ is unimodular in $\mathbb{R}_p^{p \times p}(s)$, that is, its inverse is a proper rational matrix. The unimodularity of $\hat{U}$ is equivalent to

$$\lim_{s \to \infty} \hat{U}(s) = K_u \quad \text{nonsingular}$$

Morse [1979] showed that, with some slight modifications, one could *uniquely* define the *Hermite normal form* as follows

Definition Hermite Normal Form

The *Hermite normal form* of a nonsingular plant $\hat{P}(s) \in \mathbb{R}_p^{p \times p}(s)$ is the unique matrix $\hat{H} \in \mathbb{R}_p^{p \times p}(s)$ such that

$$\hat{P}(s) = \hat{H}(s) \cdot \hat{U}(s) \qquad \hat{U} \text{ unimodular in } \mathbb{R}_p^{p \times p}(s)$$

$$\hat{H} = \begin{bmatrix} \dfrac{1}{(s+a)^{r_1}} & 0 & \cdot & \cdot \\ \dfrac{\hat{h}_{21}(s)}{(s+a)^{r_2 - 1}} & \dfrac{1}{(s+a)^{r_2}} & \cdot & \cdot \\ \cdot & \cdot & \cdot & \cdot \\ \cdot & \cdot & \cdot & \dfrac{1}{(s+a)^{r_p}} \end{bmatrix}$$

where $\partial \hat{h}_{ij}(s) \le r_i - 1$ and a is arbitrary, but fixed *a priori*.

As shown in the following proposition, the interactor matrix and the Hermite normal form are completely equivalent.

Proposition 6.3.2 Interactor Matrix and Hermite Normal Form Equivalence

Let $\hat{\xi}(s)$ be the interactor matrix of $\hat{P}(s) \in \mathbb{R}_p^{p \times p}(s)$. Let $\hat{H}(s)$ be the Hermite normal form of $\hat{P}(s)$ for $a = 0$.

Then $\hat{\xi}(s) = \hat{H}^{-1}(s)$

Proof of Proposition 6.3.2

We let $\hat{\xi} = \hat{H}^{-1}$ and show that it satisfies the conditions in the definition of the interactor matrix. First, note that $\hat{\xi}\hat{P} = \hat{H}^{-1}\hat{P} = \hat{U}$. Since $\hat{U}$ is unimodular, $\lim_{s \to \infty} \hat{\xi}\hat{P} = K_u$ nonsingular. Next, decompose

$$\hat{H} = \begin{bmatrix} s^{-r_1} & 0 & \cdot & 0 \\ 0 & s^{-r_2} & \cdot & \cdot \\ \cdot & \cdot & \cdot & \cdot \\ 0 & 0 & \cdot & s^{-r_p} \end{bmatrix} \begin{bmatrix} 1 & 0 & \cdot & 0 \\ s\hat{h}_{21}(s) & 1 & \cdot & 0 \\ \cdot & \cdot & \cdot & \cdot \\ s\hat{h}_{p1}(s) & \cdot & \cdot & \cdot \end{bmatrix}$$

$$= \hat{\Delta}^{-1}\hat{\Sigma}^{-1} = (\hat{\Sigma}\hat{\Delta})^{-1}$$

Clearly, $\hat{\Delta}$ so defined satisfies the required properties, and $\hat{\Sigma}$ is a lower triangular matrix with 1's on the diagonal. The off-diagonal terms satisfy

$$\begin{aligned} \hat{\sigma}_{21}(s) &= -s\hat{h}_{21}(s) \\ \hat{\sigma}_{31}(s) &= s(\hat{h}_{31}(s) - \hat{h}_{32}(s)\hat{h}_{31}(s)) \\ \hat{\sigma}_{32}(s) &= -s\hat{h}_{32}(s) \\ &\cdots \end{aligned}$$

so that $\hat{\Sigma}(s)$ also satisfies the required conditions. □

Hermite Form and Model Reference Control

The significance of the Hermite normal form in model reference adaptive control may be understood from the following discussion. An arbitrary linear time-invariant (LTI) controller may be represented as

$$u = \hat{C}_{FF}(r) + \hat{C}_{FB}(y_p)$$

where $\hat{C}_{FF}$ is a *feedforward* controller and $\hat{C}_{FB}$ is a *feedback* controller, so that the closed-loop transfer function is given by

$$\begin{aligned} y_p &= (I - \hat{P}\hat{C}_{FB})^{-1}\hat{P}\hat{C}_{FF}(u) \\ &= \hat{P}(I - \hat{C}_{FB}\hat{P})^{-1}\hat{C}_{FF}(u) \end{aligned}$$

$$= \hat{H}\hat{U}(I - \hat{C}_{FB}\hat{P})^{-1}\hat{C}_{FF}(u)$$

The transfer function is equal to the reference model transfer function $\hat{M}$ if

$$\hat{M} = \hat{H}\hat{U}(I - \hat{C}_{FB}\hat{P})^{-1}\hat{C}_{FF}$$

Assume now that the plant is *strictly proper*, and restrict the controller to be proper. Then, the transfer function $\hat{U}(I - \hat{C}_{FB}\hat{P})^{-1}\hat{C}_{FF}$ is proper. In other words, the reference model must be the product of the plant's Hermite form times an (arbitrary) proper transfer function. For SISO systems, this is equivalent to saying that a proper compensator cannot reduce the relative degree of a strictly proper plant.

6.3.3 Model Reference Adaptive Control—Controller Structure

With the foregoing preliminaries, the assumptions required for multi-input multi-output (MIMO) model reference adaptive control will look fairly similar to the assumptions in the SISO case.

Assumptions

(A1) Plant Assumptions

The plant is a strictly proper MIMO LTI system, described by a square, nonsingular, transfer function matrix

$$\hat{P} = \hat{H}\hat{U} \in \mathbb{R}_{p,o}^{p \times p}(s)$$

where $\hat{H}$ is a stable Hermite normal form of $\hat{P}$, obtained by setting $a > 0$. $\hat{H}$ is assumed known. The plant is minimum phase, and the observability index ν is known (an upper bound on the order of the system is, therefore, νp).

(A2) Reference Model Assumptions

The reference model is described by

$$\hat{M} = \hat{H}\hat{M}_0 \in \mathbb{R}_{p,o}^{p \times p}(s)$$

where $\hat{M}_0$ is a proper, stable transfer function matrix and $\hat{H}$ is the Hermite normal form of the plant.

(A3) Reference Input Assumptions

The reference input $r(\cdot) \in \mathbb{R}^m$ is piecewise continuous and bounded on $\mathbb{R}_+$.

Controller Structure

First, note that all the dynamics of $\hat{M}_0$ may be realized by prefiltering the reference input r. Therefore, we define

$$\bar{r} = \hat{M}_0(r)$$

so that

$$y_m = \hat{H}(\bar{r}) \tag{6.3.1}$$

In this manner, the model reference adaptive control problem is replaced by a problem where the reference model $\hat{M}$ is equal to the Hermite normal form of the plant.

The controller structure used for multivariable adaptive control is similar to the SISO structure, provided that adequate transpositions are made. Let $\hat{\lambda}(s)$ be an arbitrary, monic, Hurwitz polynomial of degree $\nu - 1$ (where ν is the observability index of $\hat{P}$). Define $\hat{\Lambda}(s) \in \mathbb{R}^{p \times p}(s)$ such that

$$\hat{\Lambda}(s) = \operatorname{diag}[\hat{\lambda}(s)] \tag{6.3.2}$$

The controller is represented in Figure 6.6.

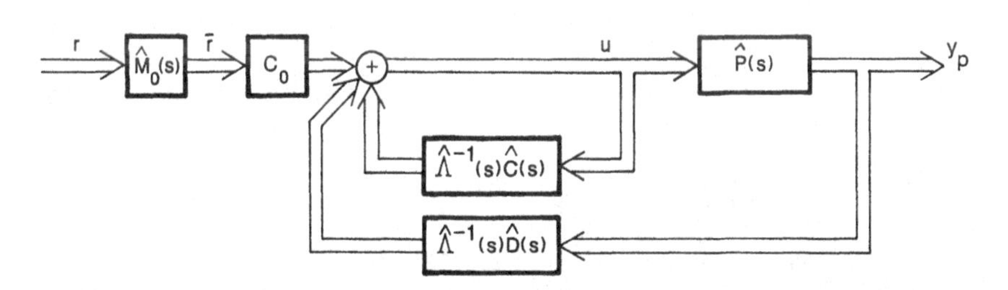

Figure 6.6 MIMO Controller Structure

It is defined by

$$\bar{r} = \hat{M}_0(r)$$

$$u = C_0\bar{r} + \hat{\Lambda}^{-1}(s)\hat{C}(s)(u) + \hat{\Lambda}^{-1}(s)\hat{D}(s)(y_p) \tag{6.3.3}$$

where $C_0 \in \mathbb{R}^{p \times p}$, $\hat{C}(s), \hat{D}(s) \in \mathbb{R}^{p \times p}(s)$. By the foregoing choice of $\hat{\Lambda}$, $\hat{\Lambda}^{-1}\hat{C} = \hat{C}\hat{\Lambda}^{-1}$, and $\hat{\Lambda}^{-1}\hat{D} = \hat{D}\hat{\Lambda}^{-1}$. Further, let $\partial\hat{C} \le \nu - 2$, $\partial\hat{D} \le \nu - 1$ (where $\partial\hat{C}$ denotes the maximum degree of all elements of $\hat{C}$).

Now, consider a right coprime fraction $(\hat{N}_R, \hat{D}_R)$ of $\hat{P}$. Therefore

$$\hat{D}_R(x) = u \qquad y_p = \hat{N}_R(x) \tag{6.3.4}$$

where x is a pseudo-state of $\hat{P}$. Combining with the expression of the controller

$$\hat{D}_R(x) = C_0\bar{r} + \hat{\Lambda}^{-1}\hat{C}(u) + \hat{\Lambda}^{-1}\hat{D}(y_p) \tag{6.3.5}$$

so that

$$\hat{\Lambda}\hat{D}_R x = \hat{\Lambda}C_0\bar{r} + \hat{C}\hat{D}_R x + \hat{D}\hat{N}_R x \tag{6.3.6}$$

$$[(\hat{\Lambda} - \hat{C})\hat{D}_R - \hat{D}\hat{N}_R]x = \hat{\Lambda}C_0\bar{r} \tag{6.3.7}$$

Therefore, the output y_p is given by

$$y_p = \hat{N}_R[(\hat{\Lambda} - \hat{C})\hat{D}_R - \hat{D}\hat{N}_R]^{-1}\hat{\Lambda}C_0(\bar{r}) \tag{6.3.8}$$

As in the SISO case, this leads us to a proposition guaranteeing that there exists a nominal controller of *prescribed degree* such that the closed-loop transfer function matches the reference model ($\hat{H}$) transfer function.

Proposition 6.3.3 MIMO Matching Equality

There exist C_0^*, $\hat{C}^*$, $\hat{D}^*$ such that the transfer function from $\bar{r} \to y_p$ is $\hat{H}$.

Proof of Proposition 6.3.3

The transfer function from $\bar{r}$ to y_p is $\hat{H}$ if and only if the following equality is satisfied

$$\hat{N}_R[(\hat{\Lambda} - \hat{C}^*)\hat{D}_R - \hat{D}^*\hat{N}_R]^{-1}\hat{\Lambda}C_0^* = \hat{H} \tag{6.3.9}$$

Since $\hat{H}^{-1}$ is a polynomial matrix, the foregoing equality may be transformed into a polynomial matrix equality, reminiscent of the *matching equality* of Chapter 3

$$(\hat{\Lambda} - \hat{C}^*)\hat{D}_R - \hat{D}^*\hat{N}_R = \hat{\Lambda}C_0^*\hat{H}^{-1}\hat{N}_R \tag{6.3.10}$$

First, we determine C_0^*. Multiply both sides by $\hat{D}_R^{-1}$ on the right and $\hat{\Lambda}^{-1}$ on the left. Then

$$(I - \hat{\Lambda}^{-1}\hat{C}^*) - \hat{\Lambda}^{-1}\hat{D}^*\hat{N}_R\hat{D}_R^{-1} = C_0^*\hat{H}^{-1}\hat{N}_R\hat{D}_R^{-1} \tag{6.3.11}$$

Taking the limit as $s \to \infty$

$$I = C_0^*K_p \to C_0^* = K_p^{-1} \tag{6.3.12}$$

The polynomial matrices $\hat{C}^*$, $\hat{D}^*$ are obtained almost as in the SISO case. Let $(\hat{D}_L, \hat{N}_L)$ be a left coprime fraction of $\hat{P}$, with $\hat{D}_L$ row reduced. Divide $\hat{\Lambda}K_p^{-1}\hat{H}^{-1}$ *on the right* by $\hat{D}_L$, so that

$$\hat{\Lambda}\hat{K}_p^{-1}\hat{H}^{-1} = \hat{Q}_L\hat{D}_L + \hat{R}_L \tag{6.3.13}$$

where $\partial_{ci}(\hat{R}_L) < \partial_{ci}(\hat{D}_L) \le \nu$. Then, let

$$\hat{D}^* = -\hat{R}_L = \hat{Q}_L\hat{D}_L - \hat{\Lambda}K_P^{-1}\hat{H}^{-1}$$

$$\hat{C}^* = \hat{\Lambda} - \hat{Q}_L\hat{N}_L \tag{6.3.14}$$

Since $\hat{D}_L\hat{N}_R = \hat{N}_L\hat{D}_R$, it is easy to show that the given C_0^*, $\hat{C}^*$, $\hat{D}^*$ solve the matching equality. Since $\partial_{ci}(\hat{D}_L) \le \nu$, $\partial\hat{D}^* = \partial(\hat{R}_L) \le \nu - 1$. On the other hand

$$\lim_{s\to\infty}(I - \hat{\Lambda}^{-1}\hat{C}^*) = \lim_{s\to\infty}(\hat{\Lambda}^{-1}\hat{D}^*\hat{P} + K_p^{-1}\hat{H}^{-1}\hat{P}) = I \tag{6.3.15}$$

so that $\hat{\Lambda}^{-1}\hat{C}^*$ is strictly proper and $\partial\hat{C}^* \le \nu - 2$. □

State-Space Representation

A state-space representation is obtained by defining the matrices $C_1, \ldots, C_{\nu-1}, D_0, D_1, \ldots, D_{\nu-1} \in \mathbb{R}^{p\times p}$ such that

$$\hat{C}(s)\hat{\Lambda}^{-1}(s) := C_1\frac{1}{\hat{\lambda}} + C_2\frac{s}{\hat{\lambda}} + \cdots + C_{\nu-1}\frac{s^{\nu-2}}{\hat{\lambda}}$$

$$\hat{D}(s)\hat{\Lambda}^{-1}(s) := D_0 + D_1\frac{1}{\hat{\lambda}} + D_2\frac{s}{\hat{\lambda}} + \cdots + D_{\nu-1}\frac{s^{\nu-2}}{\hat{\lambda}} \tag{6.3.16}$$

Consequently, the vectors $w_i^{(1)}$ and $w_i^{(2)}$ are defined by

$$w_i^{(1)} := \frac{s^{i-1}}{\hat{\lambda}}(u) \qquad w_i^{(2)} := \frac{s^{i-1}}{\hat{\lambda}}(y_p) \quad i = 1\ldots\nu-1 \tag{6.3.17}$$

The regressor vector w is defined as

$$w^T := (\bar{r}^T, w_1^{(1)^T}, \ldots, w_{\nu-1}^{(1)^T}, y_p^T, w_1^{(2)^T}, \ldots, w_{\nu-1}^{(2)^T}) \in \mathbb{R}^{p\times 2\nu p} \tag{6.3.18}$$

and the matrix of controller parameters is

$$\Theta^T := (C_0, C_1, \ldots, C_{\nu-1}, D_0, D_1, \ldots, D_{\nu-1}) \in \mathbb{R}^{p\times 2\nu p} \tag{6.3.19}$$

so that the control input is given by

$$u = \Theta^T w \in \mathbb{R}^p \tag{6.3.20}$$

The controller structure is represented in Figure 6.7. By letting the controller parameter Θ, i.e., $C_0, \ldots, C_{\nu-1}, D_0, \ldots, D_{\nu-1}$ vary with time, the scheme will be made adaptive. We define the parameter error

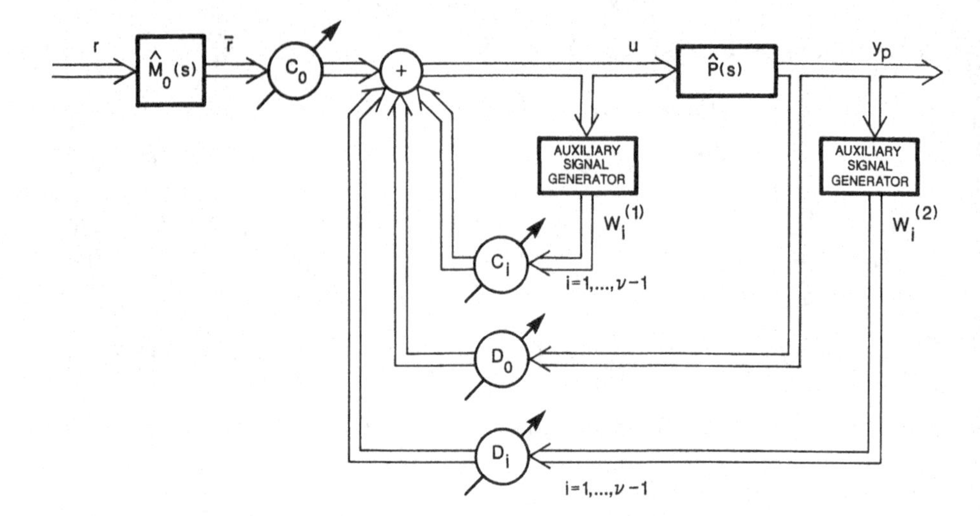

Figure 6.7 MIMO Controller Structure—Adaptive Form

$$\Phi(t) := \Theta(t) - \Theta^* \in \mathbb{R}^{p \times 2\nu p} \tag{6.3.21}$$

and

$$\bar{\Theta}^T := (C_1, \ldots, C_{\nu-1}, D_0, D_1, \ldots, D_{\nu-1})$$

$$\bar{w} := (w_1^{(1)^T}, \ldots, w_{\nu-1}^{(1)^T}, y_p^T, w_1^{(2)^T}, \ldots, w_{\nu-1}^{(2)^T})$$

$$\bar{\Phi} := \bar{\Theta} - \bar{\Theta}^* \tag{6.3.22}$$

6.3.4 Model Reference Adaptive Control—Input Error Scheme

For simplicity in the following derivations, we make the following assumption

(A4) High Frequency Gain Assumption
The high-frequency gain matrix K_p is known.

Consequently, we let

$$C_0 = C_0^* = K_p^{-1}$$

$$u = C_0^* \bar{r} + \bar{\Theta}^T \bar{w} \tag{6.3.23}$$

and we look for an update law for $\bar{\Theta}(t)$.

Consider the matching equality

$$(\hat{\Lambda} - \hat{C}^*)\hat{D}_R - \hat{D}^*\hat{N}_R = \hat{\Lambda} C_0^* \hat{H}^{-1}\hat{N}_R \tag{6.3.24}$$

and multiply both sides by $\hat{\Lambda}^{-1}$ on the left and $\hat{D}_R^{-1}$ on the right so that

$$I - (\hat{\Lambda}^{-1}\hat{C}^*) - (\hat{\Lambda}^{-1}\hat{D}^*)\hat{P} = C_0^*\hat{H}^{-1}\hat{P} \tag{6.3.25}$$

Now, define

$$\hat{L}(s) := \text{diag}\ [\hat{l}(s)] \in \mathbb{R}^{p\times p}(s) \tag{6.3.26}$$

where $\hat{l}(s)$ is a monic, Hurwitz polynomial such that $\partial\hat{l} = \partial(\hat{H}^{-1})$ (the maximum degree of all elements of $\hat{H}^{-1}$). Since $\hat{\Lambda}$ and $\hat{L}$ are given by (6.3.2), (6.3.26), they commute with any matrix. Multiplying both sides of (6.3.25) by $\hat{L}^{-1}$ and applying both transfer function matrices to u leads to

$$\begin{aligned}\hat{L}^{-1}(u) &= C_0^*(\hat{H}\hat{L})^{-1}(y_p) + \hat{L}^{-1}(\hat{C}^*\hat{\Lambda}^{-1}(u) + \hat{D}^*\hat{\Lambda}^{-1}(y_p)) \\ &= C_0^*(\hat{H}\hat{L})^{-1}(y_p) + \hat{L}^{-1}(\bar{\Theta}^{*^T}\bar{w}) \\ &= C_0^*(\hat{H}\hat{L})^{-1}(y_p) + \bar{\Theta}^{*^T}\hat{L}^{-1}(\bar{w})\end{aligned} \tag{6.3.27}$$

As in the SISO case, we define

$$\begin{aligned}v^T &:= ((\hat{H}\hat{L})^{-1}y_p^T, \hat{L}^{-1}(\bar{w}^T)) \\ \bar{v}^T &:= \hat{L}^{-1}(\bar{w}^T) \\ e_2 &:= \Theta^T v - \hat{L}^{-1}(u)\end{aligned} \tag{6.3.28}$$

so that, using (6.3.27)

$$\begin{aligned}e_2 &= \Phi^T v \\ &= \bar{\Phi}^T \bar{v}\end{aligned} \tag{6.3.29}$$

where the last equality follows because we assumed that K_p is known, that is, $C_0 = C_0^*$. The error equation is a linear equation, but a multivariable one, with $e_2 \in \mathbb{R}^p$, $\bar{\Phi}^T \in \mathbb{R}^{p\times(2\nu-1)p}$. However, standard update laws are easily extended to the multivariable case, with similar properties. For example, the normalized gradient algorithm for (6.3.29) becomes

$$\dot{\bar{\Theta}} = -g\,\frac{\bar{v}e_2^T}{1+\gamma\bar{v}^T\bar{v}} \qquad g, \gamma > 0 \tag{6.3.30}$$

This equation may be obtained by considering each component of e_2, forming p scalar error equations, and collecting the standard SISO update laws. Similarly, a least-squares algorithm may also be defined.

The scheme may also be extended to the unknown high-frequency gain case. However, some procedure must be devised to prevent C_0 from becoming singular (cf. $c_0 \neq 0$ in SISO).

6.3.5 Alternate Schemes

The multivariable input error scheme presented here is equivalent to the scheme presented by Elliott & Wolovich [1982]. In discrete time, a similar scheme is obtained by Goodwin & Long [1980]. An output error version of the model reference adaptive control scheme is found in Singh & Narendra [1984].

It is fairly straight forward to derive an indirect scheme, based on the solution of the matching equality given in the proof of proposition 6.3.3. An interesting contribution of the proposition is to show that the solution of the matching equality requires only one polynomial matrix division. This is to be contrasted with the situation for pole placement, where the general Diophantine equation needs to be solved. In fact, the matrix polynomial division itself is simpler than it looks at first, and may be calculated without matrix polynomial inversion (see Wolovich [1984]).

A possible advantage of the indirect approach is that the interactor matrix may be estimated on-line (Elliott & Wolovich [1984]). Indeed, the requirement of the knowledge of the Hermite form may be too much to ask for, unless $\hat{H}$ is diagonal (cf. Singh & Narendra [1982], [1984]). When $\hat{H}$ is not diagonal, the off-diagonal elements depend on the unknown plant parameters.

The model reference adaptive control objective with $\hat{M} = \hat{H}$ is somewhat restrictive: Indeed, all desired dynamics are generated by *prefiltering* the input signal. In continuous time, all zeros are cancelled by poles, and the remaining poles are defined by $\hat{H}$. In discrete time, the remaining poles are all at the origin (d-step ahead control). This is not very desirable for implementation, as discussed in Chapter 3. In the SISO case, a more adequate scheme was presented such that the internal dynamics of the closed-loop system are actually those of the reference model.

Stability proofs for MIMO adaptive control follow similar paths as for SISO (Goodwin & Long [1980], Singh & Narendra [1982]). Convergence properties have not been established. Indeed, the *uniqueness* of the controller parameter is not guaranteed by proposition 6.3.3, as it was in the SISO case. More research will be needed for the dynamics of

MIMO adaptive control systems to be well-understood.

6.4 CONCLUSIONS

In this chapter, we first discussed how prior information may be used in the context of identification. Then, we presented flexible indirect adaptive control schemes, and proved their global stability under a richness condition on the exogeneous reference input. Applications to pole placement and the factorization approach were discussed. We then turned to the extension of model reference adaptive control schemes to MIMO systems. After some preliminaries, we established a parameterization of the adaptive controller, following lines parallel to the SISO case. An input error scheme was finally presented. More research will be needed to better understand the dynamics of MIMO adaptive control schemes. The combination of modern control theories based on factorization approaches and of MIMO recursive identification algorithms in a flexible indirect adaptive controller scheme is a promising area for further developments and applications.

CHAPTER 7

ADAPTIVE CONTROL OF A CLASS OF NONLINEAR SYSTEMS

7.1 INTRODUCTION

In recent years there has been a great deal of interest in the use of state feedback to exactly linearize the input-output behavior of nonlinear control systems, for example of the form

$$\dot{x} = f(x) + \sum_{i=1}^{p} g_i(x)\, u_i$$

$$y_1 = h_1(x) \ldots y_p = h_p(x) \tag{7.1.1}$$

In (7.1.1), $x \in \mathbb{R}^n$, $u \in \mathbb{R}^p$, $y \in \mathbb{R}^p$ and $f, g_i : \mathbb{R}^n \rightarrow \mathbb{R}^n$. Also the h_j's are scalar valued functions from $\mathbb{R}^n$ to $\mathbb{R}$. The theory of linearization by exact feedback was developed through the efforts of several researchers, such as Singh and Rugh [1972], Freund [1975], Meyer & Cicolani [1980], Isidori, Krener, Gori-Giorgi & Monaco [1981] in the continuous time case. Good surveys are available in Claude [1986], Isidori [1985, 1986] and Byrnes & Isidori [1984]. The discrete time case and sampled data cases are more involved and are developed in Monaco & Normand-Cyrot [1986]. A number of applications of these techniques have been made. Their chief drawback however seems to be in the fact that they rely on exact cancellation of nonlinear terms in order to get linear input-output behavior. Consequently, if there are errors in the model of the nonlinear terms, the cancellation is no longer exact and the input-output behavior no longer linear. In this chapter, we suggest the use of parameter adaptive control to help make more robust the cancellation of the nonlinear terms when the uncertainty in the nonlinear terms is parametric. The results of this chapter are based on Sastry & Isidori [1987].

The remainder of the chapter is organized as follows: we give a brief review of linearization theory along with the concept of a minimum phase nonlinear system in Section 7.2. We discuss the adaptive version of this control strategy in Section 7.3 along with its application to the adaptive control of rigid robot manipulators, based on Craig, Hsu & Sastry [1987]. In Section 7.4, we collect some suggestions for future work.

7.2 LINEARIZING CONTROL FOR A CLASS OF NONLINEAR SYSTEMS—A REVIEW

7.2.1 Basic Theory

SISO Case

A large class of nonlinear control systems can be made to have linear input-output behavior through a choice of *nonlinear state feedback* control law. Consider, at first, the single-input single-output system

$$\begin{aligned} \dot{x} &= f(x) + g(x)u \\ y &= h(x) \end{aligned} \tag{7.2.1}$$

with $x \in \mathbb{R}^n$, f, g, h all smooth nonlinear functions. In this chapter, a *smooth function* will mean an infinitely differentiable function. Differentiating y with respect to time, one obtains

$$\begin{aligned} \dot{y} &= \frac{\partial h}{\partial x} f(x) + \frac{\partial h}{\partial x} g(x)u \\ &:= L_f h(x) + L_g h(x)u \end{aligned} \tag{7.2.2}$$

where $L_f h(x) : \mathbb{R}^n \to \mathbb{R}$ and $L_g h(x) : \mathbb{R}^n \to \mathbb{R}$ stand for the *Lie derivatives* of h with respect to f, g respectively. If $L_g h(x)$ is bounded away from zero for all x, the state feedback control law (of the form $u = \alpha(x) + \beta(x)v$)

$$u = \frac{1}{L_g h}(-L_f h + v) \tag{7.2.3}$$

yields the linear system (linear from the new input v to y)

$$\dot{y} = v$$

The control law (7.2.3) has the effect of rendering $(n-1)$ of the states of

(7.2.1) unobservable through appropriate choice of state feedback.

In the instance that $L_g h(x) \equiv 0$ meaning $L_g h(x) = 0$ for all x, one differentiates (7.2.2) further to get

$$\ddot{y} = L_f^2 h(x) + (L_g L_f h)(x) u \tag{7.2.4}$$

In (7.2.4), $L_f^2 h(x)$ stands for $L_f(L_f h)(x)$ and $L_g L_f h(x) = L_g(L_f h(x))$. As before, if $L_g L_f h(x)$ is bounded away from zero for all x, the control law

$$u = \frac{1}{L_g L_f h(x)} (-L_f^2 h(x) + v) \tag{7.2.5}$$

linearizes the system (7.2.4) to yield

$$\ddot{y} = v$$

More generally, if γ is the smallest integer such that $L_g L_f^i h \equiv 0$ for $i = 0, \ldots, \gamma - 2$ and $L_g L_f^{\gamma - 1} h$ is bounded away from zero, then the control law

$$u = \frac{1}{L_g L_f^{\gamma - 1} h} (-L_f^{\gamma} h + v) \tag{7.2.6}$$

yields

$$y^{\gamma} = v$$

The procedure described above terminates at some finite γ if the row vectors $\{\frac{dh}{dx}(x), \frac{d}{dx} L_f h(x), \ldots, \frac{d}{dx} L_f^{n-1} h(x)\}$ are linearly independent for all x.

Note that the theory is considerably more complicated and incomplete if $L_g L_f^{\gamma - 1} h$ is not identically zero, but is equal to zero for some values of x.

MIMO Case—Static State Feedback

For the multi-input multi-output case, we consider *square* systems (that is systems with as many inputs as outputs) of the form

$$\begin{aligned} \dot{x} &= f(x) + g_1(x) u_1 + \cdots + g_p(x) u_p \\ y_1 &= h_1(x) \end{aligned}$$

$$\vdots$$

$$y_p = h_p(x) \tag{7.2.7}$$

Here $x \in \mathbb{R}^n$, $u \in \mathbb{R}^p$, $y \in \mathbb{R}^p$ and f, g_i, h_j, are assumed smooth. Now, differentiate the j th output y_j with respect to time to get

$$\dot{y}_j = L_f h_j + \sum_{i=1}^{p} (L_{g_i} h_j) u_i \tag{7.2.8}$$

In (7.2.8), note that if each of the $(L_{g_i} h_j)(x) \equiv 0$, then the inputs do not appear in (7.2.8). Define γ_j to be the smallest integer such that at least one of the inputs appears in $y_j^{\gamma_j}$, that is,

$$y_j^{\gamma_j} = L_f^{\gamma_j} h_j + \sum_{i=1}^{p} L_{g_i} (L_f^{\gamma_j - 1} h_j) u_i \tag{7.2.9}$$

with at least one of the $L_{g_i} (L_f^{\gamma_j - 1} h_j) \neq 0$, for some x. Define the $p \times p$ matrix $A(x)$ as

$$A(x) = \begin{bmatrix} L_{g_1} L_f^{\gamma_1 - 1} h_1 & \cdots & L_{g_p} L_f^{\gamma_1 - 1} h_1 \\ \cdot & & \cdot \\ \cdot & & \cdot \\ \cdot & & \cdot \\ L_{g_1} L_f^{\gamma_p - 1} h_p & \cdots & L_{g_p} L_f^{\gamma_p - 1} h_p \end{bmatrix} \tag{7.2.10}$$

Then, (7.2.9) may be written as

$$\begin{bmatrix} y_1^{\gamma_1} \\ \cdot \\ \cdot \\ \cdot \\ y_p^{\gamma_p} \end{bmatrix} = \begin{bmatrix} L_f^{\gamma_1} h_1 \\ \cdot \\ \cdot \\ \cdot \\ L_f^{\gamma_p} h_p \end{bmatrix} + A(x) \begin{bmatrix} u_1 \\ \cdot \\ \cdot \\ \cdot \\ u_p \end{bmatrix} \tag{7.2.11}$$

If $A(x) \in \mathbb{R}^{p \times p}$ is bounded away from singularity (meaning that $A^{-1}(x)$ exists for all x and has bounded norm), the state feedback control law

$$u = -A(x)^{-1} \begin{bmatrix} L_f^{\gamma_1} h_1 \\ \cdot \\ \cdot \\ \cdot \\ L_f^{\gamma_p} h_p \end{bmatrix} + A(x)^{-1} v \tag{7.2.12}$$

yields the *linear* closed loop system

$$\begin{bmatrix} y_1^{\gamma_1} \\ \cdot \\ \cdot \\ \cdot \\ y_p^{\gamma_p} \end{bmatrix} = \begin{bmatrix} v_1 \\ \cdot \\ \cdot \\ \cdot \\ v_p \end{bmatrix} \tag{7.2.13}$$

Note that the system of (7.2.13) is in addition decoupled. Thus, decoupling is achieved as a by-product of linearization. A happy consequence of this is that a large number of SISO results are easily extended to this class of MIMO systems. Thus, for example, once linearization has been achieved, any further control objective such as model matching, pole placement or tracking can be easily met. The feedback law (7.2.12) is referred to as a *static state feedback linearizing control law.*

MIMO Case—Dynamic State Feedback

If $A(x)$ as defined in (7.2.10) is singular, and the drift term in (7.2.11) (i.e. the first term on the right-hand side) is not in the range of $A(x)$, linearization may still be achieved by using dynamic state feedback. To keep the notation from proliferating, we review the methods in the case when $p = 2$ (two inputs, two outputs). $A(x)$ then has rank 1 for all x. Using elementary column operations, we may compress $A(x)$ to one column, i.e.,

$$A(x)\,T(x) = \begin{bmatrix} \tilde{a}_{11}(x) & 0 \\ \tilde{a}_{21}(x) & 0 \end{bmatrix}$$

with $T(x) \in \mathbb{R}^{2\times 2}$ a nonsingular matrix. Now, defining the new inputs $w = T^{-1}(x)u$, (7.2.11) reads

$$\begin{bmatrix} y_1^{\gamma_1} \\ \\ y_2^{\gamma_2} \end{bmatrix} = \begin{bmatrix} L_f^{\gamma_1} h_1 \\ \\ L_f^{\gamma_2} h_2 \end{bmatrix} + \begin{bmatrix} \tilde{a}_{11}(x) \\ \\ \tilde{a}_{21}(x) \end{bmatrix} w_1 \tag{7.2.14}$$

Also, (7.2.7) now reads as

$$\dot{x} = f(x) + \tilde{g}_1(x)\,w_1 + \tilde{g}_2(x)\,w_2 \tag{7.2.15}$$

where

$$[\tilde{g}_1(x)\ \tilde{g}_2(x)] = [g_1(x)\ g_2(x)]\,T(x)$$

Differentiating (7.2.14), and using (7.2.15), we get

$$\begin{bmatrix} y_1^{\gamma_1+1} \\ y_2^{\gamma_2+1} \end{bmatrix} = \begin{bmatrix} L_f^{\gamma_1+1}h_1 + L_{\tilde{g}_1}L_f^{\gamma_1}h_1 w_1 + L_f\tilde{a}_{11}w_1 + L_{\tilde{g}_1}\tilde{a}_{11}w_1^2 \\ L_f^{\gamma_2+1}h_2 + L_{\tilde{g}_1}L_f^{\gamma_2}h_2 w_1 + L_f\tilde{a}_{21}w_1 + L_{\tilde{g}_1}\tilde{a}_{21}w_1^2 \end{bmatrix}$$

$$+ \begin{bmatrix} \tilde{a}_{11} & L_{\tilde{g}_2}L_f^{\gamma_1}h_1 + L_{\tilde{g}_2}\tilde{a}_{11}w_1 \\ \tilde{a}_{21} & L_{\tilde{g}_2}L_f^{\gamma_2}h_2 + L_{\tilde{g}_2}\tilde{a}_{21}w_1 \end{bmatrix} \begin{bmatrix} \dot{w}_1 \\ w_2 \end{bmatrix}$$

The two large blocks in the equation can be condensed to yield:

$$\begin{bmatrix} y_1^{\gamma_1+1} \\ y_2^{\gamma_2+1} \end{bmatrix} = C(x, w_1) + B(x, w_1) \begin{bmatrix} \dot{w}_1 \\ w_2 \end{bmatrix} \tag{7.2.16}$$

Note the appearance of the control term $\dot{w}_1$. Specifying $\dot{w}_1$ is equivalent to the placement of an integrator before w_1, that is to the addition of dynamics to the controller. Now, note that if $B(x, w_1)$ is bounded away from singularity, then the control law

$$\begin{bmatrix} \dot{w}_1 \\ w_2 \end{bmatrix} = -B^{-1}(x, w_1)C(x, w_1) + B^{-1}(x, w_1) \begin{bmatrix} v_1 \\ v_2 \end{bmatrix} \tag{7.2.17}$$

yields the linearized system

$$\begin{bmatrix} y_1^{\gamma_1+1} \\ y_2^{\gamma_2+1} \end{bmatrix} = \begin{bmatrix} v_1 \\ v_2 \end{bmatrix} \tag{7.2.18}$$

The control law (7.2.17) is a *dynamic* state feedback, linearizing, and decoupling control law. If $B(x, w_1)$ is singular, the foregoing procedure may be repeated on $B(x, w_1)$. The procedure ends in finitely many steps if and only if the system is right invertible (for details, see Descusse & Moog [1985]).

7.2.2 Minimum Phase Nonlinear Systems

The linearizing control law (7.2.3) when applied to the system (7.2.1) results in a first order input output system. Consequently, $(n-1)$ of the original n states are rendered unobservable by the state feedback. To see this more clearly, consider the linear case, i.e., $f(x) = Ax$, $g(x) = b$, and $h(x) = c^T x$. Then, the condition

$$L_g h(x) \neq 0 \iff c^T b \neq 0$$

and the control law of (7.2.3) is

$$u = \frac{1}{c^T b}(-c^T A x + v) \tag{7.2.19}$$

resulting in the closed loop system

$$\dot{x} = \left[I - \frac{bc^T}{c^T b} \right] Ax + \frac{b}{c^T b} v$$

$$y = c^T x \tag{7.2.20}$$

From the fact that the control law (7.2.19) yields a transfer function of $\frac{1}{s}$ from v to y it follows that $(n-1)$ of the eigenvalues of the closed loop matrix $(I - bc^T / c^T b)A$ are located at the zeros of the original system, and the last at the origin. Thus, the linearizing control laws may be thought of as being the nonlinear counterpart of this specific pole placement control law. The dynamics of the states rendered unobservable are indeed the so-called *zero-dynamics* of the system (see (7.2.23)). Clearly, in order to have internal stability (and boundedness of the states), it is important to have the closed loop pole-zero cancellation be stable, i.e. the system be minimum phase. This motivates the understanding and definitions of minimum phase nonlinear systems. We start with the single-input single-output case.

7.2.2.1 The Single-Input Single-Output Case

The first definition to be made is that of relative degree (or pole-zero excess).

Definition Strong Relative Degree

The system (7.2.1) is said to have *strong relative degree* γ if

$$L_g h(x) = L_g L_f h(x) = \cdots = L_g L_f^{\gamma-2} h(x) = 0 \quad \text{for all } x$$

and for all x, $L_g L_f^{\gamma-1} h(x)$ is bounded away from zero.

Comments

a) The system (7.2.1) is said to have strong relative degree γ if the output y needs to be differentiated γ times before terms involving the input appear.

b) In the instance that the system has strong relative degree γ, it is possible to verify that, for each $x^\circ \in \mathbb{R}^n$, there exists a neighborhood U° of x° such that the mapping

$$T : U^\circ \to \mathbb{R}^n$$

defined as

$$\begin{aligned} T_1(x) &= z_{11} = h(x) \\ T_2(x) &= z_{12} = L_f h(x) \\ &\ \ \vdots \\ T_\gamma(x) &= z_{1\gamma} = L_f^{\gamma-1} h(x) \end{aligned} \tag{7.2.21}$$

and $T_{\gamma+1}, \ldots, T_n$ chosen such that

$$\frac{\partial}{\partial x}\Big[T_i(x)\Big]\, g(x) = 0 \quad \text{for } i = \gamma+1, \ldots, n$$

is a diffeomorphism onto its image (see Isidori [1986]).

If we denote by $z_1 \in \mathbb{R}^\gamma$ the vector $(z_{11}, \ldots, z_{1\gamma})^T$ and by $z_2 \in \mathbb{R}^{n-\gamma}$ the vector $(T_{\gamma+1}, \ldots, T_n)^T$, it follows that the equations (7.2.1) may be replaced by

$$\begin{aligned} \dot z_{11} &= z_{12} \\ &\ \ \vdots \\ \dot z_{1\gamma-1} &= z_{1\gamma} \\ \dot z_{1\gamma} &= f_1(z_1, z_2) + g_1(z_1, z_2)u \\ \dot z_2 &= \Psi(z_1, z_2) \end{aligned}$$

$$y = z_{11} \tag{7.2.22}$$

In the equations above, $f_1(z_1, z_2)$ represents $L_f^\gamma h(x)$ and $g_1(z_1, z_2)$ represents $L_g L_f^{\gamma-1} h(x)$ in the new coordinates. $\Psi_i(z_1, z_2)$ represents $L_f T_i(x)$ for $i = \gamma + 1, \ldots, n$. Also note that the input does not directly influence the z_2 states. The representation of the system (7.2.1) through (7.2.22) is called a *normal form.*

If $x = 0$ is an equilibrium point of the undriven system, that is, $f(0) = 0$ and $h(0) = 0$ (without loss of generality), then the dynamics

$$\dot{z}_2 = \Psi(0, z_2) \tag{7.2.23}$$

are referred to as the *zero-dynamics.* Note that the subset

$$L_0 = \{x \in U^\circ \mid h(x) = \cdots = L_f^{\gamma-1} h(x) = 0\}$$

can be made invariant by choosing

$$u = \frac{1}{g_1(z_1, z_2)} \left[-f_1(z_1, z_2) + v \right] \tag{7.2.24}$$

The dynamics of (7.2.23) are the dynamics on this subspace.

Definition Minimum Phase

The nonlinear system (7.2.1) is said to be *globally (locally) minimum phase* if the zero-dynamics are globally (locally) asymptotically stable.

Comments

a) This definition may be strengthened to exponentially stable, in which case we call the system exponentially minimum phase.

b) The previous analysis identifies the normal form (7.2.22) and the zero-dynamics of (7.2.23) only *locally* around any point x° of $\mathbb{R}^n$. Recent work of Byrnes & Isidori [1988] has identified necessary and sufficient conditions for the existence of a globally defined normal form. They have shown that a global version of the notion of zero-dynamics is that of a dynamical system evolving on the smooth manifold of $\mathbb{R}^n$

$$L_0 = \{x \in \mathbb{R}^n : h(x) = L_f h(x) = \cdots = L_f^{\gamma-1} h(x) = 0\}$$

and hereby defined the vector field

$$\bar{f}(x) = f(x) - \frac{L_f^\gamma h(x)}{L_g L_f^{\gamma-1} h(x)} g(x) \qquad x \in L_0$$

Note that this is a vector field on L_0 because $\bar{f}(x)$ is tangent to L_0. If

L_0 is connected and the zero-dynamics are globally asymptotically stable (i.e. if the system is *globally minimum phase*), then the normal forms of (7.2.22) are globally defined if and only if the vector fields

$$\bar{g}(x),\ ad_{\bar{f}}\bar{g}(x), \ldots,\ ad_{\bar{f}}^{\gamma-1}\bar{g}(x)$$

are complete (i.e. have no finite escape time), where

$$\bar{g}(x) = \frac{1}{L_g L_f^{\gamma-1} h(x)} g(x) \qquad \bar{f}(x)\ , \quad \text{as above}$$

while

$$ad_{\bar{f}}\bar{g} = \frac{\partial \bar{g}}{\partial x}\bar{f}(x) - \frac{\partial \bar{f}}{\partial x}\bar{g}(x)$$

is the so-called *Lie bracket* of $\bar{f}$, $\bar{g}$ and $ad_{\bar{f}}^i\,\bar{g} = ad_{\bar{f}} \ldots ad_{\bar{f}}\,\bar{g}$ iterated i times. This is in turn guaranteed by requiring that the vector fields in question be globally Lipschitz continuous, for example.

The utility of the definition of minimum phase zero-dynamics arises in the context of tracking: if the control objective is for the output $y(t)$ to track a pre-specified reference trajectory $y_m(t)$, then the control input

$$v = y_m^{\gamma} + \alpha_{\gamma}(y_m^{\gamma-1} - y^{\gamma-1}) + \cdots + \alpha_1(y_m - y) \qquad (7.2.25)$$

results in the following equation for the tracking error $e_0 := y - y_m$

$$e_0^{\gamma} + \alpha_{\gamma} e_0^{\gamma-1} + \cdots + \alpha_1 e_0 = 0 \qquad (7.2.26)$$

It is important to note that the control law of (7.2.25) is not implemented by differentiating y repeatedly but rather as a *state feedback law* since

$$\dot{y} = L_f h$$

$$\ddot{y} = L_f^2 h$$

$$\vdots$$

$$y^{\gamma-1} = L_f^{\gamma-1} h$$

If $\alpha_1, \ldots, \alpha_\gamma$ are chosen so that $s^\gamma + \alpha_\gamma s^{\gamma-1} + \cdots + \alpha_1$ is a Hurwitz polynomial, then it is easy to see that $e_0, \dot{e}_0, \ldots, e_0^{\gamma-1}$ go to zero as t tends to ∞. Further, if $y_m, \dot{y}_m, \ldots, y_m^{\gamma-1}$ are bounded, then $y, \dot{y}, \ldots, y^{\gamma-1}$ are also bounded and so is z_1. The following proposition guarantees bounded tracking, that is tracking with bounded states.

Proposition 7.2.1 Bounded Tracking in Minimum Phase Nonlinear Systems

If the zero-dynamics of the nonlinear system (7.2.1) as defined in (7.2.23) are globally exponentially stable. Further, $\Psi(z_1, z_2)$ in (7.2.22) has continuous and bounded partial derivatives in z_1, z_2 and $y_m, \dot{y}_m, \ldots, y_m^{\gamma-1}$ are bounded

Then the control law (7.2.24)–(7.2.25) results in bounded tracking, that is, $x \in \mathbb{R}^n$ is bounded and $y(t)$ converges to $y_m(t)$.

Proof of Proposition 7.2.1

From the foregoing discussion, it only remains to show that z_2 is bounded. We accomplish this by using the converse theorem of Lyapunov of theorem 1.5.1 (as we did for theorem 5.3.1).

Since (7.2.23) is (globally) exponentially stable and Ψ has bounded derivatives, there exists $v_1(z_2)$ such that

$$\alpha_1 |z_2|^2 \le v_1(z_2) \le \alpha_2 |z_2|^2$$

$$\frac{dv_1}{dz_2} \Psi(0, z_2) \le -\alpha_3 |z_2|^2$$

$$\left| \frac{dv_1}{dz_2} \right| \le \alpha_4 |z_2| \tag{7.2.27}$$

By assumption, the control law (7.2.24)–(7.2.25) yields bounded z_1, i.e.

$$|z_1(t)| \le k \qquad \text{for all } t \tag{7.2.28}$$

Using (7.2.27) for the system (7.2.22) yields

$$\dot{v}_1(t) = \frac{dv_1}{dz_2} \Psi(z_1, z_2) \le -\alpha_3 |z_2|^2 + \frac{dv_1}{dz_2} (\Psi(z_1, z_2) - \Psi(0, z_2))$$

$$\le -\alpha_3 |z_2|^2 + \alpha_4 kl\, |z_2| \tag{7.2.29}$$

where l is the Lipschitz constant of $\Psi(z_1, z_2)$ in z_1 (Ψ is globally Lipschitz since it has bounded partial derivatives). It is now easy to see that

$$\dot{v}_1 \leq 0 \quad \text{for} \quad |z_2| \geq \left[\frac{\alpha_4 kl}{\alpha_3} \right]$$

Using this along with the bounds in (7.2.27), it is easy to establish that z_2 is bounded. □

Comments

a) Proposition 7.2.1 is a global proposition. If the zero-dynamics were only locally exponentially stable, the proposition would yield that z_2 is bounded for small enough z_1, that is, for small enough $y_m, \dot{y}_m, \ldots, y_m^{\gamma-1}$.

b) The assumptions of proposition 7.2.1 call for a strong form of stability—exponential stability. In fact, counter-examples to the proposition exist if the zero-dynamics are not exponentially stable—for example, if some of the eigenvalues of $d\Psi(0, z_2)/dz_2$ evaluated at $z_2 = 0$ lie on the $j\omega$-axis.

c) However, the hypothesis of proposition 7.2.1 can be weakened considerably without affecting the conclusion. In particular, it is sufficient to ask only that the zero-dynamics of (7.2.23) converge asymptotically to a bounded set (a form of exponential attractivity). To be concrete, the exponential minimum phase hypothesis can be replaced by the condition

$$z_2^T \Psi(0, z_2) \leq -\alpha_3 |z_2|^2 \quad \text{for} \quad |z_2| \geq k \tag{7.2.30}$$

for some k (large). Condition (7.2.30) is similar to (7.2.27) for the Lyapunov function $|z_2|^2$, except that it holds outside a ball of radius k. It is then easy to verify that all trajectories of the undriven zero-dynamics (7.2.23) eventually converge to a ball and that the proof of proposition 7.2.1 can be repeated to yield bounded tracking. This remark is especially useful in the adaptive context where the assumption of minimum phase zero-dynamics may be replaced by exponential attractivity.

7.2.2.2 The Multi-Input Multi-Output Case

Definitions of zero-dynamics for the square multi-input multi-output case are more subtle, as pointed out in Isidori & Moog [1987]. There are three different ways of defining them, depending on which definition of the zeros of an LTI system one chooses to generalize

a) the dynamics of the maximal controlled invariant manifold in the kernel of the output map, or

b) the output constrained dynamics (with output constrained to zero), or

c) the dynamics of the inverse system.

It is also pointed out that the three different definitions coincide if the nonlinear system can be decoupled by static state feedback, in which case the definition parallels the development of the SISO case above. More specifically, if $A(x)$ as defined in (7.2.10) is non-singular, then we proceed as follows. Define

$$\gamma_1 + \cdots + \gamma_p = m$$

and $z_1 \in \mathbb{R}^m$ by

$$z_1^T = (h_1, L_f h_1, \ldots, L_f^{\gamma_1 - 1} h_1, h_2, \ldots, L_f^{\gamma_2 - 1} h_2,$$
$$\ldots, h_p, \ldots, L_f^{\gamma_p - 1} h_p)$$

Also, define $z_2 \in \mathbb{R}^{n-m}$ by

$$z_{21} = T_1(x), \ldots, z_{2(n-m)} = T_{n-m}(x)$$

with $z^T = (z_1^T, z_2^T)$ representing a diffeomorphism of the state variables x. In these coordinates, the equations (7.2.1) read as

$$\begin{aligned}
\dot{z}_{11} &= z_{12} \\
&\cdot \\
&\cdot \\
&\cdot \\
\dot{z}_{1\gamma_1} &= f_1(z_1, z_2) + g_1(z_1, z_2)u \\
\dot{z}_{1(\gamma_1+1)} &= z_{1(\gamma_1+2)} \\
&\cdot \\
&\cdot \\
&\cdot \\
\dot{z}_{1m} &= f_p(z_1, z_2) + g_p(z_1, z_2)u \\
\dot{z}_2 &= \Psi(z_1, z_2) + \Phi(z_1, z_2)u \qquad (7.2.31) \\
y_1 &= z_{11}
\end{aligned}$$

$$y_2 = z_{1(\gamma_1+1)}$$

$$\vdots$$

$$y_p = z_{1(m-\gamma_p+1)} \tag{7.2.32}$$

Above, $f_1(z_1, z_2)$ represents $L_f^{\gamma_1} h_1(x)$ and $g_1(z_1, z_2)$ the first row of $A(x)$ in the (z_1, z_2) coordinates. The zero-dynamics are defined as follows. Let u be a linearizing control law, for example

$$u(z_1, z_2) = -\begin{bmatrix} g_1(z_1, z_2) \\ \vdots \\ g_p(z_1, z_2) \end{bmatrix}^{-1} \begin{bmatrix} f_1(z_1, z_2) \\ \vdots \\ f_p(z_1, z_2) \end{bmatrix} \tag{7.2.33}$$

Then, if $0 \in \mathbb{R}^n$ is an equilibrium point of the undriven system, that is $f(0) = 0$ and $h_1(0) = \cdots = h_p(0) = 0$, the zero dynamics are the dynamics of

$$\dot{z}_2 = \Psi(0, z_2) + \Phi(0, z_2)\, u(0, z_2) \tag{7.2.34}$$

It is verified in Isidori & Moog [1987] that the dynamics of (7.2.34) are independent of the choice of linearizing feedback law. Proposition 7.2.1 and the remarks following it can be verified to hold with the hypothesis being on the zero-dynamics of (7.2.34).

In the instance that the system (7.2.1) is not decouplable by static state feedback, the definition of the zero-dynamics is considerably more involved. We do not discuss it here since we will not use it.

7.2.3 Model Reference Control for Nonlinear Systems

The discussion thus far has been restricted to tracking control of linearizable nonlinear systems. For this class of systems the extension to model reference adaptive control is easy: for the single-input single-output case, consider $y_m(t)$ to be the output of a linear time invariant reference model with input $r(t)$, specified by

$$\dot{x}_m = A_m x_m + b_m r$$

$$y_m = c_m^T x_m \tag{7.2.35}$$

Then, provided that the relative degree of the reference model is *greater than or equal* to the relative degree γ of the nonlinear system, the control law (7.2.24)–(7.2.25) is easily modified to

$$u = \frac{1}{L_g L_f^{\gamma-1} h}\left[-L_f^{\gamma} h + y_m^{\gamma} + \sum_{i=0}^{\gamma-1} \alpha_{i+1}\left[y_m^i - y^i\right]\right]$$

$$= \frac{1}{L_g L_f^{\gamma-1} h}\left[-L_f^{\gamma} h + c_m^T A_m^{\gamma} x_m + c_m^T A_m^{\gamma-1} b_m r + \sum_{i=0}^{\gamma-1} \alpha_{i+1}\left[c_m^T A_m^i x_m - L_f^i h\right]\right]$$

Note that the dimensions of the model play no role. This also relates to the fact that the tracking error $e_0 := y - y_m$ satisfies the equation

$$e_0^{\gamma} + \alpha_{\gamma} e_0^{\gamma-1} + \cdots + \alpha_1 e_0 = 0 \tag{7.2.36}$$

For the multi-input multi-output case, and the model of the form

$$\dot{x}_m = A_m x_m + B_m r$$
$$y_m = C_m x_m$$

with $B_m \in \mathbb{R}^{n_m \times p}$, $C_m \in \mathbb{R}^{p \times n_m}$, the class of models that can be matched is that for which y_{m1} has relative degree γ_1, y_{m2} has relative degree γ_2 and so on. As above, the model error $e_0 := y - y_m$ satisfies

$$\hat{M}(s) e_0 = 0 \tag{7.2.37}$$

where

$$\hat{M}(s) = \operatorname{diag}\left[\frac{1}{s^{\gamma_1} + \alpha_{1\gamma_1} s^{\gamma_1 - 1} + \cdots + \alpha_{11}}, \ldots, \frac{1}{s^{\gamma_p} + \cdots + \alpha_{p1}}\right] \tag{7.2.38}$$

This is not unlike the results of Section 6.3, where linear multivariable plants are found to match their diagonal Hermite forms.

In the adaptive control sequel to this section, we will consider the tracking scenarios for compactness. Also, as we have noted above, the dimensions of the reference model and its parameters do not play much of a role except to generate $\dot{y}_m, \ddot{y}_m, \ldots, y_m^{\gamma}$.

7.3 ADAPTIVE CONTROL OF LINEARIZABLE MINIMUM PHASE SYSTEMS

In practical implementations of exactly linearizing control laws, the chief drawback is that they are based on exact cancellation of nonlinear terms. If there is any uncertainty in the knowledge of the nonlinear functions f and g, the cancellation is not exact and the resulting input-output equation is not linear. We discuss the use of parameter adaptive control to get asymptotically exact cancellation. At the outset, we will assume that h is known exactly but we will discuss how to relax this assumption later.

7.3.1 Single-Input Single-Output, Relative Degree One Case

Consider a nonlinear system of the form (7.2.1) with $L_g h(x)$ bounded away from zero. Further, let $f(x)$ and $g(x)$ have the form

$$f(x) = \sum_{i=1}^{n_1} \theta_i^{(1)^*} f_i(x) \tag{7.3.1}$$

$$g(x) = \sum_{j=1}^{n_2} \theta_j^{(2)^*} g_j(x) \tag{7.3.2}$$

where $\theta_i^{(1)^*}$, $i = 1, \ldots, n_1$; $\theta_j^{(2)^*}$, $j = 1, \ldots, n_2$ are unknown parameters and $f_i(x)$, $g_j(x)$ are *known* functions. At time t, our estimates of the functions f and g are

$$f_e(x) = \sum_{i=1}^{n_1} \theta_i^{(1)}(t) f_i(x) \tag{7.3.3}$$

$$g_e(x) = \sum_{j=1}^{n_2} \theta_j^{(2)}(t) g_j(x) \tag{7.3.4}$$

Here the subscript e stands for estimate and $\theta_i^{(1)}(t)$, $\theta_j^{(2)}(t)$ stand for the estimates of the parameters $\theta_i^{(1)^*}$, $\theta_j^{(2)^*}$ respectively at time t. Consequently, the linearizing control law (7.2.3) is replaced by

$$u = \frac{1}{(L_g h)_e} [-(L_f h)_e + v] \tag{7.3.5}$$

with $(L_f h)_e$, $(L_g h)_e$ representing the estimates of $L_f h$, $L_g h$ respectively based on (7.3.3), (7.3.4), i.e.,

$$(L_f h)_e = \sum_{i=1}^{n_1} \theta_i^{(1)}(t) L_{f_i} h \tag{7.3.6}$$

$$(L_g h)_e = \sum_{j=1}^{n_2} \theta_j^{(2)}(t) L_{g_j} h \tag{7.3.7}$$

If we define $\theta^* \in \mathbb{R}^{n_1+n_2}$ to be the nominal parameter vector $(\theta^{(1)^*}; \theta^{(2)^*})$, $\theta(t) \in \mathbb{R}^{n_1+n_2}$ the parameter estimate, and $\phi = \theta - \theta^*$ the parameter error, then using (7.3.5) in (7.2.2) yields, after some calculation

$$\dot{y} = v + \phi^{(1)^T} w^{(1)} + \phi^{(2)^T} w^{(2)} \tag{7.3.8}$$

with

$$w^{(1)} \in \mathbb{R}^{n_1} := - \begin{bmatrix} L_{f_1} h \\ \cdot \\ \cdot \\ \cdot \\ L_{f_{n_1}} h \end{bmatrix} \tag{7.3.9}$$

and

$$w^{(2)} \in \mathbb{R}^{n_2} := \begin{bmatrix} L_{g_1} h \\ \cdot \\ \cdot \\ \cdot \\ L_{g_{n_2}} h \end{bmatrix} \frac{((L_f h)_e - v)}{(L_g h)_e} \tag{7.3.10}$$

The control law for tracking is

$$v = \dot{y}_m + \alpha(y_m - y)$$

and yields the following error equation relating the tracking error $e_0 := y - y_m$ to the parameter error $\phi^T = (\phi^{(1)^T} \; \phi^{(2)^T})^T$

$$\dot{e}_0 + \alpha e_0 = \phi^T w \tag{7.3.11}$$

where $w \in \mathbb{R}^{n_1+n_2}$ is defined to be the concatenation of w_1, w_2. Equation (7.3.11) may be written

$$e_0 = \frac{1}{s + \alpha} (\phi^T w)$$

which is of the form of the SPR error equation encountered in Chapter 2. The following theorem may now be stated.

Theorem 7.3.1 Adaptive Tracking

Consider an exponentially minimum phase, nonlinear system of the form (7.2.1), with the assumptions on f, g as given in (7.3.3), (7.3.4). Define the control law

$$u = \frac{1}{(L_g h)_e}\left[-(L_f h)_e + \dot{y}_m + \alpha(y_m - y)\right] \tag{7.3.12}$$

If $(L_g h)_e$ as defined in (7.3.7) is bounded away from zero and y_m is bounded.

Then the gradient type parameter update law

$$\dot{\phi} = -e_0 w \tag{7.3.13}$$

yields bounded $y(t)$, asymptotically converging to $y_m(t)$. Further, all state variables of (7.2.1) are bounded.

Proof of Theorem 7.3.1

The Lyapunov function $v(e_0, \phi) = ½ e_0^2 + ½ \phi^T \phi$ is decreasing along the trajectories of (7.3.11), (7.3.13), with $\dot{v}(e_0, \phi) = -\alpha e_0^2 \le 0$. Therefore, e_0 and ϕ are bounded, and $e_0 \in L_2$. To establish that e_0 is uniformly continuous (to use Barbalat's lemma–lemma 1.2.1), or alternately that $\dot{e}_0$ is bounded, we need w—a continuous function of x (since $(L_g h)_e$ is bounded away from zero)—to be bounded. Now note that given a bounded e_0, y_m bounded implies y bounded. From this, and the exponentially minimum phase assumption (proposition 7.2.1), it follows that x is bounded. Hence w is bounded and e_0 is uniformly continuous, and so e_0 tends to zero as $t \to \infty$. □

Comments

a) The preceding theorem guarantees that e_0 converges to zero as $t \to \infty$. Nothing whatsoever is guaranteed about parameter convergence. It is, however, easy to see that both e_0, ϕ converge exponentially to zero if w is persistently exciting, i.e. if there exist $\alpha_1, \alpha_2, \delta > 0$ such that

$$\alpha_2 I \ge \int_{t_0}^{t_0+\delta} w\, w^T dt \ge \alpha_1 I \tag{7.3.14}$$

Unfortunately, the condition (7.3.14) is usually impossible to verify explicitly ahead of time, since w is a complicated nonlinear function of x.

b) One other popular way of dealing with parametric uncertainty is to replace the control law in (7.3.12) by the "sliding mode" control law

$$u = \frac{1}{(L_g h)_e}\{ -(L_f h)_e + \dot{y}_m + k \operatorname{sgn}(y_m - y)\} \tag{7.3.15}$$

The error equation (7.3.11) is then replaced by

$$\dot{e} + k \operatorname{sgn} e = d(t) \tag{7.3.16}$$

where $d(t)$ is a mismatch term (depending on the difference between $L_g h$ and $(L_g h)_e$, $L_f h$ and $(L_f h)_e, \ldots$). This may be bounded using bounds on f_i, g_j and the ϕ_i's above. It is then possible to see that if $k > \sup_t |d(t)|$, the error e_0 goes to zero, in fact in finite time. This philosophy is not at odds with adaptation as discussed in theorem 7.3.1. In fact, it could be used quite gainfully when the parameter error $\phi(t)$ is small. However, if $\phi(t)$ is large, the gain k needs to be large, resulting in unacceptable chatter, large control activity and other undesirable behavior.

c) An hypothesis of theorem 7.3.1 is that $(L_g h)_e$ be bounded away from zero for all x. Since $(L_g h)_e$ as defined by (7.3.7) may indeed go through zero, even if the 'true' $L_g h$ is bounded away from zero, auxiliary techniques need to be used to guarantee that $(L_g h)_e$ is bounded away from zero. One popular technique is the projection technique, in which the parameters $\theta_1^{(2)}(t), \ldots, \theta_{n_2}^{(2)}(t)$ are kept in a certain parameter range which guarantees that $(L_g h)_e$ is bounded away from zero, say by ϵ (by modifying the update law (7.3.13) as discussed in Chapter 2 and Chapter 3).

7.3.2 Extensions to Higher Relative Degree SISO Systems

We first consider the extensions of the results of the previous section to SISO systems with relative degree γ, that is, $L_g h = L_g L_f h = \cdots = L_g L_f^{\gamma-2} h \equiv 0$ with $L_g L_f^{\gamma-1} h$ bounded away from zero. The non-adaptive linearizing control law is then of the form

$$u = \frac{1}{L_g L_f^{\gamma-1} h}(-L_f^{\gamma} h + v) \tag{7.3.17}$$

If f and g are not completely known but of the form (7.3.1), (7.3.2), we need to replace $L_f^{\gamma} h$ and $L_g L_f^{\gamma-1} h$ by their estimates. We define these as follows

$$(L_f^{\gamma} h)_e := L_{f_e}^{\gamma} h \tag{7.3.18}$$

$$(L_g L_f^{\gamma-1} h)_e := L_{g_e} L_{f_e}^{\gamma-1} h \tag{7.3.19}$$

Note that for $\gamma \geq 2$, (7.3.18), (7.3.19) *are not linear in the unknown parameters* θ_i. For example,

$$(L_f^2 h)_e = \sum_{i=1}^{n_1} \sum_{j=1}^{n_1} L_{f_i}(L_{f_j} h)\theta_i^{(1)}\theta_j^{(1)} \tag{7.3.20}$$

and

$$(L_g L_f h)_e = \sum_{i=1}^{n_2} \sum_{j=1}^{n_1} L_{g_i}(L_{f_j} h)\theta_i^{(2)}\theta_j^{(1)} \tag{7.3.21}$$

and so on.

The development of the preceding section can easily be repeated if we define each of the parameter products to be a new parameter, in which case the $\theta_i^{(1)}\theta_j^{(1)}$ and $\theta_i^{(2)}\theta_j^{(1)}$ of (7.3.20) and (7.3.21) are parameters. Let $\Theta \in \mathbb{R}^k$ be the k-(large!) dimensional vector of parameters $\theta_i^{(1)}$, $\theta_j^{(2)}$, $\theta_i^{(1)}\theta_j^{(2)}$, $\theta_i^{(1)}\theta_j^{(1)}$, ... Thus, for example, if $\gamma = 3$, Θ contains $\theta_i^{(1)}$, $\theta_j^{(2)}$, $\theta_i^{(1)}\theta_j^{(1)}$, $\theta_i^{(1)}\theta_j^{(1)}\theta_k^{(1)}$, $\theta_i^{(1)}\theta_j^{(2)}$, $\theta_i^{(1)}\theta_j^{(1)}\theta_k^{(2)}$. For the purpose of tracking, the control law to be implemented is

$$v = y_m^\gamma + \alpha_\gamma(y_m^{\gamma-1} - y^{\gamma-1}) + \cdots + \alpha_1(y_m - y)$$

where $\dot{y}$, $\ddot{y}$ are obtained as *state feedback terms* using $\dot{y} = L_f h(x)$, $\ddot{y} = L_f^2 h(x)$, and so on. In the absence of precise information about $L_f h, L_f^2 h, \ldots,$ the tracking law to be implemented is

$$v_e = y_m^\gamma + \alpha_\gamma\left[y_m^{\gamma-1} - (L_f^{\gamma-1} h)_e\right] + \cdots + \alpha_1(y_m - y) \tag{7.3.22}$$

The adaptive control law then is

$$u = \frac{1}{(L_g L_f^{\gamma-1} h)_e}\left(-(L_f^\gamma h)_e + v_e\right) \tag{7.3.23}$$

This yields the error equation (with $\Phi := \Theta(t) - \Theta$ representing the parameter error)

$$\begin{aligned} & e_0^\gamma + \alpha_\gamma e_0^{\gamma-1} + \cdots + \alpha_1 e_0 \\ &= L_f^\gamma h + \frac{L_g L_f^{\gamma-1} h}{(L_g L_f^{\gamma-1} h)_e}\left(-(L_f^\gamma h)_e + v_e\right) - v \\ &= (L_f^\gamma h - (L_f^\gamma h)_e) \end{aligned}$$

$$+\left[L_g L_f^{\gamma-1}h - (L_g L_f^{\gamma-1}h)_e\right]\frac{(-(L_f^{\gamma}h)_e + v_e)}{(L_g L_f^{\gamma-1}h)_e} + v_e - v$$

$$:= \Phi^T w_1 + \Phi^T w_2 \tag{7.3.24}$$

The two terms on the right-hand side arise, respectively, from the mismatch between the ideal and actual linearizing law, and the mismatch between the ideal tracking control v and the actual tracking control v_e. For definiteness, consider the case that $\gamma = 2$ and $n_1 = n_2 = 1$. Then, with $\Theta^T = [\theta^{(1)}, \theta^{(2)}, \theta^{(1)}\theta^{(1)}, \theta^{(1)}\theta^{(2)}]$, we get

$$w_1^T = -\left[0 \quad 0 \quad L_{f_1}^2 h \quad L_{g_1}L_{f_1}h\,\frac{(-(L_f^2 h)_e + v_e)}{(L_g L_f h)_e}\right]$$

$$w_2^T = -[\,\alpha_1 L_{f_1}h \quad 0 \quad 0 \quad 0\,] \tag{7.3.25}$$

Note that w_1 and w_2 can be added to get a regressor w. It is of interest to note that $\theta^{(2)}$ cannot be explicitly identified in this case, since the terms in the regressor multiplying it are zero. Also note that w is a function of x and also $\Theta(t)$.

Consider now the form (7.3.24) of the error equation. For the purposes of adaptation, we could use an error of the form

$$e_1 = \beta_\gamma e_0^{\gamma-1} + \cdots + \beta_1 e_0 \tag{7.3.26}$$

with the transfer function

$$\frac{\beta_\gamma s^{\gamma-1} + \cdots + \beta_1}{s^\gamma + \alpha_\gamma s^{\gamma-1} + \cdots + \alpha_1} \tag{7.3.27}$$

strictly positive real. Indeed, if such a signal e_1 were measurable, the basic tracking theorem would follow easily. The difficulty with constructing the signal in (7.3.26) is that $\dot e_0, \ddot e_0, \ldots, e_0^{\gamma-1}$ are not measurable since

$$\dot e_0 = L_f h - \dot y_m$$

$$\ddot e_0 = L_f^2 h - \ddot y_m \tag{7.3.28}$$

and so on, with $L_f^i h$ not explicitly available since they may not be known functions of x. An exception is a large class of electromechanical systems of which the robot manipulator equations are a special case, see Section 7.3.3 below. In such systems, $\beta_1, \ldots, \beta_\gamma$ may be chosen so that the transfer function in (7.3.27) is strictly positive real and the adaptive law of the previous section with e_1 as given by (7.3.26) yields the desired conclusion. When the $L_f^i h$'s are not available, the following approach may be used.

Adaptive Control Using an Augmented Error

Motivated by the adaptive schemes of Chapter 3, we define the polynomial

$$\hat{L}(s) := s^\gamma + \alpha_\gamma s^{\gamma-1} + \cdots + \alpha_1 \tag{7.3.29}$$

so that equation (7.3.24) may be written as

$$e_0 = \hat{L}^{-1}(s)(\Phi^T w) \tag{7.3.30}$$

where we used the hybrid notation of previous chapters and dropped the exponentially decaying initial condition terms. Define the augmented error

$$e_1 = e_0 + \left[\Theta^T \hat{L}^{-1}(s)(w) - \hat{L}^{-1}(s)(\Theta^T w)\right] \tag{7.3.31}$$

Using the fact that constants commute with $\hat{L}^{-1}(s)$, we get

$$e_1 = e_0 + \left[\Phi^T \hat{L}^{-1}(s)(w) - \hat{L}^{-1}(s)(\Phi^T w)\right] \tag{7.3.32}$$

Note that e_1 in (7.3.31) can be obtained from available signals, unlike (7.3.32) which is used for the analysis. Using (7.3.30) in (7.3.32), we have that

$$e_1 = \Phi^T \hat{L}^{-1}(s)(w) \tag{7.3.33}$$

Equation (7.3.33) is a linear error equation. For convenience, we will denote

$$\xi := \hat{L}^{-1}(s)(w) \tag{7.3.34}$$

From the error equation (7.3.33), several parameter update laws are immediately suggested. For example, the normalized gradient type algorithm:

$$\dot{\Theta} = \dot{\Phi} = \frac{-e_1 \xi}{1 + \xi^T \xi} \tag{7.3.35}$$

As in the stability proofs of Chapter 3, we will use the following notation

(a) β is a *generic* $L_2 \cap L_\infty$ function which goes to zero as $t \to \infty$.

(b) γ is a *generic* $L_2 \cap L_\infty$ function.

(c) $\| z \|_t$ refers to the norm $\sup_{\tau \le t} | z(\tau) |$, that is the *truncated* L_∞ norm.

From the results of Chapter 2, a number of properties of $\dot{\Phi}$, e_1 follow immediately, with no assumptions on the boundedness of ξ.

Proposition 7.3.2 Properties of the Identifier
Consider the error equation

$$e_1 = \Phi^T \xi \tag{7.3.36}$$

with the update law

$$\dot{\Phi} = \frac{-e_1 \xi}{1 + \xi^T \xi} \tag{7.3.37}$$

Then $\Phi \in L_\infty$, $\dot{\Phi} \in L_2 \cap L_\infty$ and

$$| \Phi^T \xi(t) | \le \gamma (1 + \| \xi \|_t) \qquad \text{for all } t \tag{7.3.38}$$

for some $\gamma \in L_2 \cap L_\infty$.

Proof of Proposition 7.3.2: See theorem 2.4.2.

We are now ready to state and prove the main theorem.

Theorem 7.3.3 Basic Tracking Theorem for SISO Systems with Relative Degree Greater than 1
Consider the control law of (7.3.22)–(7.3.23) applied to an exponentially minimum phase nonlinear system with parameter uncertainty as given in (7.3.1)–(7.3.2).

If $y_m, \dot{y}_m, \ldots, y_m^{\gamma - 1}$ are bounded,
$(L_g L_f^{\gamma - 1} h)_e$ is bounded away from zero,
$f, g, h, L_f^k h, L_g L_f^k h$ are Lipschitz continuous functions, and $w(x, \theta)$ has bounded derivatives in x, θ.

Then the parameter update law

$$\dot{\Phi} = \frac{-e_1 \xi}{1 + \xi^T \xi} \tag{7.3.39}$$

with $\xi = \hat{L}^{-1}(s)(w)$ yields bounded tracking, i.e., $y \to y_m$ as $t \to \infty$ and x is bounded.

Remark: The proof is similar to the proof of theorem 3.7.1 in the linear case. Although the scheme is based on the output error e_0, the choice $\hat{L}^{-1} = \hat{M}$ makes it identical to the input error scheme.

Proof of Theorem 7.3.3

(a) Bounds on the Error Augmentation

Using the swapping lemma (lemma 3.6.5), we have (with notation borrowed from the lemma)

$$\Phi^T \hat{L}^{-1}(w) - \hat{L}^{-1}(\Phi^T w) = -\hat{L}_c^{-1}(\hat{L}_b^{-1}(w^T)\dot{\Phi}) \tag{7.3.40}$$

Using the fact that $\dot{\Phi} \in L_2$ and $\hat{L}_b^{-1}$ is stable (since $\hat{L}^{-1}$ is), we get

$$|(L_b^{-1}w^T)\dot{\Phi}| \le \gamma \| w \|_t + \gamma \tag{7.3.41}$$

Using lemma 3.6.4 and the fact that $\hat{L}_c^{-1}$ is strictly proper and stable, we get

$$|\Phi^T \hat{L}^{-1}(w) - \hat{L}^{-1}(\Phi^T w)| \le \beta \| w \|_t + \beta \tag{7.3.42}$$

(b) Regularity of $w, \Phi^T w$

The differential equation for $z_1 = (y, \dot{y}, \ldots, y^{\gamma-1})^T$ is

$$z_1 = \hat{M}(s) \begin{bmatrix} 1 \\ \cdot \\ \cdot \\ \cdot \\ s^{\gamma-1} \end{bmatrix} (\Phi^T w) + \begin{bmatrix} y_m \\ \cdot \\ \cdot \\ \cdot \\ y_m^{\gamma-1} \end{bmatrix} \tag{7.3.43}$$

Since Φ is bounded and $y_m, \ldots, y_m^{\gamma-1}$ are bounded by hypothesis, and $s^k \hat{M}(s)$ are all proper stable transfer functions, we have that

$$\| z_1 \|_t \le k \| w \|_t + k \tag{7.3.44}$$

Using (7.3.44) in the exponentially minimum phase zero-dynamics

$$\dot{z}_2 = \Psi(z_1, z_2) \tag{7.3.45}$$

we get

$$\| z_2 \|_t \le k \| w \|_t + k \tag{7.3.46}$$

Equations similar to (7.3.44), (7.3.46) can also be obtained for $\dot{z}_1, \dot{z}_2$ since the transfer functions $\hat{M}, \ldots, s^{\gamma-1}\hat{M}$ are *strictly* proper. Combining (7.3.44) and (7.3.46), and noting that x is a diffeomorphism of z_1, z_2 we see that

$$\| x\|_t \leq k\| w\|_t + k \tag{7.3.47}$$

and

$$\| \dot{x} \|_t \leq k\| w\|_t + k \tag{7.3.48}$$

Using the hypotheses that $\| \partial w/ \partial x \|$ and $\| \partial w/ \partial \theta \|$ are bounded and (7.3.48) we get

$$\| \dot{w} \|_t \leq k\| w \|_t + k \tag{7.3.49}$$

Thus w is regular $\Rightarrow \xi = \hat{L}^{-1} w$ is regular by corollary 3.6.3. For $\Phi^T w$, note that

$$\frac{d}{dt}(\Phi^T w) = \dot{\Phi}^T w + \Phi^T \dot{w} \tag{7.3.50}$$

Using (7.3.49), and $\Phi, \dot{\Phi} \in L_\infty$ we get

$$\| \frac{d}{dt} \Phi^T w\|_t \leq k\| w\|_t + k \tag{7.3.51}$$

But from (7.3.43) and (7.3.45) we get that

$$\| x\|_t \leq k\| \Phi^T w\|_t + k \tag{7.3.52}$$

so that

$$\| w\|_t \leq k\| \Phi^T w\|_t + k \tag{7.3.53}$$

Combining (7.3.53) with (7.3.51) yields the regularity of $\Phi^T w$.

(c) Stability Proof

From the regularity of ξ, $\Phi^T w$, one can establish that $\Phi^T \xi / 1 + \| \xi \|_t$ has bounded derivative and so is uniformly continuous. By theorem 2.4.6,

$$| e_1(t)| = | \Phi^T \xi(t)| \leq \beta(1 + \| \xi \|_t) \tag{7.3.54}$$

where $\beta \to 0$ as $t \to \infty$.

Now

$$e_0 = e_1 + \Phi^T \hat{L}^{-1}(w) - \hat{L}^{-1}(\Phi^T w) \tag{7.3.55}$$

Using (7.3.42),

$$| e_0| \leq | e_1| + \beta \| w\|_t + \beta$$

Using (7.3.53), we have

$$| e_0| \leq | e_1| + \beta \| \Phi^T w\|_t + \beta \tag{7.3.56}$$

Applying the BOBI lemma (lemma 3.6.2) to

$$e_0 = \hat{L}^{-1}(s)(\Phi^T w)$$

along with the established regularity of $\Phi^T w$, we get

$$\|\Phi^T w\|_t \;\le\; k\,\|e_0\|_t + k \tag{7.3.57}$$

Using (7.3.57) in (7.3.56)

$$|e_0| \;\le\; |e_1| + \beta\|\,e_0\|_t + \beta \tag{7.3.58}$$

and using (7.3.54) for e_1, we find

$$|e_0| \;\le\; \beta\|e_0\|_t + \beta + \beta\|\,\xi\,\|_t \tag{7.3.59}$$

Since ξ is related to w by stable filtering

$$\|\,\xi\,\|_t \;\le\; k\,\|\,w\|_t + k \tag{7.3.60}$$

Using the estimate (7.3.53), followed by (7.3.57) in (7.3.59) yields

$$|e_0| \;\le\; \beta\|e_0\|_t + \beta \tag{7.3.61}$$

Since $\beta \to 0$ as $t \to \infty$, we see from (7.3.61) that e_0 goes to zero as $t \to \infty$ (as in proof of theorem 3.7.1, using lemma 3.6.6). This in turn can be easily verified to yield bounded w, x. □

Comments

a) The parameter update law (7.3.35) appears not to take into account *prior parameter* information such as the initial existence of θ_i^*, θ_j^*, $\theta_i^*\,\theta_j^*$ and so on. It is important, however, to note that the best estimate of $\theta_i^*\,\theta_j^*$ in the transient period may not be $\theta_i(t)\,\theta_j(t)$. Since parameter convergence is not guaranteed in the proof of theorem 7.3.3, it may also not be a good idea to constrain the estimate of $\theta_i^*\,\theta_j^*$ to be close to $\theta_i\theta_j$. Note however, that the number of parameters increases very rapidly with γ.

b) In several problems, it turns out that $L_f^\gamma h$ and $L_g L_f^{\gamma-1} h$ depend linearly on some unknown parameters. It is then clear that the development of the previous theorem can be carried through.

c) Thus far, we have only assumed parameter uncertainty in f and g, but not in h. It is not hard to see that if h depends linearly on unknown parameters, then we can mimic the aforementioned procedure quite easily.

d) Parameter convergence can be guaranteed in theorem 7.3.3 above if w is persistently exciting in the usual sense (cf (7.3.14)).

7.3.3. Adaptive Control of MIMO Systems Decouplable by Static State Feedback

From the preceding discussion, it is easy to see how the linearizing, decoupling static state feedback control law for minimum, phase, square systems can be made adaptive—by replacing the control law of (7.2.12) by

$$u = (A(x))_e^{-1} \left[- \begin{bmatrix} L_f^{\gamma_1} h_1 \\ \cdot \\ \cdot \\ \cdot \\ L_f^{\gamma_p} h_p \end{bmatrix}_e + v \right] \tag{7.3.62}$$

Recall that if $A(x)$ is invertible, then the linearizing control law is also the decoupling control law. Thus, if $A(x)$ and the $L_f^{\gamma_i} h_i$'s depend linearly on certain unknown parameters, the schemes of the previous sections (those of Section 7.3.1 if $v_1 = v_2 = \cdots = v_p = 1$, and those of Section 7.3.2 in other cases) can be readily adapted. The details are more notationally cumbersome than insightful. Therefore, we choose not to discuss them here. Instead, we will illustrate our theory on an important class of such systems which partially motivated the present work (see Craig, Hsu, and Sastry [1987])—the adaptive control of rigid link robot manipulators. We sketch only a few of the details of the application relevant to our present context, the interested reader is referred to the paper referenced previously.

If $q \in \mathbb{R}^n$ represents the joint angles of a rigid link robot manipulator, its dynamics may be described by an equation of the form

$$M(q)\ddot{q} + C(q, \dot{q}) = u \tag{7.3.63}$$

In (7.3.63), $M(q) \in \mathbb{R}^{n \times n}$ is the positive definite inertia matrix, $C(q, \dot{q})$ represent the Coriolis, gravity and friction terms, and $u \in \mathbb{R}^n$ represents the control input to the joint motors (torques). In applications, $M(q)$ and $C(q, \dot{q})$ are not known exactly, but fortunately they depend linearly on unknown parameters such as payloads, frictional coefficients, ..., so that

$$M(q) = \sum_{i=1}^{n_1} \theta_i^{(2)^*} M_i(q) \tag{7.3.64}$$

$$C(q, \dot{q}) = \sum_{j=1}^{n_2} \theta_j^{(1)^*} C_j(q, \dot{q}) \tag{7.3.65}$$

Writing the equation (7.3.63) in state space form with $x^T = (q^T, \dot{q}^T)$ and $y = q$, we see that the system is decouplable in the sense of Section 7.2 with $\gamma_1 = \cdots = \gamma_n = 2$, and

$$A(x) = M^{-1}(q) \tag{7.3.66}$$

$$\begin{bmatrix} L_f^{\gamma_1} h_1 \\ \cdot \\ \cdot \\ \cdot \\ L_f^{\gamma_n} h_n \end{bmatrix} = -M^{-1}(q)C(q, \dot{q}) \tag{7.3.67}$$

while the decoupling control law is given by

$$u = C(q, \dot{q}) + M(q)v \tag{7.3.68}$$

Note that the quantities in equation (7.3.66) depend on a complicated fashion on the unknown parameters $\theta^{(1)^*}$, $\theta^{(2)^*}$ while the equation (7.3.68) depends on them *linearly*. For the sake of tracking, v is chosen to be

$$v = \ddot{q}_m + \alpha_2(\dot{q}_m - \dot{q}) + \alpha_1(q_m - q) \tag{7.3.69}$$

and the overall control law (7.3.68), (7.3.69) is referred to as the *computed torque* scheme.

To make the scheme adaptive, the law (7.3.68) is replaced by

$$u = C_e(q, \dot{q}) + M_e(q)v \tag{7.3.70}$$

Let $e_0 = q_m - q$ so that

$$\ddot{e}_0 + \alpha_2 \dot{e}_0 + \alpha_1 e_0 = M_e^{-1}(q) \sum_{j=1}^{n_1} C_j(q, \dot{q})\, \phi_j^{(1)} + M_e^{-1}(q) \sum_{i=1}^{n_2} M_i(q)\ddot{q}\, \phi_i^{(2)} \tag{7.3.71}$$

This may be abbreviated as

$$\ddot{e}_0 + \alpha_2 \dot{e}_0 + \alpha_1 e_0 = W\Phi \tag{7.3.72}$$

where $W \in \mathbb{R}^{n \times (n_1 + n_2)}$ is a function of q, $\dot{q}$, and $\ddot{q}$, and Φ is the parameter error vector. The parameter update law

$$\dot{\Phi} = -W^T e_1 \tag{7.3.73}$$

where $e_1 = \dot{e}_0 + \beta_1 e_0$ is chosen so that $(s+\beta_1)/(s^2+\alpha_2 s+\alpha_1)$ is strictly positive real. This can be shown to yield bounded tracking. *The error augmentation of Section 7.3.2 is not necessary in this application* since both $y, \dot{y}$ are available as states so that the $L_f h_i$'s do not have to be estimated. Note that the system is minimum phase—there are in fact no zero dynamics at all. It is, however, unfortunate that the signal W is a function of $\ddot{q}$—but this is caused by the form of the equations and may be avoided by modifying the scheme as in the input error approach (cf. Hsu *et al* [1987]). As in other examples, it is important to keep $M_e(q)$ from becoming singular, using prior parameter bounds.

7.4 CONCLUSIONS

We have presented some initial results on the use of parameter adaptive control for obtaining asymptotically exact cancellation in linearizing control laws. We considered the class of continuous time systems decouplable by static state feedback. The extension to continuous time systems not decouplable by static state feedback is not as obvious for two reasons

a) The different matrices involved in the development of the control laws in this case, namely, $T(x)$, $C(x, w_1)$, $B(x, w_1)$ depend in extremely complicated fashion on the unknown parameters.

b) While the "true" $A(x)$ may have rank less than p, its estimate $A_e(x)$ during the course of adaptation may well be full rank, in which case the procedure of Section 7.2.1 cannot be followed.

The discrete time and sampled data case are also not obvious for similar reasons:

a) The non-adaptive theory, as discussed in Monaco, Normand-Cyrot & Stornelli [1986] is fairly complicated since

$$y_{k+1} = h \circ (f(x_k) + g(x_k)u_k) \tag{7.4.1}$$

is not linear in u_k in the discrete time case and a formal series for (7.4.1) in u_k needs to be obtained (and inverted!) for the linearization. Consequently the parametric dependence of the control law is complex.

b) The notions of zero-dynamics are not as yet completely developed. Further, even in the linear case, the zeros of a sampled system can be outside the unit disc even when the continuous time system is minimum phase and the sampling is fast enough (Astrom, Hagander & Sternby [1984]).

Thus, the present chapter is only a first step in the development of a comprehensive theory of adaptive control for linearizable systems.

CHAPTER 8
CONCLUSIONS

8.1 General Conclusions

In this book, we have attempted to give, at a fairly advanced level of rigor, a unified treatment of current methodologies for the design and analysis of adaptive control algorithms.

First, we presented several schemes for the adaptive identification and control of linear time invariant systems. An output error scheme, an input error scheme, and an indirect scheme were derived in a unified framework. While all the schemes were shown to be globally stable, the assumptions that went into the derivation of the schemes were quite different. For instance, the input error adaptive control scheme did not require a strictly positive real (SPR) condition for the reference model. This also had implications for the transient behavior of the adaptive systems.

A major goal of this book has been the presentation of a number of recent techniques for analyzing the stability, parameter convergence and robustness of the complicated nonlinear dynamics inherent in the adaptive algorithms. For the stability proofs, we presented a sequence of lemmas drawn from the literature on input-output L_p stability. For the parameter convergence proofs, we used results from generalized harmonic analysis, and extracted frequency-domain conditions. For the study of robustness, we exploited Lyapunov and averaging methods. We feel that a complete mastery of these techniques will lay the groundwork for future studies of adaptive systems.

While we did not deal explicitly with discrete time systems, our presentation of the continuous time results may be transcribed to the discrete time case with not much difficulty. The operator relationships that were used for continuous time systems (L_p spaces) also hold true for discrete time systems (l_p spaces). In fact, many derivations may be simplified in the discrete time case because continuity conditions (such as the regularity of signals) are then automatically satisfied.

Averaging techniques have proved extremely useful and it is likely that important developments will still follow from their use. It is interesting to note that the two-time scale approximation was not only fundamental to the application of averaging methods to convergence (Chapter 4) and to robustness (Chapter 5), but was also underlying in the proofs of exponential convergence (Chapter 2), and global stability (Chapter 3). This highlights the separation between adaptation and control, and makes the connections between direct and indirect adaptive control more obvious.

Methods for the analysis of adaptive systems were a focal point of this book. As was observed in Chapter 5, algorithms that are stable for some inputs may be unstable for others. While simulations are extremely valuable to illustrate a point, they are useless to prove any global behavior of the adaptive algorithm. This is a crucial consequence of the nonlinearity of the adaptive systems, that makes rigorous analysis techniques essential to progress in the area.

8.2 Future Research

Adaptive control is a very active area of research, and there is a great deal more to be done. The area of robustness is essential to successful applications, and since the work of Rohrs *et al*, it has been understood that the questions of robustness for adaptive systems are very different from the same questions for linear time-invariant systems. This is due in great part to the *dual control* aspect of adaptive systems: the reference input plays a role in determining the convergence and robustness by providing excitation to the identification loop. A major problem remains to quantify robustness for adaptive systems. Current theory does not allow for the comparison of the robustness of different adaptive systems, and the relation to non-adaptive robustness concepts. Closer connections will probably emerge from the application of averaging methods, and from the frequency-domain results that they lead to.

Besides these fundamental questions of analysis, much remains to be done to precisely define *design methodologies* for robust adaptive systems and in particular a better understanding of which algorithms are more robust. Indeed, although the adaptive systems discussed in this book have identical stability properties in the ideal case, there is evidence that their behavior is drastically different in the presence of unmodeled dynamics. A better understanding of which algorithms are more robust will also help in deriving guidelines for the improved design of robust algorithms.

While we have extensively discussed the analysis of adaptive systems, we also feel that great strides in this area will come from experiences in implementing the algorithms on several classes of systems. With the advent of microprocessors, and of today's multi-processor environments, complicated algorithms can now be implemented at very high sample rates. The years to come will see a proliferation of techniques to effectively map these adaptive algorithms onto multiprocessor control architectures. There is a great deal of excitement in the control community at large over the emergence of such custom multiprocessor control architectures as CONDOR (Narasimhan *et al* [1988]) and NYMPH (Chen *et al* [1986]). In turn, such advances will make it possible to exploit adaptive techniques on high bandwidth systems such as flexible space structures, aircraft flight control systems, light weight robot manipulators, and the like. While past successes of adaptive control have been on systems of rather low bandwidth and benign dynamics, the future years are going to be ones of experimentation on more challenging systems.

Two other areas that promise explosive growth in the years to come are adaptive control of multi-input multi-output (MIMO) systems, and adaptive control of nonlinear systems, explicitly those linearizable by state feedback. We presented in this book what we feel is the tip of the iceberg in these areas. More needs to be understood about the sort of prior information needed for MIMO adaptive systems. Conversely, the incorporation of various forms of prior knowledge into black-box models of MIMO systems also needs to be studied. Adaptive control for MIMO systems is especially attractive because the traditional and heuristic techniques for SISO systems quickly fall apart when strong cross-couplings appear. Note also that research in the identification of MIMO systems is also relevant to nonadaptive algorithms, which are largely dependent on the knowledge of a process model, and of its uncertainty. One may hope that the recently introduced averaging techniques will help to better connect the frequency-domain properties of adaptive and nonadaptive systems.

A very large class of nonlinear systems is explicitly linearizable by state feedback. The chief difficulty with implementing the linearizing control law is the imprecise knowledge of the nonlinear functions in the dynamics, some of which are often specified in table look-up form. Adaptation then has a role in helping identify the nonlinear functions on-line to obtain asymptotically the correct linearizing control law. This approach was discussed in this book, but it is still in its early development. However, we have found it valuable in the implementation of an adaptive controller for an industrial robot (the Adept-I) and are currently working on a flight control system for a vertical take-off and landing aircraft (the Harrier).

In addition to all these exciting new directions of research in adaptive control, most of which are logical extensions and outgrowths of the developments presented in the previous chapters, we now present a few other new vistas which are not as obvious extensions.

A "Universal" Theory of Adaptive Control

While all the adaptive control algorithms developed in this book required assumptions on the plant—in the single-input single-output case, the order of the plant, the relative degree of the plant, the sign of the high-frequency gain, and the minimum phase property of the plant—it is interesting to ask if these assumptions are a minimal set of assumptions. Indeed, that these assumptions can be relaxed was established by Morse [1985, 1987], Mudgett and Morse [1985], Nussbaum [1983], and Martensson [1985] among others. Chief under the assumptions that could be relaxed was the one on the sign of the high-frequency gain.

There is a simple instance of these results which is in some sense representative of the whole family: consider the problem of adaptively *stabilizing* a first order linear plant of relative degree 1 with unknown gain k_p; i.e.,

$$\dot{y}_p = -a_p\, y_p + k_p\, u \tag{8.2.1}$$

with k_p different from zero but otherwise unknown, and a_p unknown. If the sign of k_p is known and assumed positive, the adaptive control law

$$u = d_0(t)\, y_p \tag{8.2.2}$$

and

$$\dot{d}_0 = -y_p^2 \tag{8.2.3}$$

can be shown to yield $y_p \to 0$ as $t \to \infty$. Nussbaum [1983] proposed that if the sign of k_p is *unknown*, the control law (8.2.2) can be replaced by

$$u = d_0^2(t) \cos\left(d_0(t)\right) y_p \tag{8.2.4}$$

with (8.2.3) as before. He then showed that $y_p \to 0$ as $t \to \infty$, with $d_0(t)$ remaining bounded. Heuristically, the feedback gain $d_0^2 \cos(d_0)$ of (8.2.4) alternates in sign ("searches for the correct sign") as d_0 is decreased monotonically (by (8.2.3)) until it is large enough and of the "correct sign" to stabilize the equation (8.2.1).

While the transient behavior of the algorithm (8.2.3), (8.2.4) is poor, the scheme has stimulated a great deal of interest to derive adaptive control schemes requiring a minimal set of assumptions on the plant (*universal controllers*). A further objective is to develop a unified framework which would subsume all the algorithms presented thus far. Adaptive systems may be seen as the interconnection of a plant, a parameterized controller, and adaptation law or tuner (cf. Morse [1988]). The parameterized controller is assumed to control the process, and the tuner assumed to tune the controller. Tuning is said to have taken place when a suitable tuning error goes to zero. The goal of a *universal theory* is to give a minimal set of assumptions on the process, the parameterized controller, and the tuner to guarantee global stability and asymptotic performance of the closed loop system. Further, the assumptions are to contain as special cases the algorithms presented thus far. Such a theory would be extremely valuable from a conceptual and intellectual standpoint.

Rule-Based, Expert and Learning Control Systems

As the discussions in Chapter 5 indicated, there is a great deal of work needed to implement a given adaptive algorithm, involving the use of heuristics, prior knowledge, and expertise about the system being controlled (such as the amount of noise, the order of the plant, the number of unknown parameters, the bandwidth of the parameters' variation...). This may be coded as several logic steps or rules, around the adaptive control algorithm. The resulting composite algorithm is often referred to as a *rule-based* control law, with the adaptation scheme being one of the rules. The design and evaluation of such composite systems is still an open area of research for nonadaptive as well as adaptive systems, although adaptive control algorithms form an especially attractive area of application.

One can conceive of a more complex scenario, in which the plant to be controlled cannot be easily modeled, either as a linear or nonlinear system because of the complexity of the physical processes involved. A controller then has to be built by codifying systematically into rules the experience gained from operating the system (this is referred to as querying and representation of expert knowledge). The rules then serve as a model of the plant from which the controller is constructed as a rule-based system, i.e. a conjunction of several logic steps and control algorithms. Such a composite design process is called a *rule-based expert controller design.* The sophistication and performance of the controller is dependent on the amount of detail in the model furnished by the expert knowledge. Adaptation and learning in this framework consists in refining the rule-based model on the experience gained during the course of operation of the system.

While this framework is extremely attractive from a practical point of view, it is fair to say that no more than a few case studies of expert control have been implemented, and state of the art in learning for rule-based models is rudimentary. In the context of adaptive control, a very interesting study is found in Astrom *et al* [1986]. Adapted from their work is Figure 8.1, illustrating the structure of an expert control system using an adaptive algorithm.

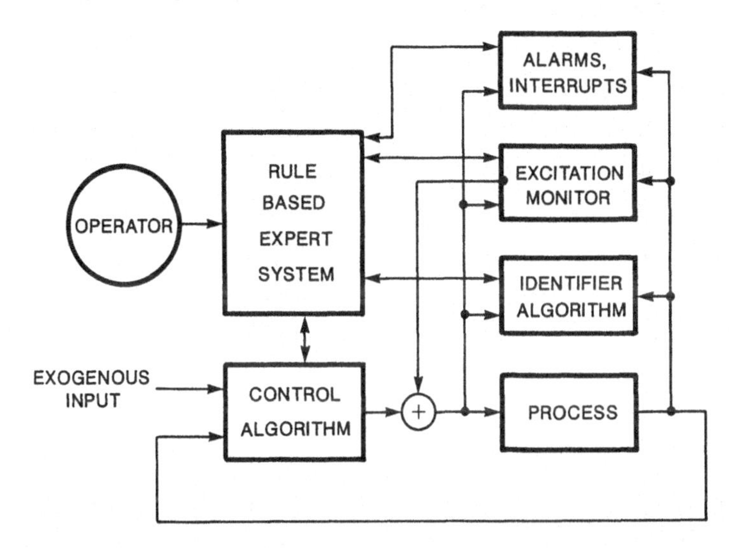

Figure 8.1: Expert Adaptive Control System

The rule-based system decides, based on the level of excitation, which of a library of identification algorithms to use and, if necessary, to inject new excitation. It also decides which of a family of control laws to use and communicates its inferencing procedures to the operator. A supervisor provides alarms and interrupts.

Adaptation, Learning, Connectionism and all those things...

While the topics in the title have the same general philosophical goals, namely, the understanding, modeling and control of a given process, the fields of identification and adaptive control have made the largest strides in becoming a design methodology by limiting their universe of discourse to a small (but practically meaningful) class of systems with linear or linearizable dynamics, and a finite dimensional state-space. Learning has, however, been merely parameter updating.

The goals of connectionism and neural networks (see for example Denker [1986] and the parallel distributed processing models) have been far more lofty: the universe of discourse includes human systems and the learning mimics our own human development. A few applications of this work have been made to problems of speech recognition, image recognition, associative memory storage elements and the like, and it is interesting to note that the 'learning' algorithms implemented in the successful algorithms are remarkably reminiscent of the gradient type and least-squares type of update laws studied in this book. We feel, consequently, that in the years to come, there will be a confluence of the theories and techniques for learning.

APPENDIX

Proof of Lemma 1.4.2

Let

$$r(t) = \int_0^t a(\tau)x(\tau)\,d\tau \tag{A1.4.1}$$

so that, by assumption

$$\dot{r}(t) = a(t)x(t) \leq a(t)r(t) + a(t)u(t) \tag{A1.4.2}$$

that is, for some positive $s(t)$

$$\dot{r}(t) - a(t)r(t) - a(t)u(t) + s(t) = 0 \tag{A1.4.3}$$

Solving the differential equation with $r(0) = 0$

$$r(t) = \int_0^t e^{\int_\tau^t a(\sigma)\,d\sigma}(a(\tau)u(\tau) - s(\tau))\,d\tau \tag{A1.4.4}$$

Since exp(.) and s(.) are positive functions

$$r(t) \leq \int_0^t e^{\int_\tau^t a(\sigma)\,d\sigma}\,a(\tau)u(\tau)\,d\tau \tag{A1.4.5}$$

By assumption $x(t) \leq r(t) + u(t)$ so that (1.4.11) follows. Inequality (1.4.12) is obtained by integrating (1.4.11) by parts. □

Proof of Lemma 2.5.2

We consider the system

$$\dot{x}(t) = A(t)x(t)$$
$$y(t) = C(t)x(t) \tag{A2.5.1}$$

and the system under output injection

$$\dot{w}(t) = \left[A(t) + K(t)C(t)\right] w(t)$$
$$z(t) = C(t)w(t) \tag{A2.5.2}$$

where $x, w \in \mathbb{R}^n$, $A \in \mathbb{R}^{n \times n}$, $C \in \mathbb{R}^{m \times n}$, $K \in \mathbb{R}^{n \times m}$, and $y, z \in \mathbb{R}^m$.

It is sufficient to derive equations the inequalities giving β_1', β_2', β_3'.

Derivation of β_1'

Consider the trajectories $x(\tau)$ and $w(\tau)$, corresponding to systems (A2.5.1) and (A2.5.2) respectively, with identical initial conditions $x(t_0) = w(t_0)$. Then

$$w(\tau) - x(\tau) = \int_{t_0}^{\tau} \Phi(\tau, \sigma) K(\sigma) C(\sigma) w(\sigma)\, d\sigma \tag{A2.5.3}$$

Let $e(\sigma) = K(\sigma)C(\sigma)w(\sigma) / |K(\sigma)C(\sigma)w(\sigma)| \in \mathbb{R}^n$, so that

$$|C(\tau)(w(\tau) - x(\tau))|^2 = \left|\int_{t_0}^{\tau} C(\tau)\Phi(\tau,\sigma)K(\sigma)C(\sigma)w(\sigma)\,d\sigma\right|^2$$

$$\leq \left[\int_{t_0}^{\tau} |C(\tau)\Phi(\tau, \sigma)e(\sigma)|\ \|K(\sigma)\|\ |C(\sigma)w(\sigma)|\ d\sigma\right]^2$$

$$\leq \int_{t_0}^{\tau} |C(\nu)w(\nu)|^2 d\nu \int_{t_0}^{\tau} |C(\tau)\Phi(\tau, \sigma)e(\sigma)|^2\ \|k(\sigma)\|^2 d\sigma \tag{A2.5.4}$$

using the definition of the induced norm and Schwartz inequality. On the other hand, using the triangular inequality

$$\left[\int_{t_0}^{t_0+\delta} |C(\tau)w(\tau)|^2 d\tau\right]^{\frac{1}{2}} \geq \left[\int_{t_0}^{t_0+\delta} |C(\tau)x(\tau)|^2 d\tau\right]^{\frac{1}{2}}$$

$$- \left[\int_{t_0}^{t_0+\delta} |C(\tau)(w(\tau) - x(\tau))|^2 d\tau \right]^{\frac{1}{2}} \tag{A2.5.5}$$

so that, using (A2.5.4), and the UCO of the original system

$$\left[\int_{t_0}^{t_0+\delta} |C(\tau) w(\tau)|^2 d\tau \right]^{\frac{1}{2}} \geq \sqrt{\beta_1}\, |w(t_0)|$$

$$- \left[\int_{t_0}^{t_0+\delta} \int_{t_0}^{\tau} |C(\nu) w(\nu)|^2 d\nu \int_{t_0}^{\tau} | C(\tau) \Phi(\tau, \sigma) e(\sigma)|^2 \| K(\sigma)\|^2 d\sigma\, d\tau \right]^{\frac{1}{2}}$$

$$\geq \sqrt{\beta_1}\, |w(t_0)| - \left[\int_{t_0}^{t_0+\delta} |C(\nu) w(\nu)|^2 d\nu \right]^{\frac{1}{2}}$$

$$\cdot \left[\int_{t_0}^{t_0+\delta} \int_{t_0}^{\tau} \| K(\sigma)\|^2 |C(\tau) \Phi(\tau, \sigma) e(\sigma)|^2 d\sigma\, d\tau \right]^{\frac{1}{2}} \tag{A2.5.6}$$

Changing the order of integration, the integral in the last parenthesis becomes

$$\int_{t_0}^{t_0+\delta} \| K(\sigma)\|^2 \int_{\sigma}^{t_0+\delta} |C(\tau) \Phi(\tau, \sigma) e(\sigma)|^2 d\tau\, d\sigma \tag{A2.5.7}$$

Note that $t_0 + \delta - \sigma \leq \delta$, $|e(\sigma)| = 1$, while $\Phi(\tau, \sigma) e(\sigma)$ is the solution of system (A2.5.1) starting at $e(\sigma)$. Therefore, using the UCO property on the original system, and the condition on $K(.)$, (A2.5.7) becomes

$$\int_{t_0}^{t_0+\delta} \| K(\sigma)\|^2 \int_{\sigma}^{t_0+\delta} |C(\tau) \Phi(\tau, \sigma) e(\sigma)|^2 d\tau\, d\sigma \leq k_\delta \beta_2 \tag{A2.5.8}$$

Inequality (2.5.7) follows directly from (A2.5.6) and (A2.5.8).

Derivation of β_2'

We use a similar procedure, using (A2.5.4)

$$|C(\tau) w(\tau)|^2 \leq |C(\tau) x(\tau)|^2$$

$$+ \left| \int_{t_0}^{\tau} C(\tau)\,\Phi(\tau,\sigma)\,K(\sigma)\,C(\sigma)\,w(\sigma)\,d\sigma \right|^2$$

$$\leq |C(\tau)x(\tau)|^2$$

$$+ \left[\int_{t_0}^{\tau} |C(\sigma)\,w(\sigma)| \;\; |C(\tau)\,\Phi(\tau,\sigma)\,e(\sigma)| \;\; \|K(\sigma)\|\,d\sigma \right]^2$$

$$\leq |C(\tau)x(\tau)|^2$$

$$+ \int_{t_0}^{\tau} |C(\nu)\,w(\nu)|^2 d\nu \int_{t_0}^{\tau} |C(\tau)\Phi(\tau,\sigma)e(\sigma)|^2 \;\|K(\sigma)\|^2 d\sigma \qquad \text{(A2.5.9)}$$

and, for all $t_0 \leq t \leq t_0 + \delta$

$$\int_{t_0}^{t} |C(\tau)w(\tau)|^2 d\tau \;\leq\; \int_{t_0}^{t_0+\delta} |C(\tau)x(\tau)|^2 d\tau + \int_{t_0}^{t}\int_{t_0}^{\tau} |C(\nu)w(\nu)|^2 d\nu$$

$$\cdot \int_{t_0}^{\tau} |C(\tau)\,\Phi(\tau,\sigma)\,e(\sigma)|^2 \;\|K(\sigma)\|^2 d\sigma\, d\tau \qquad \text{(A2.5.10)}$$

and, using the Bellman-Gronwall lemma (lemma 1.4.2), together with the UCO of the original system

$$\int_{t_0}^{t} |C(\tau)\,w(\tau)|^2 d\tau \;\leq\; \beta_2 |\, w(t_0)|^2$$

$$\exp\left[\int_{t_0}^{t}\int_{t_0}^{\tau} |C(\tau)\,\Phi(\tau,\sigma)e(\sigma)|^2 \;\|K(\sigma)\|^2 d\sigma\, d\tau \right] \qquad \text{(A2.5.11)}$$

for all t, and in particular for $t = t_0 + \delta$.

The integral in the exponential can be transformed, by changing the order of integration, as in (A2.5.8). Inequality (2.5.8) follows directly from (A2.5.8) and (A2.5.11). □

Proof of Lemma 2.6.6

We wish to prove that for some δ, α_1, $\alpha_2 > 0$, and for all x with $|x| = 1$

$$\alpha_2 \ge \int_{t_0}^{t_0+\delta} \left[(w^T + e^T)x\right]^2 d\tau \ \ge\ \alpha_1 \qquad \text{for all } t_0 \ge 0 \tag{A2.6.1}$$

By assumption, $e \in L_2$, so that $\int_0^\infty (e^T x)^2 d\tau \le m$ for some $m \ge 0$. Since w is PE, there exist σ, β_1, $\beta_2 > 0$ such that

$$\beta_2 \ge \int_{t_0}^{t_0+\sigma} (w^T x)^2 d\tau \ \ge\ \beta_1 \qquad \text{for all } t_0 \ge 0 \tag{A2.6.2}$$

Let $\delta \ge \sigma \left[1 + \frac{m}{\beta_1}\right]$, $\alpha_1 = \beta_1$, $\alpha_2 = m + \beta_2 \left[1 + \frac{m}{\beta_1}\right]$ so that

$$\int_{t_0}^{t_0+\delta} \left[(w^T + e^T)x\right]^2 d\tau \ \ge\ \int_{t_0}^{t_0+\delta} (w^T x)^2 d\tau - \int_{t_0}^{t_0+\delta} (e^T x)^2 d\tau$$

$$\ge \beta_1 \left[1 + \frac{m}{\beta_1}\right] - m = \alpha_1 \tag{A2.6.3}$$

and

$$\int_{t_0}^{t_0+\delta} \left[(w^T + e^T)x\right]^2 d\tau \ \le\ \int_{t_0}^{t_0+\delta} (w^T x)^2 d\tau + \int_{t_0}^{t_0+\delta} (e^T x)^2 d\tau$$

$$\le \beta_2 \left[1 + \frac{m}{\beta_1}\right] + m = \alpha_2 \tag{A2.6.4}$$

□

Proof of Lemma 2.6.7

We wish to prove that for some δ, α_1, $\alpha_2 > 0$, and for all x with $|x| = 1$

$$\alpha_2 \ge \int_{t_0}^{t_0+\delta} \left[\hat{H}(w^T)x\right]^2 d\tau \ \ge\ \alpha_1 \qquad \text{for all } t_0 \ge 0 \tag{A2.6.5}$$

Denote $u = w^T x$ and $y = \hat{H}(u) = \hat{H}(w^T x) = \hat{H}(w^T)x$ (where the last inequality is true because x does not depend on t). We thus wish to show that

$$\alpha_2 \ge \int_{t_0}^{t_0+\delta} y^2(\tau)\, d\tau \ \ge\ \alpha_1 \qquad \text{for all } t_0 \ge 0 \tag{A2.6.6}$$

Since w is PE, there exists σ, β_1, $\beta_2 > 0$ such that

$$\beta_2 \geq \int_{t_0}^{t_0+\sigma} u^2(\tau)\, d\tau \geq \beta_1 \qquad \text{for all } t_0 \geq 0 \tag{A2.6.7}$$

In this form, the problem appears on the relationship between truncated L_2 norms of the input and output of a stable, minimum phase LTI system. Similar problems are addressed in Section 3.6, and we will therefore use results from lemmas in that section.

Let $\delta = m\sigma$, where m is an integer to be defined later. Since u is bounded, and $y = \hat{H}(u)$, it follows that y is bounded (lemma 3.6.1) and the upper bound in (A2.6.6) is satisfied. The lower bound is obtained now, by inverting $\hat{H}$ in a similar way as is used in the proof of lemma 3.6.2. We let

$$\hat{z}(s) = \frac{a^r}{(s+a)^r}\, \hat{u}(s) \tag{A2.6.8}$$

where $a > 0$ will be defined later, and r is the relative degree of $\hat{H}(s)$. Thus

$$\hat{y}(s) = \frac{(s+a)^r}{a^r}\, \hat{H}(s)\, \hat{z}(s) \tag{A2.6.9}$$

The transfer function from $\hat{z}(s)$ to $\hat{y}(s)$ has relative degree 0. Being minimum phase, it has a proper and stable inverse. By lemma 3.6.1, there exist $k_1, k_2 \geq 0$ such that

$$\int_{t_0}^{t_0+\delta} z^2(\tau)\, d\tau \leq k_1 \int_{t_0}^{t_0+\delta} y^2(\tau)\, d\tau + k_2 \tag{A2.6.10}$$

Since $\dot{u}$ is bounded

$$\int_{t_0}^{t_0+\delta} \dot{u}^2(\tau)\, d\tau \leq k_3 \delta \tag{A2.6.11}$$

for some $k_3 \geq 0$. Using the results in the proof of lemma 3.6.2 ((A3.6.14)), we can also show that, with the properties of the transfer function $a^r / (s+a)^r$

$$\int_{t_0}^{t_0+\delta} u^2(\tau)\, d\tau \leq \int_{t_0}^{t_0+\delta} z^2(\tau)\, d\tau + \frac{r}{a} k_3 \delta + k_4 \tag{A2.6.12}$$

where k_4 is another constant due to initial conditions. It follows that

$$\int_{t_0}^{t_0+\delta} y^2(\tau)\,d\tau \ \geq \ \frac{1}{k_1}\left[\int_{t_0}^{t_0+\delta} u^2(\tau)\,d\tau - \frac{r}{a}\,k_3\,\delta - k_2 - k_4\right]$$

$$\geq \ \frac{1}{k_1}\left[m\,(\beta_1 - \frac{r}{a}\,k_3\,\sigma) - k_2 - k_4\right] \qquad \text{(A2.6.13)}$$

Note that r/a is arbitrary, and although k_1 depends on r/a, the constants β_1, k_3, and σ do not. Consequently, we can let r/a sufficiently small that $\beta_1 - (r/a)k_3\sigma \geq \beta_1/2$. We can also let m be sufficiently large that $m\beta_1/2 - k_2 - k_4 \geq \beta_1$. Then the lower bound in (A2.6.6) is satisfied with

$$\alpha_1 \ = \ \frac{\beta_1}{k_1} \qquad \text{(A2.6.14)}$$

□

Proof of Lemma 3.6.2

The proof of lemma 3.6.2 relies on the auxiliary lemma presented hereafter.

Auxiliary Lemma

Consider the transfer function

$$\hat{K}(s) \ = \ \frac{a^r}{(s+a)^r} \qquad a>0 \qquad \text{(A3.6.1)}$$

where r is a positive integer.

Let $k(t)$ be the corresponding impulse response and define

$$g(t-\tau) \ = \ \int_{t-\tau}^{\infty} k(\sigma)d\sigma \ = \ \int_{-\infty}^{\tau} k(t-\sigma)d\sigma \qquad t-\tau \geq 0 \qquad \text{(A3.6.2)}$$

Then

$$k(t) \ = \ \frac{a^r}{(r-1)!}\,t^{r-1}e^{-at} \qquad t \geq 0 \qquad \text{(A3.6.3)}$$

and $k(t) = 0$ for $t<0$. It follows that $k(t) \geq 0$ for all t, and

$$\| k \|_1 \ = \ \int_0^{\infty} k(\sigma)\,d\sigma \ = \ \int_{-\infty}^{t} k(t-\sigma)\,d\sigma \ = \ 1 \qquad \text{(A3.6.4)}$$

Similarly

$$g(t) = e^{-at} \sum_{k=1}^{r} \frac{t^{r-k}}{(r-k)!} a^{r-k} \qquad t \geq 0 \tag{A3.6.5}$$

and $g(t) = 0$ for $t < 0$. It follows that $g(t) \geq 0$ for all t, and

$$\| g \|_1 = \int_0^\infty g(\sigma) d\sigma = \int_{-\infty}^{t} g(t-\sigma) d\sigma = \frac{r}{a} \tag{A3.6.6}$$

□

We are now ready to prove lemma 3.6.2. Let r be the relative degree of $\hat{H}$, and

$$\hat{z}(s) = \frac{a^r}{(s+a)^r} \hat{u}(s) \tag{A3.6.7}$$

where $a > 0$ is an arbitrary constant to be defined later. Using (A3.6.7)

$$\hat{y}(s) = \frac{(s+a)^r}{a^r} \hat{H}(s) \hat{z}(s) \tag{A3.6.8}$$

Since the transfer function from $\hat{z}(s)$ to $\hat{y}(s)$ has relative degree 0 and is minimum phase, it has a proper and stable inverse. By lemma 3.6.1

$$\| z_t \|_p \leq b_1 \| y_t \|_p + b_2 \tag{A3.6.9}$$

We will prove that

$$\| u_t \|_p \leq c_1 \| z_t \|_p + c_2 \tag{A3.6.10}$$

so that the lemma will be verified with $a_1 = c_1 b_1$, $a_2 = c_1 b_2 + c_2$.

Derivation of (A3.6.10)

We have that

$$z(t) = \epsilon(t) + \int_0^t k(t-\tau) u(\tau) d\tau \tag{A3.6.11}$$

where $\epsilon(t)$ is an exponentially decaying term due to the initial conditions, and $k(t)$ is the impulse response corresponding to the transfer function in (A3.6.7) (derived in the auxiliary lemma). Integrate (A3.6.11) by parts to obtain

$$z(t) = \epsilon(t) + u(t) \int_{-\infty}^{t} k(t-\sigma) d\sigma - u(0) \int_{-\infty}^{0} k(t-\sigma) d\sigma$$

$$- \int_{-\infty}^{t} \left[\int_{-\infty}^{\tau} k(t-\sigma)\, d\sigma \right] \dot{u}(\tau)\, d\tau \qquad \text{(A3.6.12)}$$

Using the results of the auxiliary lemma

$$z(t) = \epsilon(t) + u(t) - u(0)g(t) - \int_{0}^{t} g(t-\tau)\dot{u}(\tau)\, d\tau \qquad \text{(A3.6.13)}$$

Since $g(t)$ is exponentially decaying, $u(0)g(t)$ can be included in $\epsilon(t)$. Also, using again the auxiliary lemma, together with lemma 3.6.1, and then the assumption on $\dot{u}$, it follows that

$$\begin{aligned} \| u_t \|_p &\le \| z_t \|_p + \| \epsilon_t \|_p + \frac{r}{a} \| \dot{u}_t \|_p \\ &\le \| z_t \|_p + \| \epsilon_t \|_p + \frac{r}{a} k_1 \| u_t \|_p + \frac{r}{a} k_2 \end{aligned} \qquad \text{(A3.6.14)}$$

Since a is arbitrary, let it be sufficiently large that $\frac{r}{a} k_1 < 1$. Consequently,

$$\begin{aligned} \| u_t \|_p &\le \frac{1}{1 - \frac{r}{a} k_1} \| z_t \|_p + \frac{\| \epsilon \|_p + \frac{r}{a} k_2}{1 - \frac{r}{a} k_1} \\ &:= c_1 \| z_t \|_p + c_2 \end{aligned} \qquad \text{(A3.6.15)}$$

□

Proof of Corollary 3.6.3

(a) From lemma 3.6.2.

(b) Since $\hat{H}$ is strictly proper, both y and $\dot{y}$ are bounded.

(c) We have that $y = \hat{H}(u)$ and $\dot{y} = \hat{H}(\dot{u})$. Using succesively lemma 3.6.1, the regularity of u, and lemma 3.6.2, it follows that for some constants $k_1, \ldots, k_6$

$$\begin{aligned} |\dot{y}| &\le k_1 \| \dot{u}_t \|_\infty + k_2 \\ &\le k_3 \| u_t \|_\infty + k_4 \\ &\le k_5 \| y_t \|_\infty + k_6 \end{aligned} \qquad \text{(A3.6.16)}$$

The proof can easily be extended to the vector case. □

Proof of Lemma 3.6.4

Let

$$\hat{H}(s) = h_0 + \hat{H}_1(s) \tag{A3.6.17}$$

where $\hat{H}_1$ is strictly proper (and stable). Let h_1 be the impulse response corresponding to $\hat{H}_1$. The output $y(t)$ is given by

$$y(t) = \epsilon(t) + h_0 u(t) + \int_0^t h_1(t-\tau)\, u(\tau)\, d\tau \tag{A3.6.18}$$

where $\epsilon(t)$ is due to the initial conditions. Inequality (3.6.9) follows, if we define

$$\gamma_1(t) := |h_0|\, \beta_1(t) + \int_0^t |h_1(t-\tau)|\, \beta_1(\tau)\, d\tau \tag{A3.6.19}$$

and

$$\gamma_2(t) := |\epsilon(t)| + |h_0|\, \beta_2(t) + \int_0^t |h_1(t-\tau)|\, \beta_2(\tau)\, d\tau \tag{A3.6.20}$$

Since $\epsilon \in L_2$ and $h_1 \in L_1 \cap L_\infty$, we also have that $|\epsilon| \in L_2$, $|h_1| \in L_1 \cap L_\infty$. Since $\beta_1, \beta_2 \in L_2$, it follows that the last term of (A3.6.19) and similarly the last term of (A3.6.20) belong to $L_2 \cap L_\infty$, and go to zero as $t \to \infty$ (see e.g., Desoer & Vidyasagar [1975], exercise 5, p. 242). The conclusions follow directly from this observation. □

Proof of Lemma 3.6.5

Let $[A, b, c^T, d]$ be a minimal realization of $\hat{H}$, with $A \in \mathbb{R}^{m \times m}$, $b \in \mathbb{R}^m$, $c \in \mathbb{R}^m$, and $d \in \mathbb{R}$. Let $x: \mathbb{R}_+ \to \mathbb{R}^m$, and $y_1: \mathbb{R}_+ \to \mathbb{R}$ such that

$$\begin{aligned} \dot{x} &= Ax + b\,(w^T \phi) \\ y_1 &= c^T x \end{aligned} \tag{A3.6.21}$$

and $W: \mathbb{R}_+ \to \mathbb{R}^{m \times n}$, $y_2: \mathbb{R}_+ \to \mathbb{R}$ such that

$$\begin{aligned} \dot{W} &= AW + bw^T \\ y_2 &= c^T W \phi \end{aligned} \tag{A3.6.22}$$

Thus

$$\hat{H}(w^T \phi) = y_1 + d\,(w^T \phi) \qquad \hat{H}(w^T)\phi = y_2 + (d\, w^T)\phi \tag{A3.6.23}$$

Since

$$\frac{d}{dt}(W\phi) = \dot{W}\phi + W\dot{\phi} = AW\phi + bw^T\phi + W\dot{\phi} \tag{A3.6.24}$$

it follows that

$$\frac{d}{dt}(x - W\phi) = A(x - W\phi) - W\dot{\phi}$$

$$y_1 - y_2 = c^T(x - W\phi) \tag{A3.6.25}$$

The result then follows since

$$\hat{H}(w^T\phi) - \hat{H}(w^T)\phi = y_1 - y_2 = \hat{H}_c(W\dot{\phi}) = \hat{H}_c(\hat{H}_b(w^T)\dot{\phi}) \tag{A3.6.26}$$

□

Proof of Theorem 3.7.3

The proof follows the steps of the proof of theorem 3.7.1 and is only sketched here.

(a) *Derive properties of the identifier that are independent of the boundedness of the regressor*

The properties of the identifier are the standard properties obtained in theorems 2.4.1–2.4.4

$$|\psi^T(t)\tilde{w}(t)| = \beta(t)\|\tilde{w}_t\|_\infty + \beta(t)$$

$$\beta \in L_2 \cap L_\infty$$

$$\psi \in L_\infty \qquad \dot{\psi} \in L_2 \cap L_\infty$$

$$a_{m+1}(t) \geq k_{min} > 0 \qquad \text{for all } t \geq 0 \tag{A3.7.1}$$

The inequality for $a_{m+1}(t)$ follows from the use of the projection in the update law.

We also noted, in Section 3.3, that if π is bounded and a_{m+1} is bounded away from zero, then θ is also bounded, and the transformation has bounded derivatives. The vector q of coefficients of the polynomial $\hat{q}$ is also bounded. By definition of the transformation, $\theta(\pi^*) = \theta^*$. Therefore, $\psi \in L_\infty$, $\dot{\psi} \in L_2 \cap L_\infty$ implies that $\phi \in L_\infty$, $\dot{\phi} \in L_2 \cap L_\infty$. Also, we have that $(k_m / \|a_{m+1}\|_\infty) \leq c_0(t) \leq k_m / k_{min}$, for all $t \geq 0$.

(b) *Express the system states and inputs in term of the control error*

As in theorem 3.7.1.

(c) *Relate the identifier error to the control error*

We first establish an equality of ratios of polynomials, then we transform it to an operator equality. Using a similar approach as in the comments before the proof, we have that

$$\begin{aligned} \hat{q}\hat{a} - \hat{q}\hat{a}^* &= a_{m+1}(\hat{\lambda} - \hat{c}) - \hat{q}k_p\hat{n}_p \\ &= -a_{m+1}(\hat{c} - \hat{c}^*) + a_{m+1}(\hat{\lambda} - \hat{c}^*) - k_p\hat{q}\hat{n}_p \\ &= -a_{m+1}(\hat{c} - \hat{c}^*) + (a_{m+1}\hat{q}^* - k_p\hat{q})\,\hat{n}_p \end{aligned} \tag{A3.7.2}$$

$$\begin{aligned} \hat{q}\hat{b} - \hat{q}\hat{b}^* &= \hat{q}\hat{\lambda} - a_{m+1}\hat{d} - \hat{\lambda}_0\hat{d}_m - \hat{q}\hat{\lambda} + \hat{q}\hat{d}_p \\ &= -a_{m+1}(\hat{d} - \hat{d}^*) + (-a_{m+1}\hat{d}^* - \hat{\lambda}_0\hat{d}_m + \hat{q}\hat{d}_p) \\ &= -a_{m+1}(\hat{d} - \hat{d}^*) \\ &\quad + \left[-\frac{a_{m+1}}{k_p}\hat{q}^* + \frac{a_{m+1}}{k_p}\hat{\lambda}_0\frac{\hat{d}_m}{\hat{d}_p} + \hat{q} - \hat{\lambda}_0\frac{\hat{d}_m}{\hat{d}_p} \right] \hat{d}_p \end{aligned} \tag{A3.7.3}$$

Therefore

$$\begin{aligned} &\frac{\hat{q}}{\hat{\lambda}_0}\left[\frac{\hat{a} - \hat{a}^*}{\hat{\lambda}}\right] + \frac{\hat{q}}{\hat{\lambda}_0}\left[\frac{\hat{b} - \hat{b}^*}{\hat{\lambda}}\right]\frac{k_p\hat{n}_p}{\hat{d}_p} \\ &\quad = -a_{m+1}\left[\frac{\hat{c} - \hat{c}^*}{\hat{\lambda}\hat{\lambda}_0} + \frac{\hat{d} - \hat{d}^*}{\hat{\lambda}\hat{\lambda}_0}\hat{P}\right] - (k_p - a_{m+1})\frac{\hat{\lambda}_0}{\hat{\lambda}}\,\frac{\hat{d}_m\hat{n}_p}{\hat{d}_p}\,\frac{1}{\hat{\lambda}_0} \\ &\quad = -\frac{k_m}{c_0}\left[\frac{\hat{c} - \hat{c}^*}{\hat{\lambda}\hat{\lambda}_0} + \frac{\hat{d} - \hat{d}^*}{\hat{\lambda}\hat{\lambda}_0}\hat{P} + (c_0 - c_0^*)\hat{M}^{-1}\frac{\hat{P}}{\hat{\lambda}_0}\right] \end{aligned} \tag{A3.7.4}$$

where we divided by $\hat{\lambda}\hat{\lambda}_0$ to obtain proper stable transfer functions. The polynomial $\hat{\lambda}_0$ is Hurwitz and q is bounded, so that the operator $q^T\hat{s}_r/\hat{\lambda}_0$ is a bounded operator.

We now transform this polynomial equality into an operator equality as in the comments before the proof. Applying both sides of (A3.7.4) to u

$$-q^T\frac{\hat{s}_r}{\hat{\lambda}_0}(\tilde{w}^T)\psi = \frac{k_m}{c_0}\left[(c_0 - c_0^*)\hat{M}^{-1}\frac{\hat{P}}{\hat{\lambda}_0}(u) + \bar{\phi}^T\frac{1}{\hat{\lambda}_0}(\overline{w})\right] \tag{A3.7.5}$$

The right-hand side is very reminiscent of the signal z obtained in the input error scheme. A filtered version of the signal $\hat{M}^{-1}\hat{P}(u) = r_p$ appears, instead of r, with the error $c_0 - c_0^*$. From proposition 3.3.1, with $\hat{L} = \hat{\lambda}_0$ (cf. (3.3.17))

$$c_0^* \hat{M}^{-1} \hat{P} \frac{1}{\hat{\lambda}_0}(u) = \frac{1}{\hat{\lambda}_0}(u) - \frac{1}{\hat{\lambda}_0}(\bar{\theta}^{*T} \bar{w}) \tag{A3.7.6}$$

and since $u = c_0 r + \bar{\theta}^T \bar{w}$, it follows that

$$\hat{M}^{-1} \frac{\hat{P}}{\hat{\lambda}_0}(u) = \frac{1}{c_0^*} \left[\frac{1}{\hat{\lambda}_0}(c_0 r) + \frac{1}{\hat{\lambda}_0}(\bar{\phi}^T \bar{w}) \right] \tag{A3.7.7}$$

The right-hand side of (A3.7.5) becomes, using (A3.7.7) followed by the swapping lemma (and using the notation of the swapping lemma)

$$\frac{k_m}{c_0^*} \left[\frac{c_0 - c_0^*}{c_0} \frac{1}{\hat{\lambda}_0}(c_0 r) + \frac{1}{\hat{\lambda}_0}(\bar{\phi}^T \bar{w}) + \frac{c_0^*}{c_0} \left[\bar{\phi}^T \frac{1}{\hat{\lambda}_0}(\bar{w}) - \frac{1}{\hat{\lambda}_0}(\bar{\phi}^T \bar{w}) \right] \right]$$

$$= \frac{k_m}{c_0^*} \left[\frac{1}{\hat{\lambda}_0}((c_0 - c_0^*) r) - \hat{\Lambda}_{oc} \left[\hat{\Lambda}_{oc}(c_0 r) \left[\frac{c_0 - c_0^*}{c_0} \right] \right] \right.$$

$$\left. + \frac{1}{\hat{\lambda}_0}(\bar{\phi}^T \bar{w}) - \frac{c_0^*}{c_0} \hat{\Lambda}_{oc} \left[\hat{\Lambda}_{ob}(\bar{w}^T) \dot{\bar{\phi}} \right] \right] \tag{A3.7.8}$$

On the other hand, using again the swapping lemma, the left-hand side of (A3.7.5) becomes

$$q^T \frac{\hat{S}_r}{\hat{\lambda}_0}(\tilde{w}^T) \psi = q^T \frac{\hat{S}_r}{\hat{\lambda}_0}(\tilde{w}^T \psi) - q^T \hat{S}_{rc}(\hat{S}_{rb}(\tilde{w}^T) \dot{\psi}) \tag{A3.7.9}$$

where the transfer functions $\hat{\Lambda}_{ob}$, $\hat{\Lambda}_{oc}$, $\hat{S}_{rb}$, and $\hat{S}_{rc}$ result from the application of the swapping lemma. The output error is then equal to (using (3.7.2), (A3.7.5), (A3.7.8), (A3.7.9))

$$y_p - y_m = \frac{1}{c_0^*} \hat{M} \left[(c_0 - c_0^*) r + \bar{\phi}^T \bar{w} \right]$$

$$= \frac{1}{k_m} \hat{M} \hat{\lambda}_0 \left[\frac{k_m}{c_0^*} \frac{1}{\hat{\lambda}_0}((c_0 - c_0^*) r) + \frac{1}{\hat{\lambda}_0}(\bar{\phi}^T \bar{w}) \right]$$

$$= \frac{1}{k_m} \hat{M} \hat{\lambda}_0 \left[- q^T \frac{\hat{s}_r}{\hat{\lambda}_0} (\tilde{w}^T \psi) + q^T \hat{S}_{rc} \left[\hat{S}_{rb} (\tilde{w}^T) \dot{\psi} \right] \right.$$

$$+ \frac{k_m}{c_0^*} \hat{\Lambda}_{oc} \left[\hat{\Lambda}_{ob} (c_0 r) \left[\frac{c_0 - c_0^*}{c_0} \right] \right]$$

$$\left. + \frac{k_m}{c_0} \hat{\Lambda}_{oc} \left[\hat{\Lambda}_{ob} (\overline{w}^T) \dot{\overline{\phi}} \right] \right] \tag{A3.7.10}$$

(d) *Establish the regularity of the signals*
As in theorem 3.7.1.
(e) *Stability Proof*
$\hat{M} \hat{\lambda}_0$ is a stable transfer function and since q^T is bounded, $q^T \hat{s}_r / \hat{\lambda}_0$ is a bounded operator. We showed that $\dot{\psi}$, $\dot{\overline{\phi}}$, $\dot{c}_0 \in L_2$, so that, from (A3.7.10) and part (a), an inequality such as (3.7.19) can be obtained. As before $\tilde{w}$ regular implies that $\beta \to 0$ as $t \to \infty$. The boundedness of all signals in the adaptive system then follows as in theorem 3.7.1. Similarly, $y_p - y_m \in L_2$ and tends to zero as $t \to \infty$. Since the relative degree of the transfer function from $u \to \tilde{w}$ is the same as the relative degree of $\hat{P}$, $\hat{M}$, and therefore $\hat{L}^{-1}$, the same result is true for $\tilde{w} - \tilde{w}_m$.

Proof of Lemma 4.2.1
Define

$$w_\epsilon(t, x) = \int_0^t d(\tau, x) e^{-\epsilon(t-\tau)} d\tau \tag{A4.2.1}$$

and

$$w_0(t, x) = \int_0^t d(\tau, x) d\tau \tag{A4.2.2}$$

From the assumptions

$$| w_0(t + t_0, x) - w_0(t_0, x) | \le \gamma(t) \cdot t \tag{A4.2.3}$$

for all t, $t_0 \ge 0$, $x \in B_h$. Integrating (A4.2.1) by parts

$$w_\epsilon(t,x) = w_0(t,x) - \epsilon \int_0^t e^{-\epsilon(t-\tau)} w_0(\tau,x)\,d\tau \tag{A4.2.4}$$

Using the fact that

$$\epsilon \int_0^t e^{-\epsilon(t-\tau)} w_0(t,x)\,d\tau = w_0(t,x) - w_0(t,x)e^{-\epsilon t} \tag{A4.2.5}$$

(A4.2.4) can be rewritten as

$$w_\epsilon(t,x) = w_0(t,x)e^{-\epsilon t} + \epsilon \int_0^t e^{-\epsilon(t-\tau)}(w_0(t,x) - w_0(\tau,x))\,d\tau \tag{A4.2.6}$$

and, using (A4.2.3) and the fact that $w_0(0,x) = 0$,

$$|w_\epsilon(t,x)| \le \gamma(t)te^{-\epsilon t} + \epsilon \int_0^t e^{-\epsilon(t-\tau)}(t-\tau)\gamma(t-\tau)d\tau \tag{A4.2.7}$$

Consequently,

$$|\epsilon w_\epsilon(t,x)| \le \sup_{t' \ge 0} \gamma\left[\frac{t'}{\epsilon}\right] t'e^{-t'} + \int_0^\infty \gamma\left[\frac{\tau'}{\epsilon}\right] \tau'e^{-\tau'}d\tau' \tag{A4.2.8}$$

Since, for some β, $|d(t,x)| \le \beta$, we also have that $\gamma(t) \le \beta$. Note that, for all $t' \ge 0$, $t'e^{-t'} \le e^{-1}$, and $t'e^{-t'} \le t'$, so that

$$|\epsilon w_\epsilon(t,x)| \le \sup_{t' \in [0,\sqrt{\epsilon}]} \left[\gamma\left[\frac{t'}{\epsilon}\right] t'e^{-t'}\right] + \sup_{t' \ge \sqrt{\epsilon}} \left[\gamma\left[\frac{t'}{\epsilon}\right] t'e^{-t'}\right]$$

$$+ \int_0^{\sqrt{\epsilon}} \gamma\left[\frac{\tau'}{\epsilon}\right] \tau'e^{-\tau'}d\tau' + \int_{\sqrt{\epsilon}}^\infty \gamma\left[\frac{\tau'}{\epsilon}\right] \tau'e^{-\tau'}d\tau' \tag{A4.2.9}$$

This, in turn, implies that

$$|\epsilon w_\epsilon(t,x)| \le \beta\sqrt{\epsilon} + \gamma\left[\frac{1}{\sqrt{\epsilon}}\right] e^{-1} + \beta\frac{\epsilon}{2} + \gamma\left[\frac{1}{\sqrt{\epsilon}}\right](1+\sqrt{\epsilon})e^{-\sqrt{\epsilon}}$$

$$:= \xi(\epsilon) \tag{A4.2.10}$$

From the assumption on γ, it follows that $\xi(\epsilon) \in K$. From (A4.2.1)

$$\frac{\partial w_\epsilon(t,x)}{\partial t} - d(t,x) = -\epsilon w_\epsilon(t,x) \tag{A4.2.11}$$

so that the first part of the lemma is verified.

If $\gamma(T) = a / T^r$, then the right-hand side of (A4.2.8) can be computed explicitly

$$\sup_{t' \geq 0} a\,\epsilon^r (t')^{1-r} e^{-t'} = a\,\epsilon^r (1-r)^{1-r} e^{r-1} \leq a\,\epsilon^r \quad \text{(A4.2.12)}$$

and, with Γ denoting the standard gamma function,

$$\int_0^\infty a\,\epsilon^r (\tau')^{1-r} e^{-\tau'} d\tau' = a\,\epsilon^r\,\Gamma(2-r) \leq a\,\epsilon^r \quad \text{(A4.2.13)}$$

Defining $\xi(\epsilon) = 2\,a\,\epsilon^r$, the second part of the lemma is verified. □

Proof of Lemma 4.2.2

Define $w_\epsilon(t,x)$ as in lemma 4.2.1. Consequently,

$$\frac{\partial w_\epsilon(t,x)}{\partial x} = \frac{\partial}{\partial x}\left[\int_0^t d(\tau, x) e^{-\epsilon(t-\tau)} d\tau\right]$$

$$= \int_0^t \left[\frac{\partial}{\partial x} d(\tau, x)\right] e^{-\epsilon(t-\tau)} d\tau \quad \text{(A4.2.14)}$$

Since $\dfrac{\partial d(t,x)}{\partial x}$ is zero mean, and is bounded, lemma 4.2.1 can be applied to $\dfrac{\partial d(t,x)}{\partial x}$, and inequality (4.2.6) of lemma 4.2.1 becomes inequality (4.2.10) of lemma 4.2.2. Note that since $\dfrac{\partial d(t,x)}{\partial x}$ is bounded, and $d(t, 0) = 0$ for all $t \geq 0$, $d(t,x)$ is Lipschitz.

Since $d(t,x)$ is zero mean, with convergence function $\gamma(T)|x|$, the proof of lemma 4.2.1 can be extended, with an additional factor $|x|$. This leads directly to (4.2.8) and (4.2.9) (although the function $\xi(\epsilon)$ may be different from that obtained with $\dfrac{\partial d(t,x)}{\partial x}$, these functions can be replaced by a single $\xi(\epsilon)$). □

Proof of Lemma 4.2.3

The proof proceeds in two steps.

(a) For ϵ sufficiently small, and for t fixed, the transformation is a homeomorphism.

Apply lemma 4.2.2, and let ϵ_1 such that $\xi(\epsilon_1) < 1$. Let $\epsilon \leq \epsilon_1$. Given $z \in B_h$, the corresponding x such that

$$x = z + \epsilon w_\epsilon(t, z) \tag{A4.2.15}$$

may not belong to B_h. Similarly, given $x \in B_h$, the solution z of (A4.2.15) may not exist in B_h. However, for any x, z satisfying (A4.2.15), inequality (4.2.8) implies (4.2.16) and

$$(1 - \xi(\epsilon))\,|z| \;\leq\; |x| \;\leq\; (1 + \xi(\epsilon))\,|z| \tag{A4.2.16}$$

Define

$$h'(\epsilon) = \min\left[h(1 - \xi(\epsilon)),\ \frac{h}{1 + \xi(\epsilon)}\right] = h(1 - \xi(\epsilon)) \tag{A4.2.17}$$

and note that $h'(\epsilon) \to h$ as $\epsilon \to 0$.

We now show that

- for all $z \in B_{h'}$, there exists a unique $x \in B_h$ such that (A4.2.15) is satisfied,
- for all $x \in B_{h'}$, there exists a unique $z \in B_h$ such that (A4.2.15) is satisfied.

In both cases, $|x - z| \leq \xi(\epsilon)h$.

The first part follows directly from (A4.2.15), (A4.2.16). The fact that $|x - z| \leq \xi(\epsilon)h$ also follows from (A4.2.15), (4.2.8) and implies that, if a solution z exists to (A4.2.15), it must lie in the closed ball U of radius $\xi(\epsilon)h$ around x. It can be checked, using (4.2.10), that the mapping $F_x(z) = x - \epsilon w_\epsilon(t, z)$ is a contraction mapping in U, provided that $\xi(\epsilon) < 1$. Consequently, F has a unique fixed point z in U. This solution is also a solution of (A4.2.15), and since it is unique in U, it is also unique in B_h (and actually in R^n). For $x \in B_h$, but outside $B_{h'}$, there is no guarantee that a solution z exists in B_h, but if it exists, it is again unique in B_h. Consequently, the map $x \to z$ defined by (A4.2.15) is well defined. From the smoothness of $w_\epsilon(t, z)$ with respect to z, it follows that the map is a homeomorphism.

(b) The transformation of variable leads to the differential equation (4.2.17)

Applying (A4.2.15) to the system (4.2.1)

$$\left[I + \epsilon \frac{\partial w_\epsilon}{\partial z}\right] \dot{z} = \epsilon f_{av}(z) + \epsilon\left[f(t, z, 0) - f_{av}(z) - \frac{\partial w_\epsilon}{\partial t}\right]$$

$$+ \epsilon\left[f(t, z + \epsilon w_\epsilon, \epsilon) - f(t, z, \epsilon)\right]$$

$$+ \epsilon \left[f(t, z, \epsilon) - f(t, z, 0) \right]$$

$$:= \epsilon f_{av}(z) + \epsilon p'(t, x, z, \epsilon) \qquad \text{(A4.2.18)}$$

where, using the assumptions, and the results of lemma 4.2.2

$$|p'(t, z, \epsilon)| \le \xi(\epsilon)|z| + \xi(\epsilon)l_1|z| + \epsilon l_2|z| \qquad \text{(A4.2.19)}$$

For $\epsilon \le \epsilon_1$, (4.2.10) implies that $\left[I + \epsilon \frac{\partial w_\epsilon}{\partial z} \right]$ has a bounded inverse for all $t \ge 0$, $z \in B_h$. Consequently, z satisfies the differential equation

$$\dot{z} = \left[I + \epsilon \frac{\partial w_\epsilon}{\partial z} \right]^{-1} \left[\epsilon f_{av}(z) + \epsilon p'(t, z, \epsilon) \right]$$

$$= \epsilon f_{av}(z) + \epsilon p(t, z, \epsilon) \qquad z(0) = x_0 \qquad \text{(A4.2.20)}$$

where

$$p(t, z, \epsilon) = \left[I + \epsilon \frac{\partial w_\epsilon}{\partial z} \right]^{-1} \left[p'(t, z, \epsilon) - \epsilon \frac{\partial w_\epsilon}{\partial z} f_{av}(z) \right] \qquad \text{(A4.2.21)}$$

and

$$|p(t, z, \epsilon)| \le \frac{1}{1 - \xi(\epsilon_1)} \left[\xi(\epsilon) + \xi(\epsilon)l_1 + \epsilon l_2 + \xi(\epsilon)l_{av} \right] |z|$$

$$:= \psi(\epsilon)|z| \qquad \text{(A4.2.22)}$$

for all $t \ge 0$, $\epsilon \le \epsilon_1$, $z \in B_h$. □

Proof of Lemma 4.4.1

The proof is similar to the proof of lemma 4.2.3. We consider the transformation of variable

$$x = z + \epsilon w_\epsilon(t, z) \qquad \text{(A4.4.1)}$$

with $\epsilon \le \epsilon_1$, such that $\xi(\epsilon_1) < 1$. (4.4.1) becomes

$$\dot{z} = \left[I + \epsilon \frac{\partial w_\epsilon}{\partial z} \right]^{-1} \epsilon \left\{ f_{av}(z) + \left[f(t, z, 0) - f_{av}(z) - \frac{\partial w_\epsilon}{\partial t} \right] \right.$$

$$\left. + \left[f(t, z + \epsilon w_\epsilon, 0) - f(t, z, 0) \right] \right.$$

$$+ \left[f(t, z + \epsilon w_\epsilon, y) - f(t, z + \epsilon w_\epsilon, 0) \right] \Bigg\} \tag{A4.4.2}$$

or

$$\dot{z} = \epsilon f_{av}(z) + \epsilon p_1(t, z, \epsilon) + \epsilon p_2(t, z, y, \epsilon) \quad z(0) = x_0 \tag{A4.4.3}$$

where

$$|p_1(t, z, \epsilon)| \leq \frac{1}{1 - \xi(\epsilon_1)} \left[\xi(\epsilon) l_{av} + \xi(\epsilon) + \xi(\epsilon) l_1 \right] |z|$$

$$:= \xi(\epsilon) k_1 |z| \tag{A4.4.4}$$

and

$$|p_2(t, z, y, \epsilon)| \leq \frac{1}{1 - \xi(\epsilon_1)} l_2 |y| := k_2 |y| \tag{A4.4.5}$$

□

Proof of Theorem 4.4.2

The proof assumes that for all $t \in [0, T/\epsilon]$, the solutions $x(t)$, $y(t)$, and $z(t)$ (to be defined) remain in B_h. Since this is not guaranteed a priori, the steps of the proof are only valid for as long as the condition is verified. By assumption, $x_{av}(t) \in B_{h'}$, with $h' < h$. We will show that by letting ϵ and h_0 sufficiently small, we can let $x(t)$ be arbitrarily close to $x_{av}(t)$ and $y(t)$ arbitrarily small. It then follows, from a contradiction argument, that $x(t), y(t) \in B_h$ for all $t \in [0, T/\epsilon]$, provided that ϵ and h_0 are sufficiently small.

Using lemma 4.4.1, we transform the original system (4.4.1), (4.4.2) into the system (4.4.11), (4.4.2). A bound on the error $|z(t) - x_{av}(t)|$ can be calculated by integrating the difference (4.4.11)–(4.4.4), and by using (4.4.7) and (4.4.12)

$$|z(t) - x_{av}(t)| \leq \epsilon l_{av} \int_0^t |z(\tau) - x_{av}(\tau)| \, d\tau + \epsilon \xi(\epsilon) k_1 \int_0^t |z(\tau)| \, d\tau$$

$$+ \epsilon k_2 \int_0^t |y(\tau)| \, d\tau \tag{A4.4.6}$$

Bound on | y(t)|

To obtain a bound on $|y(t)|$, we divide the interval $[0, T/\epsilon]$ in intervals $[t_i, t_{i+1}]$ of length ΔT (the last interval may be of smaller length, and ΔT will be defined later). The differential equation for y is

$$\dot{y} = A(x)y + \epsilon g(t, x, y) \tag{A4.4.7}$$

and is rewritten on the time interval $[t_i, t_{i+1}]$ as follows

$$\dot{y} = A_{x_i} y + \epsilon g(t, x, y) + (A_{x_t} - A_{x_i}) y \tag{A4.4.8}$$

where $A_{x_t} = A(x(t))$, and $A_{x_i} = A(x(t_i))$, so that the solution $y(t)$, for $t \in [t_i, t_{i+1}]$, is given by

$$y(t) = e^{A_{x_i}(t - t_i)} y_i + \epsilon \int_{t_i}^{t} e^{A_{x_i}(t - \tau)} g(\tau, x, y) d\tau$$

$$+ \int_{t_i}^{t} e^{A_{x_i}(t - \tau)} (A_{x_\tau} - A_{x_i}) y(\tau) d\tau \tag{A4.4.9}$$

where $y_i = y(t_i)$. From the assumptions, it follows that

$$\| A_{x_\tau} - A_{x_i} \| \leq k_a |\dot{x}| (\tau - t_i) \leq \epsilon (l_1 + l_2) h\, k_a \Delta \tag{A4.4.10}$$

and, using the uniform exponential stability assumption on $A(x)$

$$|y(t)| \leq m |y_i| e^{-\lambda(t - t_i)} + \epsilon \frac{m}{\lambda} h \left[(l_3 + l_4) + (l_1 + l_2) k_a \Delta T \right] \tag{A4.4.11}$$

Let the last term in (A4.4.11) be denoted by ϵk_b, and use (A4.4.11) as a recursion formula for y_i, so that

$$|y_i| \leq \left[m e^{-\lambda \Delta T} \right]^i |y_0| + \epsilon k_b \sum_{j=0}^{i-1} \left[m e^{-\lambda \Delta T} \right]^j \tag{A4.4.12}$$

Choose ΔT sufficiently large that

$$m e^{-\lambda \Delta T} \leq e^{-\lambda \Delta T / 2} \quad \text{or} \quad \Delta T \geq \frac{2}{\lambda} \ln m \tag{A4.4.13}$$

It follows that

$$\sum_{j=0}^{i-1} \left[m e^{-\lambda \Delta T} \right]^j \leq \sum_{j=0}^{\infty} \left[e^{-\lambda \Delta T / 2} \right]^j = \frac{1}{1 - e^{-\lambda \Delta T / 2}} \tag{A4.4.14}$$

Combining (A4.4.12)–(A4.4.14) and using the assumption $y_0 \in B_{h_0}$,

$$|y_i| \leq e^{-\lambda \Delta T\, i/2} h_0 + \frac{\epsilon k_b}{1 - e^{-\lambda \Delta T / 2}} := e^{-\lambda t_i / 2} h_0 + \epsilon k_c \tag{A4.4.15}$$

Using this result in (A4.4.11), it follows that for all $t \in [t_i, t_{i+1}]$

$$|y(t)| \leq m e^{-\lambda t_i / 2} h_0 e^{-\lambda(t - t_i)} + m \epsilon k_c e^{-\lambda(t - t_i)} + \epsilon k_b$$

$$\leq mh_0 e^{-\lambda t/2} + \epsilon(mk_c + k_b) \tag{A4.4.16}$$

Since the last inequality does not depend on i, it gives a bound on $|y(t)|$ for all $t \in [0, T/\epsilon]$.

Bound on z(t) – x$_{av}$(t)

We now return to (A4.4.6), and to the approximation error, using the bound on $|y(t)|$

$$|z(t) - x_{av}(t)| \leq \epsilon l_{av} \int_0^t |z(\tau) - x_{av}(\tau)|\, d\tau + \epsilon \xi(\epsilon) k_1 \int_0^t h\, d\tau$$

$$+ \epsilon k_2 \int_0^t \left[mh_0 e^{-\lambda\tau/2} + \epsilon(mk_c + k_b) \right] d\tau \tag{A4.4.17}$$

so that, using the Bellman-Gronwall lemma (lemma 1.4.2)

$$|z(t) - x_{av}(t)|$$

$$\leq \int_0^t \left[\xi(\epsilon) k_1 h + k_2 mh_0 e^{-\lambda\tau/2} + k_2 \epsilon(mk_c + k_b) \right] \epsilon e^{\epsilon l_{av}(t-\tau)} d\tau$$

$$\leq (\epsilon + \xi(\epsilon)) \left[k_1 h + \frac{k_2 mh_0 l_{av}}{\lambda/2 + \epsilon l_{av}} + k_2(mk_c + k_b) \right] \frac{e^{l_{av}T}}{l_{av}}$$

$$:= \psi(\epsilon) a_T \tag{A4.4.18}$$

and, using (4.4.10)

$$|x(t) - x_{av}(t)| \leq \psi(\epsilon) b_T \tag{A4.4.19}$$

for some b_T.

Assumptions

We assumed in the proof that all signals remained in B_h. By assumption, $x_{av}(t) \in B_{h'}$, for some $h' < h$. Let h_0, and ϵ_T be sufficiently small so that, for all $\epsilon \leq \epsilon_T \leq \epsilon_1$, we have that $mh_0 + \epsilon(mk_c + k_b) \leq h$ (cf. (A4.4.16)), and that $\psi(\epsilon) b_T \leq h - h'$ (cf. (4.4.27)). It follows, from a simple contradiction argument, that the solutions $x(t)$, $y(t)$ and $z(t)$ remain in B_h for all $t \in [0, T/\epsilon]$, so that all steps of the proof are valid, and (A4.4.19) is in fact satisfied over the whole time interval. □

Proof of Theorem 4.4.3

The proof relies on the converse theorem of Lyapunov for exponentially stable systems (theorem 1.4.3). Under the hypotheses, there exists a function $v(x_{av}) : \mathbb{R}^n \to \mathbb{R}_+$, and strictly positive constants $\alpha_1, \alpha_2, \alpha_3, \alpha_4$ such that, for all $x_{av} \in B_h$,

$$\alpha_1 |x_{av}|^2 \le v(x_{av}) \le \alpha_2 |x_{av}|^2 \tag{A4.4.20}$$

$$\dot{v}(x_{av})\Big|_{(4.4.4)} \le -\epsilon\alpha_3 |x_{av}|^2 \tag{A4.4.21}$$

$$\left|\frac{\partial v}{\partial x_{av}}\right| \le \alpha_4 |x_{av}| \tag{A4.4.22}$$

The derivative in (A4.4.21) is to be taken along the trajectories of the averaged system (4.4.4).

We now study the stability of the original system (4.4.1), (4.4.2), through the transformed system (4.4.11), (4.4.2), where $x(z)$ is defined in (4.4.9). Consider the following Lyapunov function

$$v_1(z, y) = v(z) + \frac{\alpha_2}{p_2} y^T P(x(z)) y \tag{A4.4.23}$$

where $P(x)$, p_2 are defined in the comments after the definition of uniform exponential stability of $A(x)$. Defining $\alpha'_1 = \min(\alpha_1, \frac{\alpha_2}{p_2} p_1)$, it follows that

$$\alpha'_1 (|z|^2 + y|^2) \le v_1(z, y) \le \alpha_2 (|z|^2 + y|^2) \tag{A4.4.24}$$

The derivative of v_1 along the trajectories of (4.4.11)–(4.4.2) can be bounded, using the foregoing inequalities

$$\dot{v}_1(z, y) \le -\epsilon\alpha_3 |z|^2 + \epsilon\xi(\epsilon) k_1 \alpha_4 |z|^2 + \epsilon k_2 \alpha_4 |z|\,|y|$$
$$+ \frac{\alpha_2}{p_2} \left\|\frac{\partial P(x)}{\partial x}\right\| \left\|\frac{\partial x}{\partial z}\right\| |\dot{z}|\,|y|^2$$
$$- \frac{\alpha_2}{p_2} q_1 |y|^2 + 4\epsilon l_3 \alpha_2 |z|\,|y| + 2\epsilon l_4 \alpha_2 |y|^2 \tag{A4.4.25}$$

for $\epsilon \le \epsilon_1$ (so that the transformation $x \to z$ is well defined and $|x| \le 2|z|$). We now calculate bounds on the terms in (A4.4.25).

Bound on $|\,\partial P/\partial x\,|$

Note that $P(x)$ can be defined by

$$P(x) = \int_0^{\infty} e^{A^T(x)t} Q e^{A(x)t} dt \tag{A4.4.26}$$

so that

$$\frac{\partial P(x)}{\partial x_i} = \int_0^{\infty} \left\{ \left[\frac{\partial}{\partial x_i} e^{A^T(x)t} \right] Q e^{A(x)t} + e^{A^T(x)t} Q \left[\frac{\partial}{\partial x_i} e^{A(x)t} \right] \right\} dt \tag{A4.4.27}$$

The partial derivatives in parentheses solve the differential equation

$$\frac{d}{dt} \left[\frac{\partial}{\partial x_i} e^{A(x)t} \right] = A(x) \left[\frac{\partial}{\partial x_i} e^{A(x)t} \right] + \frac{\partial A(x)}{\partial x_i} e^{A(x)t} \tag{A4.4.28}$$

with zero initial conditions, so that

$$\frac{\partial}{\partial x_i} e^{A(x)} t = \int_0^t e^{A(x)(t-\tau)} \frac{\partial A(x)}{\partial x_i} e^{A(x)\tau} d\tau \tag{A4.4.29}$$

From the boundedness of $\frac{\partial A(x)}{\partial x_i}$, and from the exponential stability of $A(x)$, it follows that

$$\left\| \frac{\partial}{\partial x} e^{A(x)t} \right\| \le m^2 k_a t e^{-\lambda t} \tag{A4.4.30}$$

With (A4.4.27), this implies that $\| \partial P(x)/\partial x \|$ is bounded by some $k_p \ge 0$.

Bound on $\|\partial x/\partial z\|$ and $|z|$

On the other hand, using (4.4.9), (4.2.8) and (4.4.12)

$$\left\| \frac{\partial x}{\partial z} \right\| < 1 + \xi(\epsilon) < 2$$

$$\text{and} \quad |\dot{z}| \le \epsilon h(l_{av} + \xi(\epsilon) k_1 + k_2) \tag{A4.4.31}$$

Using these results in (A4.4.25), and noting the fact that, for all y, $z \in \mathbb{R}$

$$\epsilon|z|\;|y| \;\leq\; \frac{1}{2}\,(\epsilon^{4/3}|z|^2 + \epsilon^{2/3}|y|^2) \tag{A4.4.32}$$

it follows that

$$\begin{aligned}
\dot{v}_1(z, y) \;\leq\; & -\epsilon\left[\alpha_3 - \xi(\epsilon)\,k_1\,\alpha_4 - \epsilon^{1/3}\frac{k_2\,\alpha_4}{2} - 2\,\epsilon^{1/3}\,l_3\,\alpha_2\right]|z|^2 \\
& -\left[\frac{\alpha_2}{p_2}\,q_1 - 2\,\epsilon\,l_4\,\alpha_2 - \epsilon^{2/3}\frac{k_2\,\alpha_4}{2} - 2\,\epsilon^{2/3}\,l_3\,\alpha_2\right. \\
& \left. + 2\,\epsilon\,\frac{\alpha_2}{p_2}\,k_p h\left[l_{av} + \xi(\epsilon)\,k_1 + k_2\right]\right]|y|^2 \\
:= \; & -2\,\epsilon\,\alpha_2\,\alpha(\epsilon)\,|z|^2 - q(\epsilon)\,|y|^2
\end{aligned} \tag{A4.4.33}$$

Note that, with this definition, $\alpha(\epsilon) \rightarrow \dfrac{1}{2}\dfrac{\alpha_3}{\alpha_2}$ as $\epsilon \rightarrow 0$, while $q(\epsilon) \rightarrow \dfrac{\alpha_2}{p_2}\,q_1$.

Let $\epsilon \leq \epsilon_1$ be sufficiently small that $\alpha(\epsilon) > 0$ and $2\,\epsilon\,\alpha_2\,\alpha(\epsilon) \leq q(\epsilon)$. Then

$$\dot{v}_1(z, y) \;\leq\; -2\,\epsilon\,\alpha(\epsilon)\,v_1(z, y) \tag{A4.4.34}$$

so that the z, y system is exponentially stable with rate of convergence $\epsilon\,\alpha(\epsilon)$ (v_1 being bounded above and below by the *square* of the norm of the state). The same conclusion holds for the x, y system, given the transformation (4.4.9), with (4.4.10). Also, for ϵ, h_0 sufficiently small, all signals are actually guaranteed to remain in B_h so that all assumptions are valid. □

Auxiliary Lemmas for Section 6.2

Lemma A6.2.1

Consider the least squares identification algorithm described by (6.2.8), (6.2.9) with the sequence of resetting times $\{0, t_1, t_2, \ldots\}$, that is

$$\dot{\phi} \;=\; -P w w^T \phi \tag{A6.2.1}$$

$$\frac{d}{dt}(P^{-1}) \;=\; w w^T \qquad t \neq t_i \tag{A6.2.2}$$

$$P^{-1}(t_i^+) = k_0 I \qquad t_i = 0, t_1, t_2, \ldots \tag{A6.2.3}$$

If w satisfies

$$\int_{t_i}^{t_{i+1}} w\, w^T\, dt \geq \alpha_1 I \qquad \text{for all } t_i \tag{A6.2.4}$$

Then

$$|\phi(t_i)| \leq \left[\frac{k_0}{k_0 + \alpha_1}\right]^i |\phi(0)| \tag{A6.2.5}$$

Proof of Lemma A6.2.1

Note that for $t \neq 0, t_1, t_2, \ldots$

$$\frac{d}{dt}(P^{-1}\phi) = 0 \tag{A6.2.6}$$

Thus

$$P^{-1}(t_i^-)\,\phi(t_i) = P^{-1}(t_{i-1}^+)\,\phi(t_{i-1}) \tag{A6.2.7}$$

so that

$$\phi(t_i) = k_0 P(t_i^-)\,\phi(t_{i-1}) \tag{A6.2.8}$$

and, with (A6.2.4)

$$|\phi(t_i)| \leq \left[\frac{k_0}{k_0 + \alpha_1}\right] |\phi(t_{i-1})| \tag{A6.2.9}$$

Recursion on (A6.2.9) yields (A6.2.5).

Comments

If $\alpha_1 = 0$, the lemma shows that $\phi(t_i)$ is bounded. If $\alpha_1 > 0$ and the sequence t_i is infinite, $\phi(t_i) \to 0$ as $i \to \infty$. Further, if the intervals $t_{i+1} - t_i$ are bounded, then $\phi(t_i) \to 0$ exponentially. □

Lemma A6.2.2

Consider the following linear systems

$$\dot{z}_0 = A z_0 + b r \tag{A6.2.10}$$

$$\dot{z} = (A + \Delta A(t))z + (b + \Delta b(t))r \tag{A6.2.11}$$

with A stable and ΔA, Δb both bounded and converging to zero as

$t \to \infty$.

If the input r is bounded

Then given $\epsilon > 0$, there exists $k > 0$ (independent of ϵ) and a $T(\epsilon)$ such that

$$| z(t) - z_0(t) | \leq \epsilon k \qquad \text{for all} \quad t \geq T \tag{A6.2.12}$$

Proof of Lemma A6.2.2

From lemma 6.2.1, it follows that $A + \Delta A(t)$ is asymptotically stable and that there exists T_1 such that the state transition matrix of $A + \Delta A(t)$ satisfies

$$\| \phi(t, \tau) \| \leq m (\exp - \alpha(t - \tau)) \tag{A6.2.13}$$

for some m, $\alpha > 0$ and $t \geq \tau > T_1$. Using this estimate, it is easy to show that $z(t)$ is bounded. Now, defining the error $e(t) := z(t) - z_0(t)$, we have that

$$\dot{e} = Ae + \Delta A\, z + \Delta b\, r \tag{A6.2.14}$$

For T sufficiently large, ΔA and Δb are arbitrarily small, so that e may be showed to satisfy (A6.2.12). □

Lemma A6.2.3 Solution of the Pole Placement Equation

Consider two coprime polynomials: $\hat{d}_p$ monic of order n, and $\hat{n}_p$ monic of order $\leq n - 1$. Let k_p be a real number.

Then given an arbitrary polynomial $\hat{d}_{cl}(s)$ of order $2n - 1$, there exist unique polynomials $\hat{n}_c$ and $\hat{d}_c$ of order at most $n - 1$ so that

$$\hat{n}_c k_p \hat{n}_p + \hat{d}_c \hat{d}_p = \hat{d}_{cl} \tag{A6.2.15}$$

Proof of Lemma A6.2.3

Since $k_p \hat{n}_p$ and $\hat{d}_p$ are coprime and of order $n - 1$, n, respectively, there exist polynomials $\hat{u}$, $\hat{v}$ of degree at most n, $n - 1$, respectively so that

$$\hat{u}\, k_p \hat{n}_p + \hat{v}\, \hat{d}_p = 1 \tag{A6.2.16}$$

Thus, we see that

$$\hat{u}\, \hat{d}_{cl}\, k_p \hat{n}_p + \hat{v}\, \hat{d}_{cl}\, \hat{d}_p = \hat{d}_{cl} \tag{A6.2.17}$$

Further, we may modify (A6.2.17) to

$$(\hat{u}\,\hat{d}_{cl} - \hat{q}\,\hat{d}_p)k_p\,\hat{n}_p + (\hat{v}\,\hat{d}_{cl} + \hat{q}\,k_p\,\hat{n}_p)\,\hat{d}_p = \hat{d}_{cl} \qquad \text{(A6.2.18)}$$

for an arbitrary polynomial $\hat{q}$. Let

$$\begin{aligned} \hat{n}_c &:= \hat{u}\,\hat{d}_{cl} - \hat{q}\,\hat{d}_p \\ \hat{d}_c &:= \hat{v}\,\hat{d}_{cl} + \hat{q}\,k_p\,\hat{n}_p \end{aligned} \qquad \text{(A6.2.19)}$$

Since $\hat{d}_p$ is of order n, we may choose $\hat{q}$ so that $\hat{n}_c$ is of order $\leq n-1$ (for instance, as the quotient obtained by dividing $\hat{u}\,\hat{d}_{cl}$ by $\hat{d}_p$). Then, $\hat{d}_c$ is constrained to be of order $\leq n-1$, since the other two polynomials in (A6.2.18), that is $\hat{n}_c\,k_p\,\hat{n}_p$ and $\hat{d}_{cl}$, are of order $\leq 2n-1$.

It is useful to note that if

$$\begin{aligned} \hat{d}_{cl} &= d_{2n}\,s^{2n-1} + \cdots + d_1 \\ \hat{d}_p &= s^n + \beta_n\,s^{n-1} + \cdots + \beta_1 \\ k_p\,\hat{n}_p &= \alpha_n\,s^{n-1} + \cdots + \alpha_1 \\ \hat{n}_c &= a_n\,s^{n-1} + \cdots + a_1 \\ \hat{d}_c &= b_n\,s^{n-1} + \cdots + b_1 \end{aligned} \qquad \text{(A6.2.20)}$$

then the linear equation relating the coefficients of $\hat{n}_c$, $\hat{d}_c$ to those of $\hat{d}_{cl}$ is

$$\begin{bmatrix}
\alpha_1 & 0 & \cdots & 0 & 0 & \beta_1 & 0 & \cdots & 0 & 0 \\
\alpha_2 & \alpha_1 & \cdots & 0 & 0 & \beta_2 & \beta_1 & \cdots & 0 & 0 \\
\cdot & \cdot & & \cdot & \cdot & \cdot & \cdot & & \cdot & \cdot \\
\cdot & \cdot & & \cdot & \cdot & \cdot & \cdot & & \cdot & \cdot \\
\alpha_{n-1} & \alpha_{n-2} & \cdots & \alpha_1 & 0 & \beta_{n-1} & \beta_{n-2} & \cdots & \beta_1 & 0 \\
\alpha_n & \alpha_{n-1} & \cdots & \alpha_2 & \alpha_1 & \beta_n & \beta_{n-1} & \cdots & \beta_2 & \beta_1 \\
0 & \alpha_n & \cdots & \alpha_3 & \alpha_2 & 1 & \beta_n & \cdots & \beta_3 & \beta_2 \\
0 & 0 & \cdots & \alpha_4 & \alpha_3 & 0 & 1 & \cdots & \beta_4 & \beta_3 \\
\cdot & \cdot & & \cdot & \cdot & \cdot & \cdot & & \cdot & \cdot \\
\cdot & \cdot & & \cdot & \cdot & \cdot & \cdot & & \cdot & \cdot \\
0 & 0 & \cdots & 0 & \alpha_n & 0 & 0 & \cdots & 1 & \beta_n \\
0 & 0 & \cdots & 0 & 0 & 0 & 0 & \cdots & 0 & 1
\end{bmatrix}$$

$$\cdot \begin{bmatrix} a_1 \\ a_2 \\ \cdot \\ \cdot \\ a_{n-1} \\ a_n \\ b_1 \\ b_2 \\ \cdot \\ \cdot \\ b_{n-1} \\ b_n \end{bmatrix} = \begin{bmatrix} d_1 \\ d_2 \\ \cdot \\ \cdot \\ d_{n-1} \\ d_n \\ d_{n+1} \\ d_{n+2} \\ \cdot \\ \cdot \\ d_{2n-1} \\ d_{2n} \end{bmatrix} \tag{A6.2.21}$$

(A6.2.21) is of the form $A(\theta^*)\theta_c^* = d$ where θ^* is the nominal plant parameter, and θ_c^* the nominal controller parameter. □

REFERENCES

Anderson, B.D.O., "Exponential Stability of Linear Equations Arising in Adaptive Identification," *IEEE Trans. on Automatic Control,* Vol. AC-22, no. 1, pp. 83-88, 1977.

Anderson, B.D.O., "Adaptive Systems, Lack of Persistency of Excitation and Bursting Phenomena," *Automatica,* Vol. 21, no. 3, pp. 247-258, 1985.

Anderson, B.D.O., R.R. Bitmead, C.R. Johnson, P.V. Kokotovic, R.L. Kosut, I.M.Y. Mareels, L. Praly, & B.D. Riedle, *Stability of Adaptive Systems, Passivity and Averaging Analysis,* MIT Press, Cambridge, Massachusetts, 1986.

Anderson, B.D.O., S. Dasgupta, & A.C. Tsoi, "On the Convergence of a Model Reference Adaptive Control Algorithm with Unknown High Frequency Gain," *Systems & Control Letters,* Vol. 5, pp. 303-307, 1985.

Anderson, B.D.O., & R.M. Johnstone, "Adaptive Systems and Time Varying Plants," *Int. J. Control,* Vol. 37, no. 2, pp. 367-377, 1983.

Anderson, B.D.O., & R.M. Johnstone, "Global Adaptive Pole Positioning," *IEEE Trans. on Automatic Control,* Vol. AC-30, no. 1, pp. 11-22, 1985.

Anderson, B.D.O., & J.B. Moore, "New Results in Linear System Stability," *SIAM J. Control,* Vol. 7, no. 3, pp. 398-414, 1969.

Anderson, B.D.O., & S. Vongpanitlerd, *Network Analysis and Synthesis,* Prentice-Hall, Englewood Cliffs, New Jersey, 1973.

Arnold, V.I., *Geometric Methods in the Theory of Ordinary Differential Equations,* Springer-Verlag, New York, 1982.

Astrom, K.J., "Analysis of Rohrs Counterexamples to Adaptive Control," *Proc. of the 22nd IEEE Conf. on Decision and Control,* pp. 982–987, San Antonio, Texas, 1983.

Astrom, K.J., "Interactions Between Excitation and Unmodeled Dynamics in Adaptive Control," *Proc. of the 23rd IEEE Conf. on Decision and Control,* pp. 1276–1281, Las Vegas, Nevada, 1984.

Astrom, K.J., "Adaptive Feedback Control," *Proc. of the IEEE,* Vol. 75, no. 2, pp. 185–217, 1987.

Astrom, K.J., J.J. Anton, & K.E. Arzen, "Expert Control," *Automatica,* Vol. 22, no. 3, pp. 277–286, 1986.

Astrom, K.J., P. Hagander, & J. Sternby, "Zeros of Sampled Systems," *Automatica,* Vol. 20, pp. 31–38, 1984.

Astrom, K.J., & B. Wittenmark, "On Self Tuning Regulators," *Automatica,* Vol. 9, pp. 185–199, 1973.

Bai, E.W., & S. Sastry, "Parameter Identification Using Prior Information," *Int. J. Control,* Vol. 44, no. 2, pp. 455–473, 1986.

Bai, E.W., & S. Sastry, "Global Stability Proofs for Continuous-Time Indirect Adaptive Control Schemes," *IEEE Trans. on Automatic Control,* Vol. AC–32, no. 6, pp. 537–543, 1987.

Balachandra, M., & P.R. Sethna, "A Generalization of the Method of Averaging for Systems with Two-Time Scales," *Archive for Rational Mechanics and Analysis,* Vol. 58, pp. 261–283, 1975.

Bar-Shalom, Y., & E. Tse, "Dual Effect, Certainty Equivalence, and Separation in Stochastic Control," *IEEE Trans. on Automatic Control,* Vol. AC–19, no. 5, pp 494–500, 1974.

Bellman, R., "The Stability of Solutions of Linear Differential Equations," *Duke Math. J.,* Vol. 10, pp. 643–647, 1943.

Bodson, M., "Stability, Convergence, and Robustness of Adaptive Systems," Ph.D. Dissertation, Memorandum No. UCB/ERL M86/66, Electronics Research Laboratory, University of California, Berkeley, 1986.

Bodson, M., "Effect of the Choice of Error Equation on the Robustness Properties of Adaptive Control Systems," *Int. J. of Adaptive Control and Signal Processing,* to appear, 1988.

Bodson, M., & S. Sastry, "Small Signal I/O Stability of Nonlinear Control Systems: Application to the Robustness of a MRAC Scheme," *Proc. of the 23rd IEEE Conf. on Decision and Control,* pp. 1282–1285, Las Vegas, Nevada, 1984.

Bodson, M., & S. Sastry, "Input Error versus Output Error Model Reference Adaptive Control," *Proc. of the Automatic Control Conference,* pp. 224-229, Minneapolis, Minnesota, 1987.

Bodson, M., S. Sastry, B.D.O. Anderson, I. Mareels, & R.R. Bitmead, "Nonlinear Averaging Theorems, and the Determination of Parameter Convergence Rates in Adaptive Control," *Systems & Control Letters,* Vol. 7, no. 3, pp. 145-157, 1986.

Bogoliuboff, N.N., & Y.A. Mitropolskii, *Asymptotic Methods in the Theory of Nonlinear Oscillators,* Gordon & Breach, New York, 1961.

Borison, V., "Self-Tuning Regulators for a Class of Multivariable Systems," *Automatica,* Vol. 15, pp. 209-215, 1979.

Boyd, S., & S. Sastry, "On Parameter Convergence in Adaptive Control," *Systems & Control Letters,* Vol. 3, pp. 311-319, 1983.

Boyd, S., & S. Sastry, "Necessary and Sufficient Conditions for Parameter Convergence in Adaptive Control," *Automatica,* Vol. 22, no. 6, pp. 629-639, 1986.

Byrnes, C., & A. Isidori, Applications to Stabilization and Adaptive Control," *Proc. of the 23rd IEEE Conf. on Decision and Control,* pp. 1569-1573, Las Vegas, Nevada, 1984.

Caines, P.E., *Linear Stochastic Systems,* John Wiley, New York, 1988.

Callier, F.M., & C.A. Desoer, *Multivariable Feedback Systems,* Springer-Verlag, New York, 1982.

Chen, J.B., B.S. Armstrong, R.S. Fearing, & J.W. Burdick, "Satyr and the Nymph: Software Archetype for Real Time Robotics," *Proc. of the IEEE-ACM Joint Conference,* Dallas, Texas, 1988.

Chen, Z.J., & P.A. Cook, "Robustness of Model-Reference Adaptive Control Systems with Unmodelled Dynamics," *Int. J. Control,* Vol. 39, no. 1, pp. 201-214, 1984.

Clarke, D., C. Mohtadi, & P. Tuffs, "Generalized Predictive Control—Part I. The Basic Algorithm," *Automatica,* Vol. 23, no. 2, pp. 137-148, 1987.

Clary J.P., "Robust Algorithms in Adaptive Control," Ph.D. Dissertation, Information Syst. Lab., Stanford University, Stanford, California, 1984.

Claude, D., "Everything You Always Wanted to Know About Linearization," *Algebraic and Geometric Methods in Nonlinear Control Theory,* Riedel, Dordrecht, 1986.

Coddington, E.A., & N. Levinson, *Theory of Ordinary Differential Equations,* McGraw-Hill, New York, 1955.

Craig, J., P. Hsu, & S. Sastry, "Adaptive Control of Mechanical Manipulators," *Int. J. of Robotics Research*, Vol. 6, no. 2, pp. 16–28, 1987.

Dasgupta, S., "Adaptive Identification and Control," Ph.D. Dissertation, Australian National University, Canberra, Australia, 1984.

De Larminat, P.H., "On the Stabilizability Condition in Indirect Adaptive Control," *Automatica,* Vol. 20, no. 6, pp. 793–795, 1984.

Denker, J. (Ed.), " Neural Networks for Computing," *Proc. of the 1986 Conference of the American Institute of Physics,* Snowbird, Utah, 1986.

Descusse, J., and C. H. Moog, "Decoupling with Dynamic Compensation for Strongly Invertible Affine Nonlinear Systems," *Int. J. of Control,* no. 42, pp. 1387–1398, 1985.

Desoer, C.A., & M. Vidyasagar, *Feedback Systems: Input-Ouput Properties,* Academic Press, New York, 1975.

Donalson, D.D., & C.T. Leondes, "A Model Referenced Parameter Tracking Technique for Adaptive Control Systems—The Principle of Adaptation," *IEEE Trans. on Applications and Industry,* Vol. 82, pp. 241–251, 1963a.

Donalson, D.D., & C.T. Leondes, "A Model Referenced Parameter Tracking Technique for Adaptive Control Systems—Stability Analysis by the Second Method of Lyapunov," *IEEE Trans. on Applications and Industry,* Vol. 82, pp. 251–262, 1963b.

Doyle, J.C., "Advances in Multivariable Control," *Notes for the Honeywell Workshop in Multivariable Control,* Minneapolis, 1984.

Doyle, J.C., "A Review of μ—For Case Studies in Robust Control," *Proc of the 10th IFAC Congress,* Vol. 8, pp. 395–402, Munich, 1987.

Doyle, J.C., & G. Stein, "Multivariable Feedback Design: Concepts for a Classical/Modern Synthesis," *IEEE Trans. on Automatic Control,* Vol. AC–26, no. 1, pp. 4–16, 1981.

Dugard, L., & J.M. Dion, "Direct Adaptive Control for Linear Systems," *Int. J. Control,* Vol. 42, no. 6, pp. 1251–1281, 1985.

Dymock, A.J., J.F. Meredith, A. Hall, & K.M. White, "Analysis of a Type of Model Reference-Adaptive Control System," *Proc. IEE,* Vol. 112, no. 4, pp. 743–753, 1965.

Egardt, B., *Stability of Adaptive Controllers,* Springer Verlag, New York, 1979.

Elliott, H., R. Cristi, & M. Das, "Global Stability of Adaptive Pole Placement Algorithms," *IEEE Trans. on Automatic Control,* Vol. AC–30, no. 4, pp. 348–356, 1985.

Elliott, H., & W. Wolovich, "A Parameter Adaptive Control Structure for Linear Multivariable Systems," *IEEE Trans. on Automatic*

Control, Vol. AC-27, no. 2, pp. 340-352, 1982.

Elliott, H., & W. Wolovich, "Parameterization Issues in Multivariable Adaptive Control," *Automatica,* Vol. 20, no. 5, pp. 533-545, 1984.

Elliott, H., W. Wolovich, & M. Das, "Arbitrary Adaptive Pole Placement for Linear Multivariable Systems," *IEEE Trans. on Automatic Control,* Vol. AC-29, no. 3, pp. 221-229, 1984.

Eykhoff, P., *System Identification,* Wiley & Sons, New York, 1974.

Freund, E., "The Structure of Decoupled Nonlinear Systems," *Int. J. of Control,* Vol. 21, pp. 651-659, 1975.

Fu, L.-C., M. Bodson, & S. Sastry, "New Stability Theorems for Averaging and Their Application to the Convergence Analysis of Adaptive Identification and Control Schemes," *Singular Perturbations and Asymptotic Analysis in Control Systems,* Lecture Notes in Control and Information Sciences, P. Kokotovic, A. Bensoussan, & G. Blankenship (Eds.), Springer-Verlag, New York, 1986.

Fu, L.-C., & S. Sastry, "Slow Drift Instability in Model Reference Adaptive Systems—An Averaging Analysis," *Int. J. Control,* Vol. 45, no. 2, pp. 503-527, 1987.

Goodwin, G.C., & R.S. Long, "Generalization of Results on Multivariable Adaptive Control," *IEEE Trans. on Automatic Control,* Vol. AC-25, no. 6, pp. 1241-1245, 1980.

Goodwin, G.C., & D.Q. Mayne, "A Parameter Estimation Perspective of Continuous Time Model Reference Adaptive Control," *Automatica,* Vol. 23, no. 1, pp. 57-70, 1987.

Goodwin, G.C., & R.L. Payne, *Dynamic System Identification,* Academic Press, New York, 1977.

Goodwin, G.C., P.J. Ramadge, & P.E. Caines, "Discrete-Time Multivariable Adaptive Control," *IEEE Trans. on Automatic Control,* Vol. AC-25, no. 3, pp. 449-456, 1980.

Goodwin, G.C., & K.S. Sin, *Adaptive Filtering Prediction and Control,* Prentice-Hall, Englewood Cliffs, New Jersey, 1984.

Guckenheimer, J., & P. Holmes, *Nonlinear Oscillations, Dynamical Systems, and Bifurcations of Vector Fields,* Springer-Verlag, New York, 1983.

Hahn, W., *Stability of Motion,* Springer-Verlag, Berlin, 1967.

Hale, J.K., *Ordinary Differential Equations,* Krieger, Huntington, New York, 1980.

Harris, C.J., & S.A. Billings, *Self Tuning and Adaptive Control : Theory and Applications,* Peter Peregrinus, London, 1981.

Hill, D., & P. Moylan, "Connections Between Finite-Gain and Asymptotic Stability," *IEEE Trans. on Automatic Control,* Vol. AC-25, no. 5, pp. 931–936, 1980.

Horrocks, T., "Investigations into Model-Reference Adaptive Control Systems," *Proc. IEE,* Vol. 111, no. 11, pp. 1894–1906, 1964.

Hsu, P., M. Bodson, S. Sastry, & B. Paden, "Adaptive Identification and Control of Manipulators without Using Joint Accelerations," *Proc. of the IEEE Conf. on Robotics and Automation,* Raleigh, North Carolina, pp. 1210–1215, 1987.

Hunt, L.R., R. Su, & G. Meyer, "Global Transformations of Nonlinear Systems," *IEEE Trans. on Automatic Control,* Vol. AC–28, no. 1, pp. 24–31, 1983.

Ioannou, P.A., "Robust Adaptive Controller with Zero Residual Tracking Errors," *IEEE Trans. on Automatic Control,* Vol. AC–31, no. 8, pp. 773–776, 1986.

Ioannou, P.A., & P.V. Kokotovic, *Adaptive Systems with Reduced Models,* Springer Verlag, New York, 1983.

Ioannou, P.A., & P.V. Kokotovic, "Robust Redesign of Adaptive Control," *IEEE Trans. on Automatic Control,* Vol. AC–29, no. 3, pp. 202–211, 1984.

Ioannou, P.A., & G. Tao, "Frequency Domain Conditions for Strictly Positive Real Functions," *IEEE Trans. on Automatic Control,* Vol. AC–32, no. 1, pp. 53–54, 1987.

Ioannou, P.A., & K. Tsakalis, "A Robust Direct Adaptive Controller," *IEEE Trans. on Automatic Control,* Vol. AC–31, No. 11, pp. 1033–1043, 1986.

Isidori, A., "Control of Nonlinear Systems via Dynamic State Feedback," *Algebraic and Geometric Methods in Nonlinear Control Theory,* M. Fliess & M. Hazewinkel (Eds.), Riedel, Dordrecht, 1986.

Isidori, A., & C.I. Byrnes, "The Analysis and Design of Nonlinear Feedback Systems: Zero Dynamics and Global Normal Forms, " Preprint, Universita di Roma, "La Sapienza", Rome, 1988.

Isidori, A., A.J. Krener, C. Gori-Giorgi, & S. Monaco, "Nonlinear Decoupling via Feedback: a Differential Geometric Approach," *IEEE Trans. on Automatic Control,* Vol. AC–26, pp. 331–345, 1981.

Isidori, A., & C.H. Moog, "On the Nonlinear Equivalent of the Notion of Transmission Zeros," *Modeling and Adaptive Control,* C. Byrnes and A. Kurszanski (Eds.), Lecture Notes in Information and Control, Springer-Verlag, 1987.

James, D.J., "Stability of a Model Reference Control System," *AIAA Journal,* Vol. 9, no. 5, 1971.

Johansson, R., "Parametric Models of Linear Multivariable Systems for Adaptive Control," *IEEE Trans. on Automatic Control,* Vol. AC-32, no. 4, pp. 303-313, 1987.

Johnson, C.R., *Lectures on Adaptive Parameter Estimation,* Prentice-Hall, Englewood Cliffs, New Jersey, 1988.

Johnson, C.R., B.D.O. Anderson, & R.R. Bitmead, "A Robust Locally Optimal Model Reference Adaptive Controller," *Proc. of the 23rd IEEE Conf. on Decision and Control,* Las Vegas, Nevada, pp. 993-998, 1984.

Kailath, T., *Linear Systems,* Prentice-Hall, Englewood Cliffs, New Jersey, 1980.

Kalman, R.E., "Design of Self-Optimizing Control Systems," *Trans. ASME,* Vol. 80, pp. 468-478, 1958.

Kalman, R.E., & J. Bertram, "Control System Analysis and Design Via the 'Second Method' of Lyapunov," *J. of Basic Engineering, Trans. ASME,* Vol. 82, pp. 371-400, 1960.

Kalman, R.E., & R.S. Bucy, "New Results in Linear Filtering and Prediction Theory," *J. of Basic Engineering, Trans. ASME, Ser. D,* Vol. 83, pp. 95-108, 1961.

Koivo, H., "Multivariable Self-Tuning Controller," *Automatica,* Vol. 16, pp. 351-366, 1980.

Kokotovic, P.V., "Recent Trends in Feedback Design: An Overview," *Automatica.* Vol. 21, no.3, pp. 225-236, 1985.

Kokotovic, P., B. Riedle, & L. Praly, "On a Stability Criterion for Continuous Slow Adaptation," *Systems & Control Letters,* Vol. 6, no. 1, pp. 7-14, 1985.

Kosut, R.L., & C.R. Johnson, "An Input-Output View of Robustness in Adaptive Control," *Automatica,* Vol. 20, no. 5, pp. 569-581, 1984.

Krasovskii, N., *Stability of Motion,* Stanford University Press, Stanford, 1963.

Kreisselmeier, G., "Adaptive Observers with Exponential Rate of Convergence," *IEEE Trans. on Automatic Control,* Vol. AC-22, no. 1, pp. 2-8, 1977.

Kreisselmeier, G., "An Approach to Stable Indirect Adaptive Control," *Automatica,* Vol. 21, pp. 425-431, 1985.

Kreisselmeier, G., "A Robust Indirect Adaptive-Control Approach," *Int. J. Control,* Vol. 43, no. 1, pp. 161-175, 1986.

Kreisselmeier, G., & B.D.O. Anderson, "Robust Model Reference Adaptive Control," *IEEE Trans. on Automatic Control,* Vol. AC-31, no. 2, pp. 127-133, 1986.

Kreisselmeier, G., & K.S. Narendra, "Stable Model Reference Adaptive Control in the Presence of Bounded Disturbances," *IEEE Trans. on Automatic Control,* Vol. AC-27, no. 6, pp. 1169-1175, 1982.

Kumar, P.R., & P.P. Varaiya, *Stochastic Systems: Estimation, Identification and Adaptive Control,* Prentice Hall, Englewood Cliffs, New Jersey, 1986.

Landau, Y.D., "Unbiased Recursive Identification Using Model Reference Adaptive Techniques," *IEEE Trans. on Automatic Control,* Vol. AC-21, no. 2, pp. 194-202, 1976.

Landau, Y.D., *Adaptive Control–The Model Reference Approach,* Marcel Dekker, New York, 1979.

Lion, P.M., "Rapid Identification of Linear and Nonlinear Systems," *AIAA Journal,* Vol. 5, no. 10, pp. 1835-1842, 1967.

Ljung, L., *System Identification: Theory for the User,* Prentice-Hall, Englewood Cliffs, New Jersey, 1987.

Ljung, L., & T. Soderstrom, *Theory and Practice of Recursive Identification,* MIT Press, Cambridge, Massachusetts, 1983.

Lozano-Leal, R., & G.C. Goodwin, "A Globally Convergent Adaptive Pole Placement Algorithm without a Persistency of Excitation Requirement," *IEEE Trans. on Automatic Control,* Vol. AC-30, no. 8, pp. 795-798, 1985.

Mareels, I.M.Y., B.D.O. Anderson, R.R. Bitmead, M. Bodson, & S.S. Sastry, "Revisiting the MIT Rule for Adaptive Control," *Proc. of the 2nd IFAC Workshop on Adaptive Systems in Control and Signal Processing,* Lund, Sweden, 1986.

Martensson, B., " The Order of Any Stabilizing Regulator Is Sufficient A Priori Information for Adaptive Stabilization," *Systems and Control Letters,* Vol. 6, pp. 87-91, 1985.

Mason, J.E., E.W. Bai, L.-C. Fu, M. Bodson, & S. Sastry, "Analysis of Adaptive Identifiers in the Presence of Unmodeled Dynamics: Averaging and Tuned Parameters," *IEEE Trans. on Automatic Control,* Vol. AC-33, 1988.

Meyer, G., & L. Cicolani, "Applications of Nonlinear System Inverses to Automatic Flight Control Design - System Concepts and Flight Evaluations," AGARDograph 251 on Theory and Applications of Optimal Control in Aerospace Systems, P. Kent (Ed.), NATO, 1980.

Monaco, S., & D. Normand-Cyrot, "Nonlinear Systems in Discrete Time," *Algebraic and Geometric Methods in Nonlinear Control Theory,* M. Fliess & M. Hazewinkel (Eds.), Riedel, Dordrecht, 1986.

Monaco, S., D. Normand-Cyrot, & S. Stornelli, "On the Linearizing Feedback in Nonlinear Sampled Data Control Schemes," *Proc. of the 25th IEEE Conf. on Decision and Control,* Athens, Greece, pp. 2056–2060, 1986.

Monopoli, R.V., "Model Reference Adaptive Control with an Augmented Error Signal," *IEEE Trans. on Automatic Control,* Vol. AC–19, no. 5, pp. 474–484, 1974.

Morgan, A.P., & K.S. Narendra, "On the Uniform Asymptotic Stability of Certain Linear Nonautonomous Differential Equations," *SIAM J. Control and Optimization,* Vol. 15, no. 1, pp. 5–24, 1977a.

Morgan, A.P., & K.S. Narendra, "On the Stability of Nonautonomous Differential Equations $\dot{x} = [A + B(t)]x$, with Skew Symmetric Matrix $B(t)$," *SIAM J. Control and Optimization,* Vol. 15, no. 1, pp. 163–176, 1977b.

Morse, A. S., "Global Invariants Under Feedback and Cascade Control," *Proc. of the International Symposium on Mathematical System,* Lecture Notes in Economics and Mathematical Systems, Vol. 131, Springer Verlag, New York, 1976.

Morse, A.S., "Global Stability of Parameter-Adaptive Control Systems," *IEEE Trans. on Automatic Control,* Vol. AC–25, no. 3, pp. 433–439, 1980.

Morse, A.S., " A Three-Dimensional Universal Controller for the Adaptive Stabilization of Any Strictly Proper Minimum-phase System with Relative Degree Not Exceeding Two," *IEEE Trans. on Automatic Control,"* Vol. 30, pp. 1188–1191, 1985.

Morse, A.S., " A 4(n+1) Dimensional Model Reference Adaptive Stabilizer for Any Relative Degree One or Two, Minimum Phase System of Dimension nor Less," *Automatica,* Vol. 23, pp. 123–125, 1987.

Morse, A.S., " Towards a Unified Theory of Parameter Adaptive Control," Report no. 8807, Dept. of Electrical Engineering, Yale University, 1988.

Mudgett, D.R., & A.S. Morse, "Adaptive Stabilization of Linear Systems with Unknown High Frequency Gains," *IEEE Trans. on Automatic Control,* Vol. 30, no. 6, pp. 549–554, 1985.

Nam, K., & A. Arapostathis, "A Model Reference Adaptive Control Scheme for Pure Feedback Nonlinear Systems," Preprint, University of Texas, Austin, 1986.

Narasimhan, S., D.M. Siegel, & J.M. Hollerbach, "CONDOR: A Revised Architecture for Controlling the Utah-MIT Hand," *Proc. of the IEEE Conference on Robotics and Automation,* Philadelphia, Pennsylvania, pp. 446–449, 1988.

Narendra, K.S., "Correction to 'Stable Adaptive Controller Design—Part II: Proof of Stability'," *IEEE Trans. on Automatic Control,* Vol. AC–29, no. 7, pp. 640–641, 1984.

Narendra, K.S., & A.M. Annaswamy, "Robust Adaptive Control in the Presence of Bounded Disturbances," *IEEE Trans. on Automatic Control,* Vol. AC–31, no. 4, pp. 306–315, 1986.

Narendra, K.S., & A.M. Annaswamy, "A New Adaptive Law for Robust Adaptation without Persistent Excitation, " *IEEE Trans. on Automatic Control,* Vol. AC–32, no. 2, pp. 134–145, 1987.

Narendra, K.S., A.M. Annaswamy, & R.P. Singh, "A General Approach to the Stability Analysis of Adaptive Systems," *Int. J. Control,* Vol. 41, no. 1, pp. 193–216, 1985.

Narendra, K.S., I.H. Khalifa, & A.M. Annaswamy, "Error Models for Stable Hybrid Adaptive Systems," *IEEE Trans. on Automatic Control,* Vol. AC–30, No. 5, pp. 339–347, 1985.

Narendra, K.S., Y.-H. Lin, & L.S. Valavani, "Stable Adaptive Controller Design, Part II: Proof of Stability," *IEEE Trans. on Automatic Control,* Vol. AC–25, no. 3, pp. 440–448, 1980.

Narendra, K.S., & R.V. Monopoli (Eds.), *Applications of Adaptive Control,* Academic Press, New York, 1980.

Narendra, K.S., & L.S. Valavani, "Stable Adaptive Controller Design—Direct Control," *IEEE Trans. on Automatic Control,* Vol. AC–23, no. 4, pp. 570–583, 1978.

Nussbaum, R.D., "Some Remarks on a Conjecture in Parameter Adaptive Control," *Systems and Control Letters,* Vol. 3, pp. 243–246, 1983.

Ortega, R., L. Praly, & I.D. Landau, "Robustness of Discrete-Time Direct Adaptive Controllers," *IEEE Trans. on Automatic Control,* Vol. AC–30, no. 12, pp. 1179–1187, 1985.

Ortega, R., & T. Yu, "Theoretical Results on Robustness of Direct Adaptive Controllers : A Survey," *Proceedings of the 5th Yale Workshop on Applications of Adaptive Systems Theory,* pp. 1–15, 1987.

Osburn, P.V., H.P. Whitaker, & A. Kezer,, "New Developments in the Design of Model Reference Adaptive Control Systems," Paper no. 61-39, Institute of the Aerospace Sciences, 1961.

Ossman, K.A., & E.W. Kamen, "Adaptive Regulation of MIMO Linear Discrete-time Systems without Requiring a Persistent Excitation," *IEEE Trans. on Automatic Control,* Vol. AC–32, no. 5, pp. 397–404, 1987.

Parks, P.C., "Liapunov Redesign of Model Reference Adaptive Control Systems," *IEEE Trans. on Automatic Control,* Vol. AC-11, no. 3, pp. 362-367, 1966.

Peterson, B.B., & K.S. Narendra, "Bounded Error Adaptive Control," *IEEE Trans. on Automatic Control,* Vol. AC-27, no. 6, pp. 1161-1168, 1982.

Polderman, J.W., " Adaptive Control & Identification: Conflict or Conflux ?," Ph.D. dissertation, University of Groningen, Groningen, The Netherlands, 1988.

Popov, V.M., *Hyperstability of Control Systems,* Springer-Verlag, New York, 1973.

Prager, D., & P. Wellstead, "Multivariable Pole-Assignment and Self-Tuning Regulators," *Proc. IEE,* Part D, Vol. 128, pp. 9-18, 1981.

Praly, L., "Robustness of Model Reference Adaptive Control," *Proc. of the 3rd Yale Workshop on Applications of Adaptive Systems Theory,* pp. 224-226, 1983.

Riedle, B.D., B. Cyr, & P.V. Kokotovic, "Disturbance Instabilities in an Adaptive System," *IEEE Trans. on Automatic Control,* Vol. AC-29, no. 9, pp. 822-824, 1984.

Riedle, B.D., & P.V. Kokotovic, "A Stability-Instability Boundary for Disturbance-Free Slow Adaptation with Unmodeled Dynamics," *IEEE Trans. on Automatic Control,* Vol. AC-30, no. 10, pp. 1027-1030, 1985.

Riedle, B.D., & P.V. Kokotovic, "Integral Manifolds of Slow Adaptation," *IEEE Trans. on Automatic Control,* Vol. AC-31, no. 4, pp. 316-324, 1986.

Rohrs, C.E., "How the Paper 'Robustness of Model-Reference Adaptive Control Systems with Unmodelled Dynamics' Misrepresents the Results of Rohrs and his Coworkers," *Int. J. Control,* Vol. 41, no. 2, pp. 575-580, 1985.

Rohrs, C.E., L. Valavani, M. Athans, & G. Stein, "Robustness of Adaptive Control Algorithms in the Presence of Unmodeled Dynamics," *Proc. of the 21st IEEE Conference on Decision and Control,* Florida, pp. 3-11, 1982.

Rohrs, C.E., L. Valavani, M. Athans, & G. Stein, "Robustness of Continuous-Time Adaptive Control Algorithms in the Presence of Unmodeled Dynamics," *IEEE Trans. on Automatic Control,* Vol. AC-30, no. 9, pp. 881-889, 1985.

Samson, C., "Stability Analysis of an Adaptively Controlled System subject to Bounded Disturbances," *Automatica,* Vol. 19, pp. 81-86, 1983.

Sastry, S., "Model-Reference Adaptive Control—Stability, Parameter Convergence, and Robustness," *IMA Journal of Mathematical Control & Information,* Vol. 1, pp. 27-66, 1984.

Sastry, S., & A. Isidori, "Adaptive Control of Linearizable Systems," Electronics Research Laboratory Memo No. M87/53, University of California, Berkeley, California, 1987.

Sethna, P.R., "Method of Averaging for Systems Bounded for Positive Time," *Journal of Math Anal and Applications,* Vol. 41, pp. 621-631, 1973.

Singh, R.P., & K.S. Narendra, "A Globally Stable Adaptive Controller for Multivariable Systems," Tech. Rept. 8111, Center for Syst. Sci., Yale University, New Haven, Connecticut, 1982.

Singh, R.P., & K.S. Narendra, "Prior Information in the Design of Multivariable Adaptive Controllers," *IEEE Trans. on Automatic Control,* Vol. AC-29, no. 12, pp. 1108-1111, 1984.

Singh, S. N., & W.J. Rugh, "Decoupling in a Class of Nonlinear Systems by State Variable Feedback," *J. Dyn. Syst. Measur. and Control, Trans. ASME,* Vol. 94, pp. 323-329, 1972.

Sondhi, M.M., & D. Mitra, "New Results on the Performance of a Well-Known Class of Adaptive Filters," *Proc. of the IEEE,* Vol. 64, no. 11, pp. 1583-1597, 1976.

Staff of the Flight Research Center, "Experience with the X-15 Adaptive Flight Control System," Nasa Technical Note NASA TN D-6208, Washington D.C., 1971.

Tomizuka, M., "Parallel MRAS Without Compensation Block," *IEEE Trans. on Automatic Control,* Vol. AC-27, no. 2, pp. 505-506, 1982.

Tsypkin, Ya.Z., *Adaptation and Learning in Automatic Systems,* Academic Press, New York, 1971.

Tsypkin, Ya.Z., *Foundations of the Theory of Learning Systems,* Academic Press, New York, 1973.

Unbehauen, H. (Ed.), *Methods and Applications in Adaptive Control,* Springer Verlag, Berlin, 1980.

Vidyasagar, M., *Nonlinear Systems Analysis,* Prentice-Hall, Englewood Cliffs, New Jersey, 1978.

Vidyasagar, M., *Control System Synthesis: A Factorization Approach,* MIT Press, Cambridge, Massachusetts, 1985.

Vidyasagar, M., & A. Vannelli, "New Relationships Between Input-Ouput and Lyapunov Stability," *IEEE Trans. on Automatic Control,* Vol. AC-27, no. 2, pp. 481-483, 1982.

Volosov, V.M., "Averaging in Systems of Ordinary Differential Equations," *Russian Mathematical Surveys,* Vol. 17, no. 6, pp. 1–126, 1962.

Whitaker, H.P., "An Adaptive System for Control of the Dynamic Performance of Aircraft and Spacecraft," Paper no. 59-100, Institute of the Aeronautical Sciences, 1959.

White, A.J., "Analysis and Design of Model-Reference Adaptive Control Systems," *Proc. IEE,* Vol. 113, no. 1, pp. 175–184, 1966.

Widrow, B., & S. Stearns, *Adaptive Signal Processing,* Prentice-Hall, Englewood Cliffs, New Jersey, 1985.

Widder, D.V., *An Introduction to Transform Theory,* Academic Press, New York, 1971.

Wiener, N., "Generalized Harmonic Analysis," *Acta Mathematica,* Vol. 55, pp. 117–258, 1930.

Wittenmark, B., & K.J. Astrom, "Practical Issues in the Implementation of Self-Tuning Control," *Automatica,* Vol. 20, pp. 595–605, 1984.

Wolovich, W.A., "A Division Algorithm for Polynomial Matrices," *IEEE Trans. on Automatic Control,* Vol. 29, no. 7, pp. 656–658, 1984.

Wolovich, W.A., & P.L. Falb, "Invariants and Canonical Forms Under Dynamic Compensation," *SIAM J. Contr. Opt.,* Vol. 14, no. 6, pp. 996–1008, 1976.

INDEX

Adaptive control, 1
model reference, 5, 99, 103, 286
direct, 9, 111
input error, 111
output error, 118
indirect, 9, 14, 103, 123, 267, 274
multivariable, 277, 305
of nonlinear systems, 294, 307
of robot manipulators, 320
parallel, 6
parametric, 1
series, high gain, 5
universal theory of, 327
using M.I.T. rule. 8, 12
Adaptive pole placement, 129, 272
Algorithms
identifier with normalized gradient, 59
identifier with normalized least squares and covariance resetting, 63
indirect adaptive control, 125
input error direct adaptive control, 115, 290
model reference identifier, 80
output error direct adaptive control, 119
output error direct adaptive control, relative degree 1, 122
Autocovariance, 40
Autonomous, 20
Averaging, 158
Averaging, instability analysis, 236
mixed time scales, 183, 186
nonlinear, 164, 192
one time scale, 159, 166
two time scales, 162, 179

BIBS stability, 130
Bursting phenomenon, 233

Certainty equivalence principle, 8, 124, 268
Class K, 25
Column reduced, 280
Computed torque, 321
Connectionism, 329, 330
Controllable canonical form, 55
Convergence function, 167
Convergence, of gradient algorithms with SPR error equation, 85
of the identifier, 71, 75, 270
of the input error direct adaptive control, 154
of the indirect adaptive control, 155
of the output error direct adaptive control, 155
Coprime, left, 279

Coprime, right, 279
Corollaries
  Corollary 1.2.2, 19
  Corollary 3.6.3 Properties of Regular Signals, 140
Covariance, 62
  propagation equation, 62
  resetting, 63
  windup, 62
Cross correlation, 43, 189
Cross spectral measure, 44

Deadzone, 251
  relative, 251, 252
Diophantine equation, 129, 272
Divisor, common right, 278
  greatest common right, 278
Dual control, 10, 249, 325

Equation error identifier, 53
Equilibrium point, 21
  asymptotically stable, 24
  exponentially stable, 24, 305
  stable, 23
  uniformly asymptotically stable, 24
  uniformly stable, 23
  unstable, 237
Error,
  augmented, 119, 315
  control, 133
  equation, linear, 48, 58
  equation, modified strictly positive real, 84
  equation, strictly positive real, 82
  identification, 51, 57
  input, 102, 111, 114
  output, 7, 76, 100, 111, 118
  parameter, 57, 100
  positive real, 82
  tuned, 241, 242
Expert control, 328

Factorization approach to controller design, 274
Fraction, left, 278
  left coprime, 279
  right, 278
  right coprime, 279

Gain, adaptation, 7, 48, 58
  high frequency, 104, 283
  scheduling, 4, 12
Generalized harmonic analysis, 39
Gradient algorithm, 7, 48, 58, 121, 123, 261
  normalized, 58
  with projection, 59, 126

Hermite normal form, 284
Hurwitz, 52
Hyperstate, 10

Identifiability condition, 15, 73, 262
Identification, 1, 45, 100
  frequency domain approach, 45
  model reference, 50, 76
  structural, 1, 257, 258
Index, controllability, 280
  observability, 281
Information matrix, 177, 269
Initial conditions, 57, 65, 113
Input saturation, 117
Instability, slow drift, 233, 236
  due to bursting, 233
  high gain identifier, 236
Interactor matrix, 282

Kalman filter, 61

Leakage term, 253
Learning, 10, 329
Least Squares Algorithm, 48, 61, 261
  normalized, 62
  with forgetting factor, 62
  with covariance resetting, 63
Lemmas
  Lemma 1.2.1 Barbalat's Lemma, 19, 23
  Lemma 1.4.2 Bellman Gronwall Lemma, 23
  Lemma 2.5.2 Uniform Complete Observability under Output Injection, 73
  Lemma 2.6.2 Kalman-Yacubovitch-Popov Lemma, 83
  Lemma 2.6.6 PE and L_2 Signal, 86
  Lemma 2.6.7 PE through LTI Systems, 86
  Lemma 3.6.1 Input/Output L_p Stability, 139

Lemma 3.6.2 Output/Input L_p Stability, 139
Lemma 3.6.4, 141
Lemma 3.6.5 Swapping Lemma, 141
Lemma 3.6.6 Small Gain Theorem, 142
Lemma 4.2.1 Approximate Integral of a Zero Mean Function, 168
Lemma 4.2.2 Smooth Approximate Integral of a Zero Mean Function, 169
Lemma 4.2.3 Perturbation Formulation of Averaging, 170
Lemma 4.4.1 Perturbation Formulation of Averaging– Two Time Scales, 184
Lemma 6.1.2 Persistency of Excitation under Perturbation, 265
Lemma 6.1.3, 266
Lemma 6.2.1, 269
Lemma 6.2.2 Convergence of the Identifier, 270
Lie, bracket, 303
derivative, 295
Linear, 20
Linearizing control, 295
decoupling, 293
Lipschitz, 21
L_p spaces, 18
extended, 18
Lyapunov equation, 39
lemma, 38
redesign, 12
stability, 23
theory, 25

Matching equality, 106, 288
Mean value, 167
Measurement noise, 216
MIMO systems, 277, 305, 320, 326
poles and zeros, 279
Minimum phase, 52
nonlinear system, 299, 302
Model signals, 134
Monic, 52

Neural networks, 330
Nominal plant, 214
Nonlinear, 20
Normal form, for nonlinear systems, 302
Norms, 18

Observer, 56, 117

Parameter, error, 12, 48, 109
nominal, 11, 47, 51, 100, 109
tuned, 242
variation, 2, 251
Parameter, convergence, 71, 90, 159, 311
partial, 95
Parameterization, 53, 121, 277
Persistency of excitation, (PE), 15, 72, 154-155,175, 190, 221, 228, 248, 265
Piecewise continuous, 20
Polynomial matrix division, 281
Positive, semi-definite matrices, 19
definite matrices, 19
definite functions, 25
Positive real (PR), 82
Prefiltering, 127, 292
Projection, 59
Prior information, 250, 257
Proper, transfer function, 52
Propositions
Proposition 1.4.1, 21
Proposition 1.5.3 Uniform Asymptotic Stability of LTV Systems, 33
Proposition 1.5.4 Exponential Stability of LTV Systems, 34
Proposition 1.5.5 Exponential and Uniform Asymptotic Stability, 34
Proposition 1.6.2 Linear Filter Lemma, 42
Proposition 1.6.3 Linear Filter Lemma– Cross Correlation, 42
Proposition 2.4.6, 70
Proposition 2.6.1, 77
Proposition 2.7.1 PE and Autocovariance, 81
Proposition 3.2.1 Matching Equality, 106

Proposition 3.3.1 Fundamental Identity, 114
Proposition 6.1.1, 262
Proposition 6.3.1 Polynomial Matrix Division, 282
Proposition 6.3.2 Interactor Matrix and Hermite Normal Form Equivalence, 284
Proposition 6.3.3 MIMO Matching Equality, 288
Proposition 7.2.1 Bounded Tracking in Minimum Phase Nonlinear Systems, 304
Proposition 7.3.2 Properties of the Identifier, 316

Rate of convergence, 25, 75, 175, 177, 198, 205, 225, 261
Recursive, 1, 46, 49
Reference model, 50, 99, 286, 307
Regressor, 58
filtering, 253
Regular signals, 70
Relative degree, 52
strong for nonlinear systems, 300
Robust control, nonadaptive, 211
Robust identification, 248
Robustness, 209, 214
non-adaptive, 221
to output disturbances, 226
to unmodeled dynamics, 229
Rohrs examples, 215
Row reduced, 280
Rule based control, 328
expert controller design, 328

Self-tuning, 14
Sensitivity, 7, 212
Slow adaptation, 254
(see Averaging)
Slow drift instability, 233, 236
Solutions, existence, 21, 143
Spectral line, 41
Spectral measure, 41
Stability (see Equilibrium)
family of matrices, 183
indirect adaptive control, 151, 266
input error direct adaptive control, 149
of adaptive identifier using factorization approach, 277
of adaptive pole placement, 273
of the identifier, 69
output error direct adaptive control, 143
Stabilizing compensator, 275
State, fast, 162
generalized, 56, 117
slow, 162
State-variable filters, 117
State transition matrix, 33
Stationary, 40
Steepest descent, 58
Strictly positive real (SPR), 52, 82
Strictly proper, transfer functions, 52
Sufficient richness, 92

Theorems
Theorem 1.4.3 Basic Theorem of Lyapunov, 26
Theorem 1.5.1 Converse Theorem of Lyapunov, 28
Theorem 1.5.2 Exponential Stability Theorem, 31
Theorem 1.5.6 Exponential Stability of LTV Systems, 36
Theorem 1.5.7 Lyapunov Lemma, 38
Theorem 2.4.1 Linear Error Equation with Gradient Algorithm, 64
Theorem 2.4.2 Linear Error Equation with Normalized Gradient Algorithm, 64
Theorem 2.4.3 Effect of Initial Conditions and Projection, 65
Theorem 2.4.4 Linear Error Equation with Normalized LS Algorithm and Covariance Resetting, 67
Theorem 2.4.5 Stability of the Identifier, 69
Theorem 2.4.7 Stability of the Identifier—Unstable Plant, 71
Theorem 2.5.1 PE & Exponential Stability, 73
Theorem 2.5.3 Exponential Convergence of the Identifier, 75

Theorem 2.6.3 SPR Error Equation with Gradient Algorithm, 84
Theorem 2.6.4 Modified SPR Error Equation with Gradient Algorithm, 85
Theorem 2.6.5 Exponential Convergence of Gradient Algorithms and SPR Error Equations, 85
Theorem 2.7.2 PE and Sufficient Richness, 92
Theorem 2.7.3 Exponential Parameter Convergence and Sufficient Richness, 93
Theorem 3.7.1 Stability Proof Input Error Direct Adaptive Control, 143
Theorem 3.7.2 Stability Proof Output Error Direct Adaptive Control, 149
Theorem 3.7.3 Stability Proof– Indirect Adaptive Control, 151
Theorem 3.8.1, 155
Theorem 3.8.2, 155
Theorem 4.2.4 Basic Averaging Theorem, 172
Theorem 4.2.5 Exponential Stability Theorem, 173
Theorem 4.4.2 Basic Averaging Theorem, 185
Theorem 4.4.3 Exponential Stability Theorem, 185
Theorem 5.3.1 Small Signal I/O Stability, 221
Theorem 5.3.2 Robustness to Disturbances, 228
Theorem 5.5.3 Robustness to Unmodeled Dynamics, 230
Theorem 5.5.1 Instability Theorem for One Time Scale Systems, 237
Theorem 5.5.2 Instability Theorem for Two Time Scale Systems, 239
Theorem 6.2.3 Asymptotic Stability of an Indirect Adaptive Scheme, 271
Theorem 6.2.4 Asymptotic Stability of the Adaptive Pole Placement Algorithm, 273
Theorem 6.2.5 Asymptotic Stability of the Adaptive Identifier using the Factorization Approach, 277
Theorem 7.3.1 Adaptive Tracking, 310
Theorem 7.3.3 Basic Tracking Theorem for SISO Systems with Relative Degree Greater than 1, 316
Tracking, 304, 310, 316
adaptive, 310, 316

Uncertainty, 209
additive, 211
multiplicative, 211
structured, 209, 213, 263
unstructured, 209, 213
Uniformly completely observable (UCO), 35
Universal controller, 328
Unmodeled dynamics, 216
Unimodular, 279
Update law, 58
hybrid update, 254
modifications, 251
regressor vector filtering, 253
slow adaptation, 254
σ-modification, 253

X–15 aircraft, 6

Errata

p. 12, line 11: in eqn (0.3.5), replace "$(\theta\, r)$" by "(r)"

p. 13, line 17: replace "Note that (0.3.15)" by "Note that (0.3.14)"

p. 22, line 14: in eqn (1.4.9), replace the lower limit of the integral, "0", by "t_0"

p. 22, line 16: replace "(1.3.7)" by "(1.4.7)"

p. 24, line 17: replace "for all $t \geq 0$" by "for all $t \geq t_0 \geq 0$"

p. 37, line 5: in eqn (1.5.53), replace "$\phi(t+\delta, t)$" by "$\Phi(t+\delta, t)$"

p. 37, line 18: in eqn (1.5.56), replace "≤ 0" by "≥ 0"

p. 39, line 1: in eqn (1.5.61), replace "$d\tau$" by "$d\sigma$"

p. 45, line 16: in eqn (2.0.3), replace "$y(t)$" by "$y_p(t)$"

p. 48, line 18: in eqn (2.0.18), replace "$e_1^2(\tau)$" by "$(\theta^T(t)\, w(\tau) - y_p(\tau))^2$"

p. 48, line 21: in eqn (2.0.19), replace "$e_1^2(\tau)$" by "$(\theta^T(t)\, w(\tau) - y_p(\tau))^2$"

p. 50, line 5: in eqn (2.0.28), replace "$P(0) = P_0$" by "$P(0) = P(0)^T = P_0$"

p. 50, line 7: in eqn (2.0.29), replace "$w(t)\, w^T(\tau)$" by "$w(\tau)\, w^T(\tau)$"

p. 50, line 7: in eqn (2.0.29), replace "P_0" by "P_0^{-1}" at *both* places

p. 50, line 9: in eqn (2.0.30), replace "P_0" by "P_0^{-1}" at *both* places

p. 50, line 9: in eqn (2.0.30), replace "dt" by "$d\tau$"

p. 51, line 18: in eqn (2.0.38), replace "$a_m + k_m\, b_0(t)$" by "$a_m - k_m\, b_0(t)$"

p. 51, line 19: in eqn (2.0.38), replace "$(b_0(t) - b_0^*)$" by "$k_m\, (b_0(t) - b_0^*)$"

p. 52, line 3: in eqn (2.0.41), replace "$a_m\, e_1$" by "$a_m\, e_1^2$"

p. 64, line 15: replace "$-\frac{1}{2} g$" by "$-\frac{1}{2g}$"

p. 66, line 19: delete "that is the projection can only improve the convergence of the algorithm"

p. 67, line 21: in eqn (2.4.9), replace "$\frac{d(P^{-1})}{dt}$" by "$\| \frac{d(P^{-1})}{dt} \|$"

p. 68, line 27: in eqn (2.4.13), replace "$\beta(t)$" by "$|\beta(t)|$" at *both* places

p. 69, line 1: replace "β" by "$|\beta|$"

p. 74, line 4: replace "$[A, C]$" by "$[C, A]$"

p. 74, line 7: replace "$[A+KC, C]$" by "$[C, A+KC]$"

p. 74, line 16: replace "$[0, w^T(t)]$" by "$[w^T(t), 0]$"

p. 74, line 17-18: replace "$[-g\, w(t)\, w^T(t), w^T(t)]$" by "$[w^T(t), -g\, w(t)\, w^T(t)]$"

p. 83, line 2: replace "$\mathrm{Re}(\hat{M}(j\omega)$" by "$\mathrm{Re}(\hat{M}(j\omega))$"

p. 83, line 3: replace "$\mathrm{Re}(\hat{M}(j\omega)$" by "$\mathrm{Re}(\hat{M}(j\omega))$"

p. 84, line 2: replace "c^T" by "$c^T(t)$"

Errata

p. 88, line 13: replace $[A, c^T]$ by $[c^T, A]$

p. 88, line 18: replace "Using the triangle inequality" by "Using the fact that $(a-b)^2 \geq \frac{1}{2}a^2 - b^2$"

p. 88, last 3 lines: insert factors of 1/2 as follows

$$\int_{t_0}^{t_0+\delta} e_1^2(\tau)\,d\tau \geq \frac{1}{2}\int_{t_0}^{t_0+m\sigma} x_1^2(\tau)\,d\tau - \int_{t_0}^{t_0+m\sigma} x_2^2(\tau)\,d\tau$$

$$+\frac{1}{2}\int_{t_0+m\sigma}^{t_0+\delta} x_2^2(\tau)\,d\tau - \int_{t_0+m\sigma}^{t_0+\delta} x_1^2(\tau)\,d\tau$$

$$\geq \frac{1}{2}\gamma_3(m\,\sigma)\,|e_m(t_0)|^2 - m\,\alpha_2\,|\phi(t_0)|^2$$

p. 89, line 1: in eqn (2.6.38), replace "$n\,\alpha_1$" by "$\frac{1}{2}n\,\alpha_1$"

p. 89, line 3: adjust eqn (2.6.39) to read "$\frac{1}{2}\gamma_3(m\,\sigma) - \gamma_1 e^{-\frac{1}{2}m\,\sigma} \geq \gamma_3(m\,\sigma)/4$"

p. 89, line 5: in eqn (2.6.40), replace "$n\,\alpha_1$" by "$\frac{1}{2}\,n\,\alpha_1$"

p. 89, line 7: in eqn (2.6.41), replace "$\gamma_3(m\,\sigma)/2$" by "$\gamma_3(m\,\sigma)/4$"

p. 101, line 4: replace "$\frac{k_p}{s+a_m}\hat{M}(\phi_r\,r + \phi_y\,y_p)$" by "$\frac{k_p}{s+a_m}(\phi_r\,r + \phi_y\,y_p)$"

p. 114, line 10: in eqn (3.3.17), replace "$+\varepsilon(t)$" by "$-\varepsilon(t)$"

p. 138, line 5: in eqn (3.5.27), replace "e_1" by "e_0"

p. 143, line 31: in the equation giving $|\phi^T(t)\,v(t)|$, replace "$\beta(t)$" by "$|\beta(t)|$" at *both* places

p. 148, line 9: replace "theorem 2.4.6" by "proposition 2.4.6"

p. 148, line 14: replace "the same conditions as β" by "the same conditions as $|\beta(t)|$"

p. 155, line 8: replace "lemma 2.6.6" by "lemma 2.6.7"

p. 155, line 10: replace "lemma 2.6.5" by "lemma 2.6.6"

p. 161, line 12: in the title, replace "Sale" by "Scale"

p. 163, line 22: in eqn (4.1.30), replace "$\hat{P}(z)$" by "$\hat{P}(r)$"

p. 247, line 1: in eqn (5.5.49), replace "$-\varepsilon\,r(t)$" by "$-g\,r(t)$"

p. 247, line 13: in the caption for Fig. 5.15, replace "$r_1(t)$" by "$r_1(t) = \sin(5t)$"

p. 271, line 21: replace "$\phi(t) \to 0$ as $t \to 0$" by "$\phi(t) \to 0$ as $t \to \infty$"

p. 282, line 5: in proposition 6.3.1, delete "Let $\hat{D}_R$ be column reduced and $\hat{D}_L$ be row reduced"

p. 282, line 16: in the proof of proposition 6.3.1, delete "and $\hat{D}_R$ column reduced"

p. 283, line 11: replace "$\xi \in R^{pxp}(s)$" by "$\xi \in R^{pxp}[s]$"

p. 284, line 10: replace "Morse [1979]" by "Morse [1976]"

p. 288, line 27: replace "$\hat{D}_L$ row reduced" by "$\hat{D}_L$ column reduced (such a matrix fraction description always exists (*cf.* Beghelli S. & R. Guidorzi, "A New Input-Output Canonical Form for Multivariable Systems," *IEEE Trans. on Autom. Control,* vol. 21, pp. 692-696, 1976)"

p. 290, line 5: in eqn (6.3.22), replace "$\overline{w}$" by "$\overline{w}^T$"

p. 338, line 1: in eqn (A3.6.5), replace "$t^r - k$" by "t^{r-k}"

p. 338, line 15: under "Derivation of (A3.6.10)", insert "When $r = 0$, $u(t) = z(t)$, so that (A3.6.10) is trivially true. When $r > 0$," (we have that ...)

p. 339, line 1: in eqn (A3.6.12), replace "$\int_{-\infty}^{t}$" (the first integral) by "$\int_{0}^{t}$"

p. 339, line 12: in eqn (A3.6.15), replace ":=" by "$\leq$"

p. 341, line 14: replace "$\beta(t)$" by "$|\beta(t)|$" at *both* places

p. 356, line 10: in eqn (A6.2.13), replace "$m\,(\exp - \alpha - (t - \tau))$" by "$m\,e^{-\alpha(t-\tau)}$"

p. 360, line 37-38: the title of the paper should be "Exponential Convergence and Robustness Margins in Adaptive Control" instead of "Small Signal..."

p. 366, line 22: insert Luders, G., & K.S. Narendra, "An Adaptive Observer and Identifier for a Linear System," *IEEE Trans. on Automatic Control,* Vol. AC-18, no. 5, pp. 496-499, 1973.